LANDSCAPE PLANNING
Environmental Applications

LANDSCAPE PLANNING
Environmental Applications

Fourth Edition

William M. Marsh
University of British Columbia

WILEY

John Wiley & Sons, Inc.

ACQUISITIONS EDITOR   Jerry Corea
ASSISTANT EDITOR   Denise Powell
MARKETING MANAGER   Clay Stone
PRODUCTION EDITOR   Sarah Wolfman-Robichaud
PHOTO EDITOR   Teresa Romito and Hilary Newman
ILLUSTRATION EDITOR   Sandra Rigby
COVER   W. M. Marsh
ILLUSTRATION   Brian Larson

This book was set in 10/12 Garamond Light by LCI Design and printed and bound by Courier-Westford.
The cover was printed by Phoenix Color Corporation.

This book is printed on acid-free paper.

ISBN: 978-0-471-48583-4

Printed and bound in the United States of America

10 9 8 7 6 5 4 3 2

FOREWORD

Landscape Planning: Environmental Applications by William M. Marsh is a thoroughly splendid book, an exhaustive selection and amalgamation of the salient principles and processes derived from geology, hydrology, soil science and ecology, expressly oriented to the needs of landscape architecture.

It contains insightful descriptions of physiographic regions, each with characteristic geomorphology, hydrology, soils and ecologies. It presents fluvial and soil processes, watersheds, flood plains, erosion, sedimentation, each with enlightening text and diagrams. It could well be described as "How to Design with Nature." I give thanks to the author as should you.

I Commend *Landscape Planning* to your attention.

Ian L. McHarg
6th October 1997

University of Pennsylvania
The Graduate School of Fine Arts
Department of Landscape Architecture
and Regional Planning

PREFACE

One of the original objectives of this book was to help bring the landscape disciplines together around a set of environmental issues and land use problems—in short, to make a more effective integration among geography, landscape architecture, planning, and related fields. That objective still holds, and it is a credit to practitioners, teachers, researchers, and students that each year there are more participants, more dialogue, and fewer barriers among the fields involved in planning, designing, and managing the landscape. At the same time, the agenda is expanding and the problems are bigger and more complex—constant reminders of the struggle and responsibilities we all share.

The scope of the third edition was considerably broader than that of the second edition and this edition is somewhat broader yet. In the third edition several new topics were addressed as individual chapters—namely, stream channel and riparian environments and landscape ecology—and many original chapters were revised substantially to include topics such as landscape theory, hazardous waste disposal, slope stability assessment, small-flow wetlands, and soil sediment systems. In this edition, one new chapter on best management practices has been added, and several others have been extensively revised. New case studies have been added on sustainability, green infrastructure, industrial site management, flood risk, wetlands and water quality, shoreline stabilization, urban climate, and marsh restoration. Patrick Condon, Jack Goodnoe, Charlie Schlinger, Cory Helton, Jim Janecek, John Elder, Gerald Goddard, Elliott Menashe, and Patrick Mooney contributed these case studies, and their efforts are sincerely appreciated.

Frankly, the supplemental material has been a source of frustration with this book. Because of the field's enormous scope, the range of possibilities for data sets, maps, technical readings, problem sets, and so on is also enormous. In addition, much of the instruction related to this book uses local and regional examples and problems from all parts of the continent. In the first edition, we included large appendices and end-of-chapter problem sets. In the second edition, we provided an instructor's manual. This edition features a Web site (www.wiley.com/college/marsh) that includes links to data, maps, and other sources, as well as the material from earlier editions.

Many people contributed their time and talent to this edition. Scott Murdoch provided direction and material for early drafts of chapter 13. Alison Mewett, Jo Mitchell, Rachel Sim, and Katherine Best provided advice, clerical help, patience and sympathy. At Wiley, Denise Powell and Sarah Wolfman-Robichaud were especially helpful in moving the project along. Special thanks are extended to colleagues who provided reviews, all very helpful: Stacey Swearington White of University of Kansas, Cathleen Corlett of California Polytechnic University, Safei Hamed of Texas Tech University, Denyse Lemaire of Rowan University, Jana Carp of Appalachian State University, Carol Harden of University of Tennessee, Knoxville. We give thanks in memorium to Ian L. McHarg for the years of vision and leadership in the field. To the hundreds of students, teachers, and practitioners in landscape architecture, planning, geography, and related fields whom I have had the pleasure to lecture before and work with over the years, goes a final note of gratitude for their patience, questions, and suggestions.

W. M. Marsh
Comox, B.C.

CONTENTS

AN INTRODUCTION TO THE BOOK AND THE FIELD

0.1 OPENING STATEMENT

A predictable sameness is creeping over the face of the North American landscape. Since the early 1970s, highways, shopping centers, residential subdivisions, and most other forms of development have taken on a remarkable similarity from coast to coast. Not only do they look alike, but modern developments also tend to function alike, including the way they relate to the environment, that is, in the way land is cleared and graded, stormwater is drained, buildings are situated, and landscaping is arranged.

Landscape diversity

We know that this facade of development masks an inherently diverse landscape in North America. If we look a little deeper, it is apparent that **landscape diversity** is rooted in the varied physiographic and ecological character of the continent and this in turn reflects differences in the way the terrestrial environment functions. Does it not seem reasonable then that development and land use should also reflect these differences if they are to be responsive to the environment? Herein lies one of the important missions for environmental planning: to help guide development toward environmentally responsive landscape planning and design schemes that avoid mismatches between land uses and environment.

0.2 BACKGROUND CONCEPT

Virtually every modern field of science makes contributions toward resolving societal problems. In some cases the contributions are not very apparent, even to the practitioners in that field, because they are made via second and third parties. These parties are usually the applied professions, such as urban planning, landscape architecture, architecture, and engineering, which synthesize, reformat, refine, and adapt knowledge generated by scientific and technical investigations.

Applied professions and sciences

Geography, geology, hydrology, soil science, ecology, remote sensing, and many other fields hold such a relationship with the various fields of landscape planning and design. For more than a century, earth and environmental professionals have studied the world's physical features, learning about their makeup, how to measure them, the forces that change them, and how we humans interact with them. Planning is concerned with the use of resources, especially those of the landscape, and how to allocate them in a manner consistent with people's goals. Thus landscape planning and the environmentally-oriented sciences are linked together because of a mutual interest in resources, land use, and the nature and dynamics of the landscape.

0.3 CONTENT AND ORGANIZATION

This book addresses topics and problems of concern to planners, designers, scientists, and environmentalists. The focus is on environmental problems associated with land planning, landscape design, and land use. The coverage is broad, though not intended to represent the full range of existing applications; such a book would be too large and cumbersome for most users. The choice of topics was guided by three

Scope of study

considerations. First, the main components of the landscape should be represented—namely, topography, soils, hydrology, climate, vegetation, and habitat. Second, the topics should be pertinent to modern planning as articulated by urban planners, landscape architects, and related professionals including, for example, stormwater management, slope classification, and wetland interpretation. Third, the topics should not demand advanced training in analytical techniques, data collection, mapping, and field techniques.

Each edition of this book covers more topics than its predecessor because the field of environmental planning is continually growing. This reflects both the development of the field professionally and the expanding legitimacy of environmental planning in the planning and environmental arenas. For example, only 25 years ago it was barely legitimate for fields outside civil engineering to address stormwater management in a professional context. In the past decade or two, environmental planning has critically addressed not only stormwater as a runoff problem, but stormwater as a major source of water pollution, as well as stormwater infrastructure as a constraint in community design. In addition, the field has taken up wetland planning, urban climate, and groundwater management, and most recently a variety of ecological topics including biodiversity and landscape ecology.

0.4 LANDSCAPE PLANNING, ENVIRONMENTALISM, AND ENVIRONMENTAL PLANNING

The term **landscape planning** is used in this book to cover the macro environment of land use and planning activity dealing with landscape features, processes, and systems. Only three decades ago, the term **land use planning** was generally used for this sort of activity, but today, because of new knowledge, the recognition of new problems, the changing needs of society, and the modern proliferation of specialty fields, several new and alternative titles have emerged.

The rise of environmentalism

The environmental crisis of the 1960s and 1970s was brought on by a flurry of concern over the quality of the environment. Much of this concern took the form of a political movement to protect the "environment" from the onslaught of industry, government, urban sprawl, and war. Loosely translated, "environment" was taken by the movement to mean things of natural origin in the landscape, that is, air, water, forests, animals, river valleys, mountains, canyons, and the like. From this emerged the **environmentalist**, a person who believes in or works for the protection and preservation of the environment. **Environmentalism**, it follows, is a philosophy, a political or social ideology, that implies nothing in particular about a person's training, knowledge, or professional credentials in matters of the environment. Organizations such as the Sierra Club, Greenpeace, and Friends of the Earth practice environmentalism.

The emergence of new fields

The environmental crisis also paved the way for stronger and broader environmental legislation at all levels of government. New types of professional skills were needed to provide various services in connection with environmental assessments, waste disposal planning, air and water quality management, and so on. In response, several new "environmental" fields emerged, while many established fields, such as civil engineering and chemistry, developed "environmental" subfields. Taken as a whole, the resultant environmental fields fall roughly under three main headings: environmental science, environmental engineering and technology, and environmental planning.

Environmental planning

Environmental planning is a "catchall" sort of title applied to planning and management activities in which environmental rather than social, cultural, or political factors, for example, are central considerations. The term is often confused with environmentalism and with the preparation of environmental impact studies, but in reality environmental planning covers an enormous variety of topics associated with land development, land use, and environmental quality, including relatively new topics such as toxic waste disposal and the management of wetlands, as well as traditional problems, such as watershed management and planning municipal water supply systems. To some extent, landscape planning is also a term of convenience used to distinguish the activities of what we might call the landscape fields such as geography, landscape architecture, geomorphology, and urban planning from other areas of environmental planning, many of which are more closely tied to environmental engineering and public health.

0.5 THE SPATIAL CONTEXT: SITES AND REGIONS

Because we need to address topics and problems of the landscape that are pertinent to modern planning as articulated by the practicing professional, most material in this book is presented at the site or community scale. **Sites** are local parcels whose size usually ranges from less than an acre to hundreds of acres, with a simple ownership or stewardship arrangement (individuals, families, or organizations). They are the spatial units of land use planning, the building blocks of communities.

Whose region?   **Regions**, in the vocabulary of land use planning and landscape design, are variously defined as the geographic settings that house communities, either a single community and its rural hinterland, several communities and the systems connecting them (roads or streams, for example), or a metropolitan area with its inner city, industrial, and suburban sectors. This concept differs from the geographer's notion of a region, which encompasses a much larger area—for example, the Midwest, the Great Plains, or the Hudson Bay region. Many environmental problems have a regional scope (geographer's version)—for instance, acid rain in the eastern midsection of the continent and water supplies for irrigation in the Southwest—but for a variety of reasons, planning programs have generally not been very effective at this scale in North America. Most of our examples of effective or promising environmental planning are of regional (planner's version), community, or site scales. But you will also find these are often terms of convenience that may not accurately address the true scale of the problems and environment under consideration.

0.6 FINDING THE APPROPRIATE SCALE

Geographic or **spatial scale** is an essential part of all planning problems, but it is one of the least effectively used dimensions in modern land use systems. Illustrations of the misuse of scale permeate landscape and society in the developed world. Most stem from mismatches between the functions or processes that a program is supposed to address, the magnitude of the facilities and systems actually designed to serve it, and the environment in which it is placed and upon which it is dependent.

Missing the mark on scale   In the United States and Canada, "bigger is better" has become a rule of thumb for many decision makers in this century. Public education, for example, has shifted toward larger and larger schools that have lost touch with students as individuals and with the communities they serve. Cities in both developed and developing nations have expanded into regional entities with such large populations (Mexico City is expected to reach 50 million and Shanghai 100 million in this century) that they may now exceed the limits of manageability as physical and social systems. In the United States, Canada, and parts of Europe, expressway systems originally designed for national defense and interregional travel have become daily escape conduits for an urban-based workforce seeking to scale life down in smaller, more traditional communities beyond the urban region.

In the environment, the American federal government has approached flood management with the construction of massive dams and the manipulation of river channels while largely ignoring local and regional watershed management and flood-sensitive land use planning. The result is extensive loss of riparian habitat, woodland, and farmland while "protected" floodplain settlements have become more susceptible to impacts from large floods because of naive reliance on engineered structures. The rise of local watershed management groups and stream restoration associations is in many ways an attempt to address these problems at a grassroots level and counterbalance the top-down approach of federal, state, and provincial programs.

At the same time, communities have borrowed environmental policies from other, often distant, communities and applied them locally without regard for differences in the scale, fabric, and operation of the environmental systems for which they were originally designed. Instead of problems such as stormwater drainage, flooding, water quality, and habitat conservation being mitigated or resolved, many remain unchanged or worsened. Engineered stormwater systems designed for eastern North America, for example, are applied to topographic and hydroclimatic environments in the West where they are not only inappropriate and damaging to property, stream channels, and habitat but an economic burden to communities, especially small ones.

Sources of scale problems   Why these problems of scale? There are many reasons and the following examples are a few of the significant ones. Consider national policies that lead to sweeping applications of standard infrastructure systems, such as dams and expressways, irrespective of the character of land use and environment at local and regional scales. Or that economics based on the benefit–cost rationale show savings with large-scale development schemes. Another example is the rising threat of professional liability facing planners and engineers for not building bridges, dams, and other facilities large enough to withstand "all" environmental contingencies. And not the least among the reasons for scale problems is a lack of understanding by decision makers and their technical advisors about the workings of the environment; that is, the processes and systems that shape and sustain the landscape, including the scales at which they operate. From place to place the environment is different in fundamental ways and unless these differences are made part of the information base for decision making and design, we will continue to build missized and unsustainable infrastructures and land use systems.

An example   By way of example, the people of New Hampshire sensed the limitations of scale when they stopped the federal and state transportation departments from building a standard four-lane interstate highway through Franconia Notch, a narrow pass in the environmentally prized White Mountains (Fig. 0.1). The project was held up for more

Fig. 0.1 Interstate highway I-93 where it passes through Franconia Notch in the White Mountains of New Hampshire. A scaled-down section of expressway designed to fit a narrow valley.

Fig. 0.2 To better fit the scale of the valley and help preserve recreational and scenic features, access routes and ramps were scaled down along I-93 in New Hampshire.

than a decade until the highway planners agreed (at the request of the courts) to modify their standard approach based on federal rules and guidelines and adopt an alternative plan better suited to the topography, drainage, and recreational features of the pass. The result was a scaled-down stretch of interstate highway. It was not a perfect solution, but it was decidedly better than a conventional, large, limited-access expressway with huge exit and entry ramps squeezed into the center of the valley (Fig. 0.2). Franconia Notch serves to remind us that scale can be applied effectively if the makeup, fabric, and operation of local and regional landscapes, and the magnitude and potential impacts of the proposed development are understood, and if a compelling case is presented to decision makers based on documented evidence.

1

LANDSCAPE PLANNING: ROOTS, PROBLEMS, AND CONTENT

1.1 INTRODUCTION: ROOTS

People have probably engaged in some form of environmental planning as long as organized society has been around. There is ample evidence, for example, that the ancient Mesopotamians devised elaborate planning schemes for distributing irrigation water in the desert and that the Romans of Caesar's time programmatically drained wetlands to gain additional farmland and reconfigured harbors to improve navigation. It is clear, however, that most of the ancients' interests in environmental planning were purely practical, having to do with things like trade, food supplies, water, and defense with little or no regard for what we would call environmental impacts. In many respects, planning and engineering programs were seen as attempts to bring order to an otherwise disordered natural environment, even to correct defects in nature. Curiously, a great many of the measures applied to these "defects" such as irrigation of desert soils in Mesopotomia and reconfiguring harbors by the Romans, failed for reasons the builders never anticipated.

Early attitudes

Prior to the fifteenth century and with few exceptions, nature itself was given little regard as part of the environment. Arguably the lowest point in Western civilization's attitude toward nature occurred in Europe in the Middle Ages, when nature was commonly viewed with suspicion, fear, and ignorance. Forests, for example, were seen as dangerous places haunted with beasts and thieves, and a common person's total familiarity with local geography would not extend much beyond the village of his or her birthplace. However, with the Renaissance (beginning in the fifteenth century) and the Enlightenment (seventeenth and eighteenth centuries), humans and nature came to friendlier terms. The enlightened mind saw nature as having logic and order, something understandable to humans. Newton's model of the natural universe as a perfectly structured and perfectly functioning machine had a profound influence on our view of nature on Earth.

The Romantic movement

In the eighteenth and nineteenth centuries the concept of nature was extended to include pleasure and the enjoyment of natural things. This marked the beginning of a love affair with the environment, called the **Romantic movement**. With the Romantic movement we find nature given consideration for its own sake and for its beauty, spiritual meaning, and influence on the quality of life. In landscape design a new school of thought, called Landscape Gardening, emerged in England in which the landscapes of rural estates were made to look "natural" by using curved lines in gardens, field edges, and water features (Fig. 1.1). The arts of the Romantic movement also reflect the rise in environmental consciousness; nineteenth-century painting, music, and literature illustrate this awareness especially well with, among other things, bucolic scenes and pastoral moods.

In the United States the first village improvement associations, which began in the 1850s, applied Romantic concepts to communities. They beautified streets, cemeteries, and town squares and promoted laws for the protection of songbirds and the creation of parks. In cities, public parks emerged in an attempt to bring nature into urban centers such as New York as part of nineteenth-century social reform efforts. Urban parks also followed the landscape gardening concepts of the British with curving landscape forms and features reminiscent of idealistic rural scenery. Overall, the Romantic movement can be credited with elevating the concept of nature and the natural environment to the status of an important human value, which forms an important underpinning for modern environmental planning. Indeed, the environmental crisis of the 1960s and 1970s was founded largely around environmental quality and decline as moral issues.

The public health movement

Another development of the nineteenth century also provided an important underpinning for environmental planning: namely, the scientific understanding of the environment's role in **public health**. This understanding came about through the

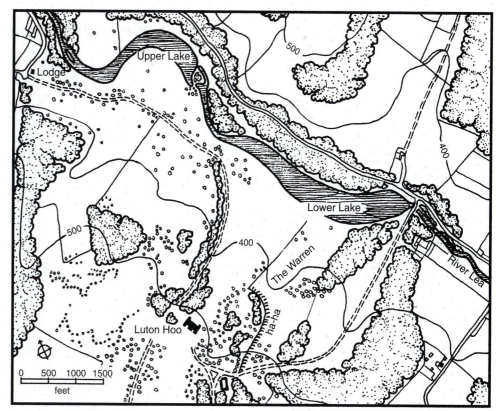

Fig. 1.1 Curved edges of this landscape design by Capability Brown at Luton Hoo, England, illustrate one of the central themes of the English Landscape Gardening School of the 1700s.

documentation of environmentally sensitive diseases, such as malaria, dysentery, and typhoid fever (Fig. 1.2). It resulted in an improved public and institutional understanding of the relationship between human impacts on the environment (such as the decline in water quality from sewage discharges) and the health and well-being of society. One manifestation of this understanding was the planning and development of municipal sanitary sewers; Chicago's system, built in 1855, was one of the first in North America. Another was the development of public water supply systems which could deliver safe drinking water to cities; water purification by filtration and chlorination was introduced in the 1930s.

The conservation movement   A third underpinning for environmental planning was the **conservation movement**. It, too, began in the 1800s, growing out of a concern for the damage and loss of land and its resources as a result of development and misuse. Tied to both Romantic and scientific thought, the conservation movement, led by environmental stalwarts such as John Muir and J. J. Audubon, initiated the national park system. Yellowstone, the first U.S. national park, was established in 1872, and Banff, the first Canadian national park, was established in 1885. The movement subsequently led to many other major conservation programs, including the U.S. Forest Service, the U.S. Natural Resources Conservation Service, and the U.S. Bureau of Land Management, as well as many state and local programs.

The conservation concept also influenced community land use planning in the United States, Canada, the United Kingdom, and elsewhere. Conservation-sensitive land use planning in the 1970s adopted an ecological perspective in which uses were assigned to the land according to its carrying capacities, environmental sensitivity, and

Fig. 1.2 An illustration from the mid-1800s in New York City, illustrating the sort of living conditions that led to the public health movement.

suitability as a human habitat. Although the conservation concept has been practiced in one form or another for many decades, its application to community development was advanced significantly by Ian L. McHarg, a landscape planner and designer, who in the 1960s, 1970s, and 1980s championed the concept of integrated landscape planning as a means of striking a balance between land use and the environment.

The environmental crisis   In the face of rapidly expanding cities, highway development, and a burgeoning industrial sector after 1940, these movements crystallized in the **environmental crisis** of the 1960s and 1970s (Fig. 1.3). Although the environmental crisis is often remembered for protest movements and social upheaval, its most lasting effects in the United States are represented by a massive body of environmental law, the National Environmental Policy Act, which addresses air quality, water quality, energy, the work environment, and many other areas. Similar bodies of environmental policy were enacted in many other developed countries as well as selected developing countries.

1.2 THE PROBLEM: CHANGE AND IMPACT

Landscape change   The rate at which North Americans have developed this continent is unprecedented in the history of the world. In scarcely 100 years, from about 1800 to 1900, they spread across the continent, probing and transforming virtually every sort of landscape. In the vast woodlands and grasslands of the continent's midsection, scarcely a whit of the

Fig. 1.3 Social protest of the environmental crisis of the late 1960s and early 1970s marking the beginning of environmentalism as a major political force in North America.

original landscape remained by the opening of the twentieth century. In its place came crop farms, ranches, towns, cities, and connecting railways and highways. Not only that, but it can be argued that since 1950 or so, a second wave of landscape change has swept across North America, especially in the region east of the Mississippi and along the Pacific Coast, changing the rural landscape of farms and small towns into the satellite communities, dormitory settlements, and playgrounds of huge urban complexes. Wholesale transformation of natural landscapes, and later the settled rural landscape, represents only part of the story, however. The introduction of synthetic materials and forms represents the rest.

Materialism and waste production

The age of materialism and economic expansionism has produced a colossal system of resource extraction which, for the United States and Canada, reaches over most of the world. At the output end of the system is the manufacture of products and residues of various compositions, many decidedly harmful to humans and other organisms. Both end up in the landscape: steel, glass, concrete, and plastics (in the form of buildings and cities), waste residues (in the form of chemical contaminants in air, water, soil, and biota), and solid and hazardous wastes (in landfills, waterbodies, and wetlands). The waste stream is enormous; in the United States, for example, the annual production of synthetic organic compounds, which includes scores of hazardous substances such as pesticides, exceeds 100 million tons; the output of solid waste from urban areas alone approaches a billion tons annually.

Built environments

The landscapes that are ultimately created are essentially new to the earth. Cities, for example, are often built of materials that are thermally and hydrologically extreme to the land and in structural forms that are geomorphically atypical in most landscapes. It is a landscape distinctly different from the one it displaced and, in many respects, decidedly inferior as a human habitat. The modern metropolitan environment that results tends to be less healthy, less safe, and less emotionally secure than most people desire. Moreover, the very existence of such environments poses a serious uncertainty to future generations, owing to the high cost of maintaining both the environment and the quality of human life within them. In addition, their relationship with the natural environment of water, air, soil, and ecological systems is a lopsided one that does not adequately fit our notion of a sustainable balance between an organism and its habitat. Herein lies much of the basis for environmental and land use planning, landscape design, and urban and regional planning.

Mismatches between land use and environment

The planning problems we are facing today are many and complex, and not all, of course, are tied directly to the landscape. For those that are, most seem to result from mismatches between land use and environment. The mismatches are of mainly four origins: (1) those that stem from *initially poor land use decisions* because of ignorance or misconceptions about the environment, as exemplified by the person who unwittingly builds a house on an active fault or ignores warnings about hurricane-prone shore property; (2) those that stem from *environmental change* after a land use has been established, as illustrated by the property owner who comes to be plagued by flooding or polluted water because of new development upstream from his or her site; (3) those that stem from *social change*, including technological change, after a land use has been established and represented, for example, by the resident living along a street initially designed for horse-drawn wagons but now used by automobiles and trucks and plagued by noise, air pollution, and safety problems; and (4) those that stem from *violations of human values* concerning the mistreatment of the environment such as the eradication of species, the degradation of streams, and the alteration of historically valued landscapes (Fig. 1.4).

Fig. 1.4 Value conflict: A modern environmental protest targeting specific issues and public policies.

1.3 THE PURPOSE OF PLANNING

The need for planning

In general, the primary objective of planning is to make decisions about the use of resources. Over the past 25 years, the need for land use and environmental planning has increased dramatically as shown by the rising competition for scarce land, water, biological and energy resources; the need to protect threatened environments; and the desire to maintain or improve the quality of human life. The problems and issues are diverse in both type and scope—ranging from worldwide issues, such as desertification and misuse of the tropical rainforests, to problems of draining and filling a 2-acre patch of wetland on the edge of a city. In North America, despite the political undertones historically associated with public planning, environmental planning has gained real legitimacy in the past two decades, though more as a reactive than a proactive process—that is, more as a system of restrictive (should not) policy than one of constructive (how to) policy.

Who makes decisions

Who does planning? Actually, professional planners probably do not do the majority of planning. Most of it is done by corporation officers, government officials and their agents, the leaders of educational and religious institutions, the military, and various other organizations, including citizen's groups. Professional planners (those with formal credentials in planning, or related areas) usually function in a technical and advisory capacity to the decision makers, providing data, defining alternative courses of actions, forecasting impacts, and structuring strategies for implementation of formal plans. The overall direction of a plan, however, always represents some sort of policy decision; one based on a formal concept of what a company, city, or neighborhood intends for itself and thus will strive to become. These concepts about the future are called *planning goals*, and they are the driving force behind the planning process.

1.4 PLANNING REALMS: DECISION MAKING, TECHNICAL, AND DESIGN

Decision-making planning

Three broad classes of activity make up modern planning: decision making, technical planning, and landscape design. The first is that activity related to the decision-making process itself, which is usually carried out in conjunction with or directly by formal bodies such as planning commissions and corporate boards. It involves building the methods and means for arriving at planning decisions, formulating plans, and then providing the directions necessary for carrying out decisions. Among the tasks commonly undertaken in **decision-making planning** are consolidation of technical studies, formulation of policies, articulation of goals, definition of alternative courses of action, and selection of preferred plans.

Technical planning

The second class of planning activity can be called **technical planning**. It involves various processes and services that are used in support of both decision-making and design activities. This includes environmental inventories, such as testing soil and vegetation mapping, engineering analysis, such as testing soil suitability for construction, and assessment of the impacts that proposed land uses may have on the environment. Technical planning is usually carried out by a variety of specialists, including cultural geographers, physical geographers, geologists, ecologists, hydrologists, wildlife biologists, archaeologists, economists, and sociologists, as well as by professional planners from landscape architecture, urban planning, and architecture and civil engineering. The line separating decision-making from technical-planning activities may be distinct or indistinct, depending on organizational arrangements and the nature of the problem.

Landscape design

Following the decision-making process and the first wave of technical support studies, we move into the arena of design. **Landscape design** entails the laying out on paper or on the computer screen the configuration of the uses, features, and facil-

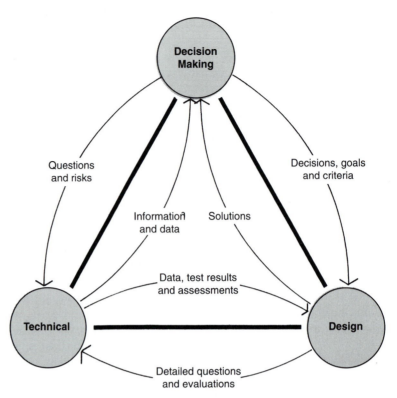

Fig. 1.5   The relationship among the technical, decision-making, and design realms of modern planning. Much of the activity of environmental planning falls within the technical sphere.

ities that are to be built, changed, or preserved by virtue of the decision maker's directives. Design may call for additional technical studies, such as soil testing, refinement of maps, and even laboratory analysis. Therefore, the planning processes and the relationship among the three areas of professional activities should not be regarded as a linear sequence—though it is often practiced that way—but more as an interrelated circuit as is depicted in Figure 1.5. Resolving a planning problem usually involves several iterations of the circuit in which a check and balance relationship often emerges among decision makers, technical planners, and designers.

1.5 ENVIRONMENTAL IMPACT ASSESSMENT

The use of technical planners or specialists in planning has increased sharply in the past three decades as a result of environmental impact legislation. Indeed, the enactment of the **National Environmental Policy Act of 1969 (NEPA)** is directly tied to the emergence of environmental planning as a formal area of professional practice. This act calls for planners to forecast and evaluate the potential impacts of a proposed action or project on the natural and human environments. Although the act covers only projects involving federal funds (such as sewage treatment systems, highways, and domestic military facilities), NEPA initiated the enactment of more broadly based environmental impact legislation at state and local levels in many parts of the United States. As a whole, the various bodies of impact legislation brought on a flurry of activity in environmental planning in the 1970s and served to establish environmental factors as legitimate considerations in urban, regional, transportation, and other types of planning.

Impact methodology   The basic **environmental impact methodology** can be summarized in five steps, or tasks, that are normally performed sequentially.

- The first is to select variables or factors that are pertinent to the problem, record them in an inventory, and identify their interrelationships.
- The second is to formulate alternative courses of action.
- The third is to forecast the effects (or impacts) of the alternatives.
- The fourth is to define the differences between the alternatives: that is, to specify what is to be gained and lost by choosing one alternative over another.
- The last is to evaluate and rank the alternatives and select the preferred one.

Environmental impact statements   The report prepared from this assessment is called an **environmental impact statement (EIS)**. It must identify the unavoidable adverse impacts of the proposed action, any irreversible and irretrievable commitments of resources as a result of the proposed action, and the relationship between short-term uses of the environment and long-term productivity. In addition, the EIS must include among its alternatives one calling for no action, and this, too, must be subjected to analysis and evaluation.

Conditions and ambiguities   One of the most important and difficult tasks in the EIS process is the third—forecasting the impacts of alternative actions. An *environmental impact* can be defined as the difference between (1) the condition or state of the environment given a proposed action and (2) the condition expected if no action were to take place. Impacts may be direct (resulting as an immediate consequence of an action) or indirect (resulting later, in a different place, and/or in different phenomena than the action). Obviously, forecasting indirect impacts, and their correlative, cumulative impacts (where many factors work in combination to produce change), can be very difficult and is often a source of much uncertainty in environmental assessment. Furthermore, since an impact represents an environmental change, the problem that also arises (as in the case of an action calling for eradicating vegetation that contains both valued and noxious plant species) is that of deciding which are desirable and undesirable impacts and how different impacts should be weighted for relative significance.

1.6 AREAS OF ACTIVITY IN LANDSCAPE PLANNING

For planning that deals with the environment, several types of activities have become conventional in the United States and Canada. One of the best known is the so-called *Environmental inventory*   **environmental inventory**, an activity designed to provide a catalog and description of the features and resources of a study area. The basic idea behind the inventory is that we must know what exists in an area before we can formulate planning alternatives for it. Among the features consistently called for in environmental inventories are water features, slopes, microclimates, floodplains, soil types, vegetation associations, and land use, as well as archaeological sites, wetlands, valued habitats, and rare and endangered species. Environmental inventories are a major part of environmental impact statements, in which the inventory must also include an evaluation of the recorded phenomena based on criteria such as relative abundance, environmental function, and local significance. This is supposed to indicate the comparative importance or value of a feature or resource (Fig. 1.6).

Opportunity and constraint   A second type of planning activity is aimed at the discovery of **opportunities and constraints**. This activity is often undertaken after a land use has been proposed for an area but the density, layout, and appropriate design of the land use program are still undetermined. The study involves searching the environment for those features and situations that would (1) facilitate a proposed land use and those that would (2) deter or

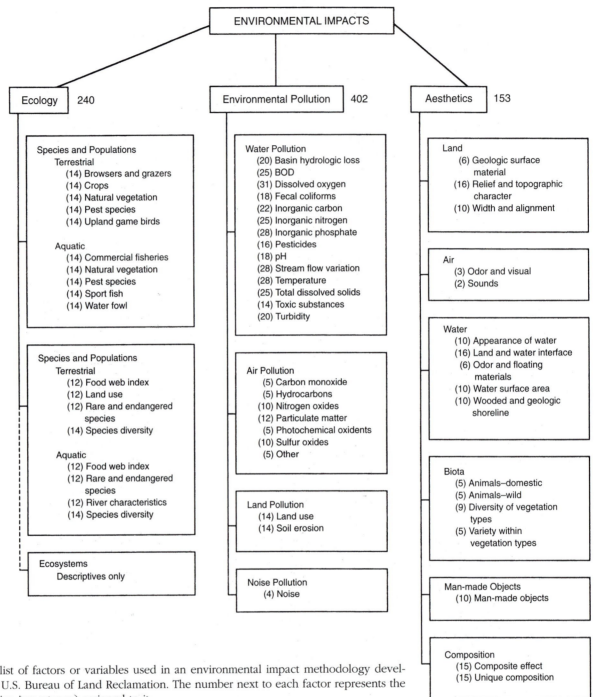

Fig. 1.6 A list of factors or variables used in an environmental impact methodology developed by the U.S. Bureau of Land Reclamation. The number next to each factor represents the weight (relative importance) assigned to it.

threaten a proposed land use. Basically, the objective is to find the potential matches and mismatches between land use and environment and recommend the most appropriate relationship between the two. This may involve a wide range of considerations including off-site ones where the site is affected by systems and actions beyond its borders, such as stormwater runoff and air pollution from development upstream and upwind.

Site assessment   Documenting opportunities and constraints is often included in **site assessments**. These are prepurchase or preplanning environmental profiles of sites highlighting whatever conditions might be germane to land value, purchase agreements, and program planning. A central part of modern site assessments, particularly on the urban/suburban fringe, are inspections for hazardous wastes such as buried storage tanks and possible safety problems such as shallow mine shafts, contaminated groundwater, potentially contentious border conditions, and unsafe soils and slopes.

Site analysis

All of the above are commonly part of **site analysis**, an activity that investigates the makeup and operation of a site vis-à-vis a proposed use program. The program serves as an initial problem statement. The analysis is tailored to the particular requirements of the program, the character of the site itself, the site's relationship to surrounding lands, waters, and facilities, community concerns, policy issues, and other factors. Site analysis may also include several of the following activities.

*Land capability,
carrying capacity,
sustainability planning*

Land capability (or suitability) studies are designed to determine what types of use and how much use the land can accommodate without degradation. For sites or study areas composed of different land types, the objective is to differentiate buildable from nonbuildable land and to define the development capacity, or **carrying capacity**, of different land units or subareas (Fig. 1.7). Capability studies may also be performed to determine best use, such as open space, agricultural, or residential,

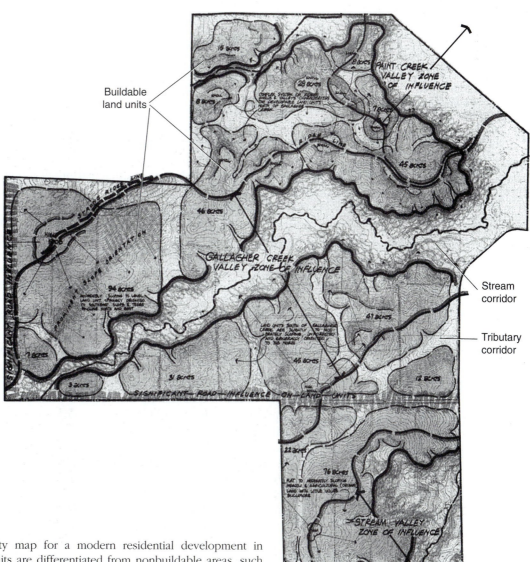

Fig. 1.7 Land capability map for a modern residential development in which buildable land units are differentiated from nonbuildable areas, such as stream corridors.

for different types of land over broad areas. Capability and capacity studies are fundamental to **sustainability planning**, in which the overriding objective is to build land use systems that are environmentally enduring and balanced for both humans and other organisms over the long term (see Case Study 1.10 "Planning for a Sustainable Landscape" at the end of this chapter).

Hazard assessment and risk management

    Hazard assessment is a specialized type of constraint study. The objective in hazard studies is to identify dangerous zones in the environment where land use is or would be in jeopardy of damage or destruction. Hazard research has been concerned with both the nature of threatening environmental phenomena, namely, floods, earthquakes, and storms, and the nature of human responses to these phenomena. Zoning and disaster relief planning for hurricane and flood-prone areas, for example, have benefited from hazard assessment at the national, state, and local levels. Another benefit is the emergence of **risk management planning** as a part of development programs, which involves building strategies and contingency plans for coping with hazards and providing emergency relief services. Both hazard assessment and risk management planning are gaining serious national attention in the United States in response to recent disasters such as the 1993 Mississippi flood, the 1994 Northridge earthquake near Los Angeles, the North Dakota flood of 1997, and the 2004 rash of hurricanes in Florida.

Forecasting impacts

    Hazard assessment, environmental impact assessment, opportunity/constraint studies and site assessments are all dependent on another activity: **forecasting impacts**. This activity involves identification of the changes called for or implied by a proposed action, followed by an evaluation of the type and magnitude of the environmental impact. The process is a tough one because of the difficulty in deriving accurate forecasts by analytical means. As a result, forecasts of impacts are usually "best estimates," and the significance assigned to them seems to be as much a matter of perspective (for example, engineer versus environmentalist) as anything else. Nevertheless, the *process* is an important one because it often leads to (1) clarification of complex issues and their environmental implications; (2) modification of a proposed action to lessen its impact; or (3) abandonment of a proposed project.

Special environments

    Analysis and evaluation of **special environments** such as wetlands, unique habitats, and archaeological sites is a rapidly rising area of planning activity. Though logically a part of impact assessment, capability studies, and most other planning activities, special environments have gained increased attention with the enforcement of wetland protection laws, rare and endangered species laws, and similar ordinances relating to prized resources in the environment. The focus of activity to date is overwhelmingly empirical, dealing mainly with field identification and mapping of the feature or organism in question. The results usually center on the question of presence or absence of, for example, a threatened species or a valued habitat, as the basis for deciding whether a proposed land use can or cannot take place in or near the area under consideration.

Restoration planning

    Other kinds of special environments are those in need of some kind of restorative action. **Restoration planning** addresses environments such as wetlands, wildflower habitats, stream channels, and shorelines that have been degraded by land use activities. Perhaps the most common restoration planning activity is associated with damaged wetlands, degraded streams, and waste disposal sites. Wetland mitigation includes both restoration of damaged wetlands and construction of essentially new wetlands.

Site selection and feasibility studies

    Site selection is a traditional planning problem that incorporates a host of planning activities. Typically, we would begin with a proposed idea for a land use program, or an actual program for a facility or enterprise in hand, and attempt to find an appropriate place to put it. Often it entails no more than an exercise in locational analysis based on economic factors, but properly done, it should also include land capability studies, hazard assessments, and various other types of environmental studies, as well as infrastructure, land use, and policy evaluations. **Feasibility studies**, on the other hand, begin with a known site and, with the aid of field studies and various forecasting techniques, attempt to determine the most appropriate use for it. Increasingly, planners and developers are interested in learning about a site's limitations based on environmentally protected areas and features, such as wetlands and habitats, as a part of feasibility studies.

Facility planning   **Facility planning** involves siting, planning, and designing installations that are dependent on structural and mechanical systems. Sewage treatment plants, industrial installations, airfields, and health complexes are typical examples. Not many years ago, little consideration was given to environmental matters in facility planning beyond water supply, sewer services, and in some cases, an environmental impact statement. Today, however, environmental analysis and site planning, both onsite and offsite, are given serious consideration as they influence environmental quality, public image and community relations, landscape management, and especially public and worker liability.

Master planning   **Master planning** may include all the planning activities mentioned above, for its overriding aim is to present a comprehensive framework to guide land use changes. Early in the master planning process, goals are formulated relating to land use, economics, environment, demographics, and transportation. Existing conditions are analyzed, and alternative plans are formulated. The alternatives are then tested against goals and existing conditions, and one is adopted. The master plan usually comprises three parts: (1) a program proposal consisting of recommendations, guidelines, and proposed land uses; (2) a physical plan, showing the recommended locations, configurations, and interrelationships of the proposed land uses; and (3) a scheme for implementing the master plan that identifies funding sources, enabling legislation, and guidelines for how the changes are to be phased over time.

Management planning   Finally, we must consider **management planning**. Although environmental planning is normally associated with the early phases of the planning process, it is becoming apparent that it must also be part of the design, construction, and operational phases of projects. In the construction phase, management plans must be formulated to minimize environmental damage from heavy equipment, material spills, soil erosion, and flooding. Similarly, once construction is over and the land use program is operational, it is often necessary to devise environmental management programs to achieve lasting stability among the landscape, built facilities, and environmental systems such as drainage, airflow, and ecosystems. As with master plans, management plans must be comprehensive to be most effective and they must meet the criteria of landscape sustainability through compatibility with larger land use-environmental systems.

1.7 METHODS AND TECHNIQUES

The methods used in environmental and landscape planning are basically no different from those in other areas of planning and design. The fact that environmental phenomena are more closely associated with the "hard," or natural, sciences does not mean that this area of planning is necessarily more rigorous than, for example, transportation or land use planning. The differences lie rather in the perspectives, particularly in what components of the plan are given greatest emphasis and in the analytic techniques used to generate data and to test the planning and design schemes.

Environmental Inventory   The questions and topics that are the focus of analysis in environmental planning originate in all phases of projects and problems and are of varying complexity and sophistication. In the early phases of a project the emphasis is generally on gathering and synthesizing data and information. Planners often refer to this listing as an **environmental inventory**, following the language of EIS methodology (see Fig. 1.6). The idea behind the inventory is to learn all we can about the character of a project site and its setting. Although this normally includes field inspection and field measurement, the generation of quantitative data usually is not the primary objective. Instead, the sources of most data are secondary (published) sources: topographic maps, soils maps, aerial imagery, climatic data, and streamflow records. When detailed field measurement is undertaken early in a project, it is usually in connection with a known or suspected engineering, safety, or health problem such as soil stability or buried waste, or in connection with policy problems such as wetlands or threatened species. For the most part, environmental analysis in the early phases of a project is typically not analytic in the scientific sense. That is, it is more concerned with defining distributions, densities, and relations among the various components of the environment than with rigorous testing of cause–effect relationships.

Analysis   Later in the project the process becomes more **analytic** as problems and questions arise relating to formulating and testing planning and design schemes. The techniques employed vary widely. For some problems, quantitative models are used, such as hydrologic models to forecast changes in streamflow and flood magnitudes in connection with land development of a watershed. For others, hardware models are called for, such as wind-tunnel analysis of building shapes and floodflow simulations in stream tanks (Fig. 1.8). For still others, a statistical analysis to test the relationships between two or more variables (such as runoff and water quality) is appropriate.

Generating results from the various data-gathering, descriptive, and analytic efforts does not mark the end of the environmental planner's or designer's responsi-
Integration   bility. Ahead lies the difficult task of **integrating** the various and sundry results in a meaningful way for decision making. No calculus has been invented that satisfactorily facilitates such a difficult integration—a dilemma faced in all planning problems. The integration almost always requires some sort of a screening and evaluation to determine the relative importance and meaning of the results. The actual integration is usually a qualitative rather than a quantitative process and typically centers on a
Displaying results visual (graphic) **display** of some sort. This may be a matrix, a flow diagram, a set of map overlays, or a gaming simulation board. Above all, it is important to understand that the final outcomes are found not in the results of specific procedures or techniques (as we might be led to believe from our experience in a college science laboratory class), but in a less exact and more eclectic process that invariably rests on a decision maker's or decision-making body's perspectives and values concerning the problem as well as on related political and financial agendas.

1.8 ENVIRONMENTAL ORDINANCES

Environmental regulations in the United States and Canada abound at all levels of government. As we noted earlier, the mother ordinance in the United States is the National Environmental Policy Act (NEPA) of 1969, which provides for a massive regulatory pro-
Policy development gram under the U.S. Environmental Protection Agency. In the 1970s, broadly similar environmental protection programs were developed by many states and provinces. In the 1980s and 1990s, as the need for environmental and land use regulation continued to rise, additional policies were legislated reflecting new and changing state and local values for groundwater protection, waste disposal management, soil erosion control, watershed management, hillslope development, and many other things.

Today there is a great multilayered umbrella of environmental rules, regulations, and guidelines covering land use activities and their interplay with the environment. Taken together, the mass of environmental ordinances borders on the incomprehensible. Like most bodies of law, the enforcement process has become highly selective and
The search for order those ordinances actually enforced tend to shift over time with changes in national, state and community interests, politics, and needs. Responding to environmental ordinances as a part of land use, engineering and related projects can be a bit of a shell game and often requires professional assistance in identifying, interpreting, and negotiating environmental policy. In most parts of Canada and the United States, local specialists, such as environmental attorneys and urban planners, have emerged to provide such services.

Fig. 1.8 A hardware model of a portion of the Mississippi River used by the U.S. Army Corps of Engineers to simulate the behavior of floodflows in a partially forested floodplain. The rows of cards produce an effect on flow similar to that of trees.

Interpreting policy   Meeting the requirements of environmental regulations is a very important part of land use and environmental planning. Countless projects have failed because of inattention to ordinances or because of judgmental errors in interpreting and negotiating regulations. Projects have also failed in some communities because of excessive and cumbersome regulatory demands placed on applicants seeking approval to make a land use change. The difficulties commonly stem from philosophical differences in what the ordinances are intended to do. For example, some communities and their officials expect strict adherence to specified measures, such as a slope requirement or setback from a stream, with little or no latitude in interpretation. **Variances**, that is, approved exceptions, are expected to be minor and few. Other communities apply the law in a more general way, intending that its spirit or intent must be satisfied and that reasonable interpretation may be applied.

Performance-based planning

Most environmental ordinances are restrictive, meaning they are aimed at what you should not do. Some, however, allow you to propose, within certain limits, your own way of meeting environmental requirements. One such approach is called **performance based planning**. Unlike restrictive ordinances that typically specify the particular measures that are to be used (such as detention ponds for stormwater management), the performance approach allows planners a certain amount of freedom to devise alternative means of coming up with the desired end. To justify an alternative approach, however, the applicant must provide reliable evidence that the measures will actually work. In the absence of such evidence, demonstration projects are often necessary before the full-scale implementation can take place. Demonstration projects are often expensive and may require several years to carry out. Thus, communities and planners are limited in using this approach and therefore usually overwhelmingly rely on conventional, restrictive type ordinances.

Problem-solving responsibility

In the end, it is important to remember that, no matter how detailed, applicable, and seriously enforced they may be, ordinances do not themselves solve planning and design problems. At best they are a safety net for society, providing some insurance that those areas, systems, features, and artifacts of value and concern are addressed as a part of land use change. The real process of landscape planning should take place at a much higher level and should be more elegant than is called for by ordinances. Among other things, it should seek **integrated solutions** which, for example, do not end with wetland and floodplain mapping, but address wetlands, floodplains, streams, groundwater, and stormwater in a comprehensive way focusing on water management, its relations to land use systems, and the creation of sustainable landscapes.

1.9 THE PLANNING PROFESSIONS AND PARTICIPATING FIELDS

Traditionally, only three fields—namely, urban planning, landscape architecture, and architecture—are recognized for training the professionals who guide the formal planning processes. These fields focus mainly on the decision-making and design aspects of planning. **Urban planning** has the broadest scope with concern for entire metropolitan areas. Most professional activity in urban planning revolves around policy development and regulatory practices in the public sector related to economic development, social programs, land use, and transportation planning. Most urban planners work for planning and related agencies in cities, townships, and counties, although the number working in the private sector for consulting firms, banks, realtors, and developers has increased in the past two decades.

Urban planning

Landscape architecture

Landscape architecture tends to be more site oriented than urban planning, with a stronger emphasis on design. Landscape architects work with both the natural and built elements of the landscape, seeking to blend the two into workable and pleasing environments. Professional activity covers the full range of settings from urban to wilderness landscapes and includes projects as small as residential site planning and those as large as national park planning. Landscape architects work in both the private and public sectors.

Architecture

Architecture has the narrowest focus in landscape planning, dealing mainly with buildings and their internal environments. Architecture is concerned with the landscape mainly as a setting for buildings and related facilities but also as a source of environmental threats (such as floods and earthquakes) to building stability. Because of today's broad environmental regulations and the high costs of energy, water, and other building resources, architecture is generally forced to pay more attention to environmental matters than has been the tradition in the field.

Each decade a growing number of scientific fields participate in planning in North America. The 1970s and 1980s saw increased participation from geography, geology, biology, chemistry, anthropology, and political science in the formal arenas of planning

Technical subfield

mainly in connection with environmental assessment and impact activities. The 1980s nurtured the development of **technical subfields** in response to the increased complexity of planning problems and the need for specialists in areas such as hazardous waste management, groundwater protection, and wetland evaluation and restoration.

Technical subfields have emerged in both traditional planning fields and the participating sciences. For many of the subfields sponsored by the scientific disciplines there are counterparts, more or less, in the traditional planning fields. Tied together by common research interests, these subfields form an important source of data and information for the decision-making and design processes. In landscape architecture, for example, there are ties with geography, botany, and ecology over issues such as watershed management, habitat planning for urban wildlife, and wetland restoration. Both architecture and geography are interested in the microclimates of building mass-

Geography

es and urban environments. **Geography** also shares an abiding interest with landscape architecture and planning in remote sensing and computer-aided mapping for land use planning, environmental assessment, wetland mitigation, stream restoration, and many other problem areas. In urban planning one of the subfields shared with political science is environmental policy, dealing with the formulation, interpretation, and applications of ordinances.

Civil engineering

Finally, we must recognize **civil engineering** for its role in environmental planning. Larger and more influential than urban planning, landscape architecture, or any of the subfields, civil engineering is involved in a major way in most private and public development projects including site assessment, environmental impact analysis, site design, risk management, stormwater management, and wetland mitigation. The field has had a major influence on environmental policy, particularly in the area of stormwater management, flood control, waste management, roadway design, and water supply planning, and has made major research contributions in the fields of hydrology, geomorphology, and pollution control.

1.10 CASE STUDY

■ ### Planning for a Sustainable Landscape

W. M. Marsh

There is a widespread belief, especially among political leaders, that growth is necessary in order for a town or city to prosper. In other words, to reach and maintain a prosperous state, a community must be continually expanding, drawing in more resources, and using more land. But there is good reason to doubt this belief. Studies reveal that the cost of growth in North American communities commonly exceeds the income from new tax revenues. Growth requires building new infrastructure, such as roads and sewers, and expanding services such as education and policing. The costs of added infrastructure and services are very high and often are not offset by the tax revenue generated by new development, especially residential development, and the community, despite outward signs of prosperity from new construction activity, often declines in quality and economic well-being.

There is another route to prosperity: building on the existing resource base. This can be achieved by upgrading existing infrastructure, improving the efficiency of services, and improving the efficiency, quality, and profitability of land use. A simple way of improving land use efficiency and profitability, for example, is by infilling vacant lands, which most communities have in abundance and are already serviced by established infrastructure. In other words, communities can move forward to a more sustainable state by creating prosperity from within through making better use of what they already have.

Building sustainable communities is fundamental to building sustainable landscapes because only when a community approaches internal sustainability is it in a position to seriously address sustainability in the larger landscape of which it is a part. How does a community go about this? What does it address in planning for a sustainable landscape? It can begin by identifying and measuring the systems upon which it depends, systems such as watersheds, ecosystems, and groundwater aquifers.

The most meaningful measure of landscape sustainability is the performance of these systems, not just one but several or more interrelated systems, both natural and manmade. For a community dependent on the forest industry, for example, this would mean finding an enduring balance in a forest ecosystem subject to lumbering activity. One approach, which is actively used by foresters, is to establish a balance between wood cut and new wood grown. By using this measure, if wood harvested is equal to wood grown, then the system would be in balance and considered sustainable.

But this is a crude and largely inaccurate measure of sustainability for it does not take into account the many other systems of the landscape that are part of the forest ecosystem and hence part of the community. These systems include runoff, groundwater, soil, animal habitat, climate, as well as land use. For instance, if we maintain a balance between wood harvested and wood grown but do it in such a way that it increases runoff rates and soil erosion, then our forest management practices are not sustainable. Increased runoff and soil erosion can reduce the capacity of the land to support forests, leading to reduced growth rates and lower quality forest. In the long run, the whole system winds down as yields fall and weedy tree species replace marketable ones, a trend which in turn can threaten the sustainability of the community. Similarly, forest management practices that reduce or eliminate other economic activities can drive a community and its landscape toward an unsustainable state. For example, cutting practices that degrade a landscape's scenic quality, damage streams, and drive away game can hurt economic activity dependent on tourism, fishing, and hunting.

Because of variations in the conditions and operations of communities and their landscapes, there is no universal formula for measuring and evaluating sustainability. Settlements and environment are different from place to place, and these differences demand that each community tailor its own version of a sustainability plan. No matter where we are, however, we have to address key systems and their performance as the starting point on the road to sustainability.

1.11 SELECTED REFERENCES FOR FURTHER READING

Catanese, Anthony J., and Snyder, James C. *Introduction to Urban Planning.* New York: McGraw–Hill, 1988.

Fodor, Eben. *Better Not Bigger.* Stony Creek, CT.: New Society Publishers, 1999.

Godschalk, David R. *Planning in America: Learning from Turbulence.* Chicago: APA Planners Press, 1974.

Hargrove, Eugene C. *Foundations of Environmental Ethics.* Englewood Cliffs, NJ: Prentice–Hall, 1989.

Holling, C. S. (ed.) *Adaptive Environmental Assessment and Management.* Hoboken, NJ: Wiley, 1978.

Leopold, Aldo. *A Sand County Almanac.* Oxford: Oxford University Press, 1949. (Reissued by Ballantine Books, 1970.)

Lynch, Kevin, and Hack, Gary. *Site Planning.* Boston: MIT Press, 1984.

Marsh, William M., and Grossa, John M. *Environmental Geography: Science, Land Use, and Earth Systems.* (3rd ed) Hoboken, NJ: Wiley, 2005.

McHarg, Ian L. *Design with Nature.* Hoboken, NJ: Wiley, 1995.

Newton, N. T. *Design on the Land: The Development of Landscape Architecture.* Cambridge, MA: Belknap Press, 1971.

Ortolano, Leonard. *Environmental Regulation and Impact Assessment.* Hoboken, NJ: Wiley, 1997.

Platt, Rutherford H. *Land Use and Society: Geography, Law, and Public Policy.* Washington, DC: Island Press, 1996.

West Publishing. *Selected Environmental Law Statutes.* St. Paul, MN: West Pub., 1995.

2

THE PHYSIOGRAPHIC FRAMEWORK OF THE UNITED STATES AND CANADA

2.1 INTRODUCTION

There is startling geographic variety to the North American landscape. We celebrate this as part of our national heritage in both Canada and the United States. For two centuries geographic diversity and dreams of economic opportunity have driven westward movement and settlement. Geographic diversity is still at the root of many of our life decisions. We carry out our lives in many different locations; the notion of interesting, picturesque, and even intriguing geographic settings is never far from our thoughts when it comes to making a move.

Geographic diversity

Yet, despite our preoccupation with the land, development in the past 50 years has largely ignored much of the inherent geographic diversity in the North American landscape. We build land use systems with little regard for place to place differences in the scale and types of landforms, drainage systems, and ecosystems. Since 1960 or so our highways, shopping centers, and residential developments are basically the same across this vast land as though one set of cookie cutter models has been used everywhere. The huge bank of knowledge we have acquired from tens of thousands of studies on the North American landscape simply has not become part of the professional culture of planning and engineering. Many community and environmental groups sense this when examining development proposals. They voice an uneasiness about what they see despite the fact that planners have usually met all the local and regional environmental rules and planning regulations. What they sense is a lack of fit with the local landscape, an unexplainable awkwardness to modern land use plans.

Fitting into the local environment

Why this state of affairs? Part is due to the "precedent" tradition in law and public policy that encourages communities to build environmental and land use ordinances according to what has worked for other communities. In other words, most ordinances and by-laws, such as those covering stormwater management, road design, and land platting, are borrowed from other jurisdictions and not tailored to fit the local environment. Part is also due to the belief among public officials that planning and design problems are solved by rules and regulations; therefore, developers are expected to go no further in their understanding landscapes than the regulations require. And, of course, part is also due to genuine lack of knowledge about local and regional environment, its essential processes and systems.

In this chapter we briefly survey the physiography of the United States and Canada. Our purpose is to gain some understanding of the different types of terrain, resources, and environmental conditions that make up the landscapes beyond our own theaters of operation. Although we are barely able to scratch the surface of this huge body of knowledge, we want at least to understand the macro-framework and maybe learn enough to ask tough questions of ourselves and others. It is important in this context to realize how mobile we are as modern professionals and that much of our project work in landscape planning and design is carried out in other places, often far beyond the home, office, or university.

Physiographic regions

Physiographic regions are defined by the composite patterns of the main elements of landscape: landforms, drainage features, soils, climate, vegetation, and land use. However, it is also important that we remember that the physiography of any region represents the product of a host of processes that operate at or near the earth's surface. These processes are arranged in various systems characterized by flows of matter driven by energy: drainage systems, climatic systems, mountain-building (geologic) systems, geomorphic systems, ecosystems, and land use systems that overlap and interact in different ways and at different rates. These systems have developed over various periods of time in North America ranging from hundreds of millions of years (in the case of the geologic systems that built the Appalachians and the Rockies) to only several centuries (in the case of the land use systems that built our settlements, farms, and highways). The physiographic patterns and features we see

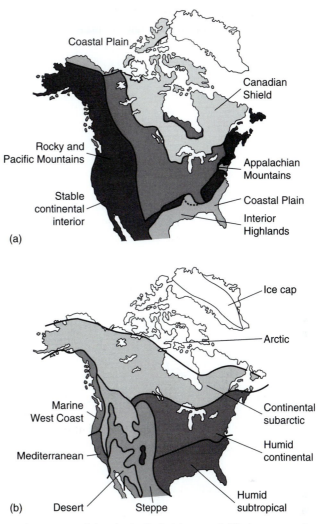

Fig. 2.1 (*a*) The major structural (geological) divisions and (*b*) the major climatic zones of the United States and Canada.

today represent an evolving picture—at this moment a mere slice of the terrestrial environment at the intersection of many time lines representing different forces and systems.

Generally, regional geology provides a useful framework for describing the gross physiography of North America. These regions are well known: the Canadian Shield, the Appalachian Mountains, the Interior Highlands, the Coastal Plain, the Interior Plains, the Rocky Mountains, and so on (Fig. 2.1*a*). The geologic structure of each region sets the drainage trends and patterns and in turn the general character of landforms. Added to this is the role of climate as it influences vegetation, soils, runoff, permafrost, and water resources (Fig. 2.1*b*). When we speak of a physiographic region, we are referring to a vast geographic entity defined by a particular combination of landforms, soils, water features, vegetation, and related resources. Ten major physiographic regions are traditionally defined for the United States and Canada (Fig. 2.2*a*), and they are broken down into smaller regions called **physiographic provinces** (Fig. 2.2*b* and Table 2.1). At the level of provinces, physiography begins to take on real meaning for many issues and problems in landscape planning. We begin with the Canadian Shield.

Fig. 2.2 (*a*) The physiographic regions of the United States and Canada and (*b*) their provinces classified as mountains, plateaus, or lowlands.

(a) Physiographic Regions

(b) Physiographic Provinces

Mountains
Plateaus
Plains

Table 2.1 Physiographic Regions and Provinces of the United States and Canada

Region	Provinces
Canadian Shield	Superior Uplands
	Laurentian Highlands
	Laborador Highlands
	Hudson Platform
Appalachian Mountains	Blue Ridge
	Piedmont
	Ridge and Valley
	Appalachian Plateaus
	Northern Appalachians
Interior Highlands	Ozark Plateaus
	Ouachita Mountains
Atlantic Coastal Plain	Outer Coastal Plain
	Inner Coastal Plain
	Mississippi Embayment
Interior Plains	Central Lowlands
	Great Plains
	St. Lawrence Lowlands
Rocky Mountain Region	Canadian Rockies
	Northern Rockies
	Central Rockies
	Southern Rockies
Intermontane Region	Colorado Plateau
	Columbia Plateau
	Basin and Range
Pacific Mountain System	Alaska Range
	Coast Mountains
	Frazier Plateau
	Cascade Mountains
	Coast Ranges
	Sierra Nevada
	Central Valley
	Puget Sound–Williamette Lowlands
Alaska–Yukon Region	Brooks Range
	Yukon Basin
Arctic Coastal Plain	North Slope
	MacKenzie Delta
	Arctic Lowlands

2.2 THE CANADIAN SHIELD

Geology

The **Canadian Shield** is a large physiographic region in the northcentral part of the continent (Fig. 2.1). It is composed of the oldest rocks in North America (older than a billion years) and is the *geologic core* of the continent. Geologically, the Canadian Shield is extraordinarily complex, with intersecting belts of highly deformed crystalline (igneous and metamorphic) rocks throughout. These rocks have been subjected to not one or two, but many, ancient episodes of tectonic deformation and metamorphism. Thus most of the rocks are hard, tightly consolidated, and diverse in mineral composition, including iron ore, nickel, silver, and gold. These minerals are the object of major mining operations, which is one of the principal economic activities of the Shield.

Stability

Most of the Canadian Shield has been *geologically stable* (relatively free of earthquakes and volcanic activity) for the past 500 million years or so. During that time, erosional forces have worn the rocks down to rough plateau surfaces, such as the Superior Uplands, the Algonquin Uplands, and the Laborador Highlands, which have the ruggedness of low mountains. In addition, large sections of the Shield lie at lower elevations and are covered with sedimentary rocks. The largest of these areas—though it does not belong to the Canadian Shield region per se—is the broad interior of the continent stretching from the U.S. Midwest through the Canadian Interior Plains northward to the Arctic Lowlands of northern Canada. Over this vast area the Shield rocks are buried under a deep cover of sedimentary rocks, thousands of feet thick in most places, and hence have essentially no influence on the surface environment.

Glaciation

A significant recent chapter in the long and complex physiographic development of the Canadian Shield was the *glaciation* of North America. Great masses of glacial ice formed in the central and eastern parts of the Shield, and in at least four different episodes in the past 1 to 2 million years, spread over all or most of the region. From the interior, the ice sheets advanced in a radial pattern, scouring the Shield's surface, removing soil cover, and rasping basins into the less resistant rocks. The last ice sheet, the Wisconsinan Glaciation, which covered the whole region 18,000 years ago, melted from the Shield's interior only 6000 to 8000 years ago when the continent was inhabited by the first North Americans.

Lakes

On the fringe of the Shield, the ice scoured large freshwater basins, called *shield-margin lakes*. On the south are the Great Lakes, and along the western margin are lakes Manitoba, Athabasca, Great Slave, and Great Bear. In the center of the Shield, marked by the Hudson Bay region, the great mass of ice depressed the earth's crust below sea level. As the ice melted back, Hudson Bay took shape. But for the past several thousand years the Bay has been shrinking as the crust rebounds from glacial unloading and pushes the water back. This process accounts for the vast, flat coastal plain that is slowly growing on the Bay's southern margin.

Landscape diversity

Over much of the Shield glaciation left the land with an irregular and generally light soil cover interspersed with low areas occupied by lakes and wetlands. When we combine this characteristic with the already diverse surface geology, it is easy to see why the Canadian Shield is one of the most *complex landscapes* in North America. From the standpoint of environmental planning and engineering, the Shield presents few easily definable patterns and trends in landforms, drainage, and soils. Considerable variation can be expected even at a local scale of observation; therefore, careful field work is called for in virtually all planning and engineering problems.

Climate and permafrost

The Canadian Shield lies principally in the *continental subarctic climate* zone. It is marked by six to seven months of freezing temperatures, fiercely cold mid winters, and short cool summers. A great proportion of the Shield north of James Bay (at the southern end of Hudson Bay) is occupied *permafrost*. Most permafrost is *discontinuous*, meaning that its distribution is irregular, usually occurring in the form of patches of different sizes and depths. The largest patches are generally found in the northern zones of the Shield and/or areas favored by cool microclimates and topography and soils conducive to cold ground (see Chapter 18 for details on permafrost).

Boreal forest

The vegetative cover is predominantly *boreal forest* except in the extreme north where it is tundra. Boreal forest is the least biologically diversified of the earth's major forest ecosystems, containing only a small fraction of the number of species in the subtropical forests 2000 miles to the south on the Atlantic Coastal Plain. In addition it is the smallest stature and least productive of the world's major forest systems. *Annual organic productivity* of the boreal forest averages 800 grams per square meter of ground and much of that is provided by the ground cover such as mosses, sedges, and various shrubs (Fig. 2.3).

Vegetation and soils

The boreal forest is dominated by white spruce, birch, aspen, and poplar in upland areas and white spruce, black spruce, and tamarack (larch) in lowlands. Trees are very slow growing and often appear in dwarf forms in wet areas and on the tundra fringe. *Bogs* are found throughout lowland areas and on lake and stream margins

Fig. 2.3 Boreal forest of the Canadian Shield, the least disturbed of the world's great forests.

in upland areas. They are characterized by zonal vegetation patterns grading from small trees on the edges to sedges and mosses toward the center. Bogs and other wetlands are dominated by organic soils (*Histosols*), whereas uplands are dominated by *Podsols*, *Brunisols*, and *Cryosols* (Canadian classification) or *Spodosols* (American classification) (see Appendix A).

Drainage   Throughout much of the Shield are extensive outcrops of resistant rock, which exert a strong influence on drainage. These outcrops vary in shape and size and in places impart to the terrain an almost chaotic fabric. Not surprisingly drainage patterns are very irregular, so much so that the term *deranged drainage* is used to describe them (Fig. 2.4). Stream systems are interrupted by topographic depressions and bedrock barriers that slow flow in some areas and hasten it in others. Lakes, streams, and wetlands are abundant, and freshwater is clearly one of the Shield's

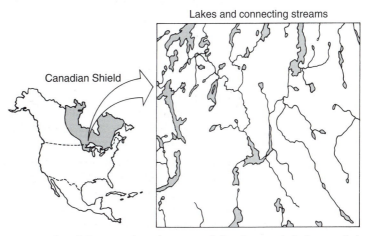

Fig. 2.4 An example of the complex character of the deranged drainage of the Canadian Shield.

greatest resources, a fact that has not gone unnoticed by the Canadian government which has promoted the development of hydroelectric dams over much of the southern part of the region.

Land use   Settlement on the Canadian Shield is light, and land use is sparse throughout most of the region, including the two sections that lie within the United States, the Adirondack Mountains and the Superior Uplands. Where settlements are found, they are usually related to an extractive economic activity, typically mining, forestry, or fishing. It is, however, an alluring recreational landscape, and it has attracted the development of parks, resorts, and summer homes along the Shield's southern margin in Ontario and Quebec. Herein lies a steadily growing challenge for planners and designers. Among other things, it requires understanding of the Shield's unique physiography, the need for careful site analysis, and the use of terrain adaptive approaches in designing sanitary, water supply, road, and energy systems.

2.3 THE APPALACHIAN REGION

Landforms   The **Appalachian Region** is an old mountain terrain that formed on the southeastern side of the Canadian Shield several hundred million years ago (Fig. 2.2). Although the Appalachians date from around the same time as the Rocky Mountains, they developed appreciably different terrains. The Appalachians are lower, less angular, and less active geologically. Whether these differences have always existed or are due to a long period of inactivity in the Appalachians during which erosional forces have worn them down, we cannot say. In any case, the Appalachians are characterized by *rounded landforms* that are generally forest-covered in their natural state and rarely exceed 6000 feet in elevation. Unlike the Canadian Shield, however, distinct topographic trends and patterns are discernible in the Appalachian Region, making it somewhat easier to deduce general physiographic conditions at the regional planning scale.

Blue Ridge   The Appalachian Region stretches from northern Alabama in the American South to Newfoundland in the Canadian Maritime provinces. It is subdivided into five provinces on the basis of landforms (Table 2.1). The highest and smallest province (in area) is the **Blue Ridge**, which stretches in a narrow band from New York to Georgia. It is composed of folded metamorphic rocks, that is, rocks hardened from the heat and pressure of mountain-building, and strongly linear mountain forms. As the backbone of the Appalachians, the Blue Ridge forms the *principal drainage divide* of the Appalachian region, marking the eastern rim of the Mississippi Basin.

Climate and vegetation   Because of its high elevation (4000 to 6000 feet), climate over much of the Blue Ridge province is distinctly wetter and cooler than lands to the east and west. This is especially pronounced in its highest reaches, the Great Smokey Mountains of North Carolina, Tennessee, and Georgia, where there is a marked *orographic effect* on the western slopes that yields 80 inches or more annual precipitation. In addition, at higher elevations the cooler climate favors a more northern forest association made up of hardwoods (birch, maple, beech, elm, oak) mixed with conifers (hemlock, white pine, spruce, and fir). And in some places mountain summits are barren (called *balds*), not because the climate is too cold for trees, but mostly because they are covered with rock rubble and soil is sparse. The province can claim only one city, Asheville, North Carolina, at an elevation around 2000 feet (600 meters). State and federal parks are common to the Blue Ridge, including the popular Great Smoky Mountain National Park in North Carolina and Tennessee.

Piedmont   East of the Blue Ridge is the **Piedmont Province** of the Appalachians (Fig. 2.5). This province is composed mainly of metamorphic rocks covered by a soil mantle of variable thickness. Although large parts of the Piedmont are fairly level, the surface is best characterized as a hilly plateau surface, especially in the south. From the Blue Ridge, the Piedmont slopes gradually eastward until it disappears under the sedimen-

Great Smokey Mountains of the Blue Ridge looking east from Grandfather Mountain in North Carolina.

Fig. 2.5 The Appalachian Region of eastern United States including the Blue Ridge, the Piedmont, the Ridge and Valley and the Appalachian Plateau. In the southeast lies the Atlantic Coastal Plain. The lower diagram shows the relationship among the Blue Ridge, Ridge and Valley, and Appalachian Plateaus provinces.

Drainage   tary rocks of the Coastal Plain, along its eastern and southeastern border. *Drainage* follows this incline toward the Atlantic. The valleys of large streams such as the Savannah and the Peedee are marked by major forest corridors that run all the way to the sea. Stream and *river gradients* are relatively steep on the Piedmont, but where they cross onto the Coastal Plain about midway between the Blue Ridge and the coast, they decline somewhat and flow becomes less irregular. For early settlers and traders moving up rivers from the Atlantic, the first fast water (rapids) they would encounter started with the eastern edge of the Piedmont. The eastern border of the Piedmont became known as the *Fall Line*, and on some rivers it became a place of settlement. Richmond, Virginia, Raleigh, North Carolina, and Macon, Georgia, are Fall Line cities (Fig. 2.5).

The Piedmont lies principally in the *humid subtropical climate* of the American South. Precipitation averages from 50 to 60 inches annually and intensive summer thun-

Climate and soils   derstorms are common. Runoff rates are high, especially where land has been cleared, and *soil erosion*, which was once very widespread, can be locally severe today on farmland and construction sites. Soils belong to the *Ultisol* order and tend to be heavily leached and generally poor in nutrients. Early in the development of Southern agriculture, the Piedmont was a favorite area for cotton and tobacco farming, but these activities waned as the soils declined and cotton farming shifted to the Mississippi Valley and later into Texas. Much of the former farmland now supports mixed oak-pine forests.

Ridge and Valley Province   The **Ridge and Valley Province** is one of the most distinctive terrains in North America. It lies west of the Blue Ridge, stretching from middle Pennsylvania to middle Alabama in a belt generally 50 to 75 miles wide. The Ridge and Valley Province topography is controlled by folded and faulted sedimentary rocks that have been eroded into long ridges separated by equally long valleys (Fig. 2.5). Ridges and valleys run parallel to each other and are more or less continuous for several hundred miles, broken only occasionally by stream valleys, called *water gaps*, or dry notches, called *wind gaps*.

Drainage   Drainage lines generally follow the trend of the landforms, with trunk streams flowing along the valley floors, forming *parallel* or *subparallel patterns*, and their tributaries draining the adjacent ridge slopes. But there are notable exceptions, for some of the large rivers in the north, specifically the Delaware, the Susquehanna, and the Potomac, drain across the grain of the ridges and flow into the Atlantic. In the southern part of the Ridge and Valley most streams drain into the Cumberland and Tennessee rivers, which are part of the Mississippi System. Settlements and farms in

Land use   the Ridge and Valley are concentrated in the valleys where soil covers are heavy and water is abundant. The upper ridges have traditionally been left mostly in forest, being too steep for much else. However, modern residential and recreation development, which is attracted to the forests, rugged terrain, and excellent vistas, has pushed its way onto some Appalachian Ridges with varying degrees of environmental disturbance related to road building, forest clearing, and slope excavation.

Appalachian Plateaus   West of the Ridge and Valley Province is a section of elevated sedimentary rocks into which rivers have cut deep valleys. This province is referred to as the **Appalachian Plateaus** (*Allegheny Plateau* in the north and *Cumberland Plateau* in the south) and it extends from western New York to northern Alabama. The plateau surface dips westward from a high elevation of 3000 feet or more along its eastern border. This border is marked by a prominent escarpment called the *Allegheny Front*, which rises 500 to 1000 feet above the adjacent valley and extends the full length of the plateaus, about 700 miles.

Landforms   The sedimentary rock formations of the Appalachian Plateaus are flat-lying for the most part. Stream valleys are incised several hundred feet into these rocks, leaving flat-topped uplands between valleys. Where the density of stream valleys is high, however, the uplands are narrow, more ridgelike, and the plateaus' topography is steeply hilly or almost mountainous, especially in the higher sections. Sandwiched among the limestone, shale, and sandstone strata within the upland areas are extensive coal deposits, generally considered to be the largest single field of bituminous reserves in the world.

Coal mining   Where the coal deposits lie close to the surface, the favored extraction method is strip mining. Where the coal is too deep for strip mining, shaft mining is used in which a horizontal tunnel is dug into the exposed end of the formation along a valley wall. Strip mining has produced some of the more serious environmental problems in the United States. Not only are topsoil, vegetation, and habitat destroyed, but also the spoils (coal debris and subsoil) are subject to erosion and weathering, resulting in sedimentation and chemical pollution of streams as well as unsightly and unproductive landscapes (Fig. 2.6).

North Appalachians   The **Northern Appalachians**, north of New York, are made up of several separate mountain ranges composed principally of crystalline rocks. The most prominent of these ranges are the *Green Mountains* and *White Mountains* of Vermont, New Hampshire, and Maine, the *Notre Dame Mountains* of the Gaspe Peninsula of Quebec, and the *Long Range Mountains* of northern Newfoundland (Fig. 2.7). The Northern

Fig. 2.6 The effects of strip mining in the Appalachian Plateaus. Today mining companies are required to restore damaged landscape, but extensive areas of badly damaged land from former operations remain in the province.

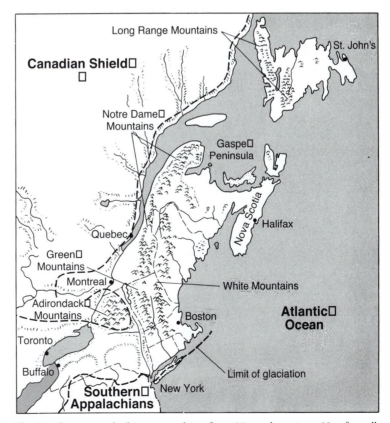

Fig. 2.7 The Northern Appalachians stretching from Massachusetts to Newfoundland.

Appalachians are *geologically complex* and made up principally of metamorphic rocks similar to those of the Blue Ridge and Piedmont provinces. The Adirondack Mountains of northern New York, despite their geographic location within the Northern Appalachian Region, are a southern extension of the Canadian Shield.

Landscape diversity   But unlike these provinces, the Northern Appalachians were glaciated, leaving it with abundant rock exposures and an irregular cover of glacial deposits. These characteristics, combined with extensive northern conifer forests and abundant lakes and wetlands, give much of the Northern Appalachian landscape a character similar to the uplands of the Canadian Shield. The lakes have been the focus of the acid rain phenomenon attributed mainly to the sulfur dioxide emissions from power plants and industrial sources fed by coal from the Appalachian Plateaus. In addition, this section of the Appalachians borders on the Atlantic Ocean, producing an especially rugged coastline of rocky headlands and deep embayments.

The comparison to the Canadian Shield extends to more than the general character of the landscape. In addition, the climate is noted for its long, cold winters, especially in the interior from Maine to Newfoundland (where there are fewer than 120 frost-free days a year), and settlement is sparse with a strong orientation toward lumbering, fishing, and tourism. Moreover, the fauna of the province, highlighted by moose, beavers, Canada geese, lynx, and black bears, are more like those of the Shield than the Appalachians to the south.

2.4 THE INTERIOR HIGHLANDS

West of the Cumberland Plateau, in southern Missouri and northern Arkansas, lies a small region of low mountainous or plateau terrain that closely resembles the Appalachians. This region is called the **Interior Highlands**, and it is made up of two main provinces: the *Ozark Plateaus*, whose landforms are similar to the Appalachian plateau and the *Ouachita Mountains*, whose landforms are very similar to those of the Ridge and Valley section except that the grain of the terrain is east–west trending (Fig.

Ozarks   2.2). The highest elevations in the **Ozarks**, as this area of plateaus is commonly called, lie between 2000 and 3000 feet; those in the Ouachita Mountains lie between 3000 and 4000 feet. Most ridges in the Ouachita Mountains are forested and too steep for settlement; in the Ozarks, however, ridges are often flat-topped and cleared for farming. Ozark stream valleys, on the other hand, are relatively deep and steep sided and, like those in the Appalachian Plateaus, are usually forested. Unlike the Appalachian Plateaus, the Ozarks are not underlain by extensive coal deposits, however.

Drainage in the Ozarks follows a radial pattern with streams on the north draining to the Missouri, those on the west draining to the Arkansas River, and those on *Drainage*   the south draining to the Black River. Limestone is the predominant bedrock throughout the Ozarks and it has given way in many areas to *karst* (dissolved rock) topography dominated by cavern systems. These caverns are tied to stream valleys, many of which have become attractive places for large reservoirs. This opportunity has not escaped the federal government, and today both the Ozarks and the Appalachian Plateaus are laced with reservoirs where forested stream valleys once existed.

2.5 THE ATLANTIC AND GULF COASTAL PLAIN

The **Atlantic and Gulf Coastal Plain** forms a broad belt along the U.S. Eastern Seaboard and the Gulf of Mexico from Cape Cod on the north to the mouth of the *General landforms*   Rio Grande in the south (Fig. 2.2). This region is composed entirely of sedimentary rock formations that dip gently seaward, disappearing at depth under the shallow

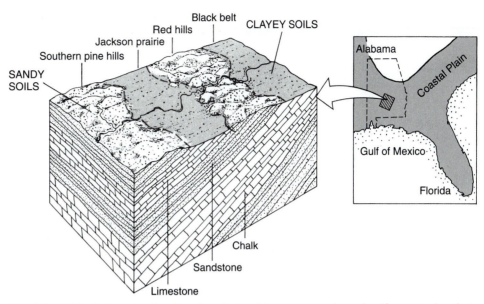

Fig. 2.8 A block diagram showing the relationship among geology, landforms, and soils in the Coastal Plain resulting in belts of lowlands and intervening hills.

waters of the continental shelf. Where resistant sedimentary rock formations such as sandstone outcrops occur within the Coastal Plain, they form belts of hilly topography with soils that are compositionally similar to the bedrock (Fig. 2.8). Conversely, weak formations, such as chalky limestones, form lowlands and deeper, richer soils. Overall, however, the *topographic relief* of the Coastal Plain is very modest, and the highest elevations [mainly along the inner (landward) edge] reach only 300 feet or so above sea level. The Coastal Plain is generally divided into three provinces: the *Inner Coastal Plain*, the *Outer Coastal Plain*, and the *Mississippi Embayment* (Fig. 2.9).

Outer Coastal Plain   Bordering the sea, in the province called the **Outer Coastal Plain**, the land is generally low, wet and heavy with stream, wetland, and shoreline sediments. Some of the largest *wetlands* in North America, such as the Everglades in Florida, the Great Dismal Swamp of Virginia and North Carolina, and the delta swamps of Louisiana, are found here. Lagoons, estuaries, and islands are abundant in addition to extensive swamps. These are subject to frequent incursion by floodwaters from streams, storm waves, and hurricane surges. Offshore islands, called *barrier islands*, which are mostly accumulations of sand from shallow water deposition by waves, are especially prone to damage from hurricanes and shore erosion. However, this has not deterred development, for each year residential and recreational land uses push further onto the barrier islands throughout the Atlantic and Gulf coasts.

Northern sections   The Outer Coastal Plain can be subdivided into at least five sections (Fig. 2.9). In the *northern section*, from Cape Hatteras north to Cape Cod, extensive lowland areas are drowned, the result of rising sea level since the last continental glaciation. The principal features of this section are the large, ecologically rich *estuaries*, such as Chesapeake Bay, which formed as the ocean flooded the lower reaches of river valleys. The next section south, the *cape-island section*, stretches from Cape Hatteras to northern Florida and is characterized by long sandy beaches (e.g., Cape Fear) and offshore islands such as those described earlier. The most impressive of these islands are the Sea Islands, which apparently formed as the land in this area slowly subsided while sea level rose over the past several thousand years.

Florida Peninsula   The *Florida Peninsula section* owes its origin to a great arch in the sedimentary rocks of the Coastal Plain. The arch is composed of limestone and its axis, which reaches about

Fig. 2.9 The principal features and sections of the Atlantic–Gulf Coastal Plain.

300 feet above sea level, and extends roughly down the center of the peninsula. The combination of Florida's wet climate and limestone bedrock has resulted in extensive karst topography. Much of the drainage is subterranean (groundwater is the state's main water supply) via *cavern systems* that discharge in springs and sinkholes. Sinkholes are common throughout the peninsula, but they are most prominent in the center of Florida where they form hundreds of large inland lakes. The outflow from Lake Okeechobee, Florida's largest lake (750 square miles in area), feeds the *Everglades* at the southern end of the peninsula. Diversion of this flow by canals coupled with water pollution from agriculture in this century has degraded this nationally valued wetland ecosystem.

Gulf Coast section Westward from the Florida Peninsula is the *Gulf Coast section*. Fed with massive amounts of river sediments, principally from the Mississippi, the Gulf Coast is conducive to barrier island formation. These low, narrow *longshore islands* parallel most of the coast and behind them are equally long lagoons with connecting wetlands, bays, and related habitats. These areas abound with waterfowl and other birds, including those that migrate between the Gulf Coast (and regions farther south) and the Arctic coastal plains in Canada.

Mississippi Delta The *Mississippi River Delta* lies roughly in the middle of the Gulf Coast section. One of the most diverse environments on the continent, the *outer delta* is actually a composite of several smaller deltas built mainly in the past 2000 years. Each finger of the delta contains many distributary channels (or bayous) bordered by natural levees behind which lie swamps, lagoons, and various types of lakes. The deep swamps support trees such as cypress and tupelo whereas the shallow ones support oaks, hickory, other hardwoods, as well as various reeds, sedges, and palmetto. Not surprisingly, *habitat diversity* is extraordinary within this waterland of fresh, brackish, and saltwater in wetlands, lakes, and streams habitats. Because of changes in the Mississippi's water quality and reductions in its sediment load, however, the delta is now showing serious signs of habitat degradation.

Mississippi Embayment   The Mississippi River brings more than a million tons of sediment to its delta every day. This process has been going on for more than 50 million years and has created a long, narrow alluvial lowland called the **Mississippi Embayment**, which begins at the southern tip of Illinois and extends 500 miles south to the modern delta (Fig. 2.10). Virtually all streams entering the Embayment as tributaries of the

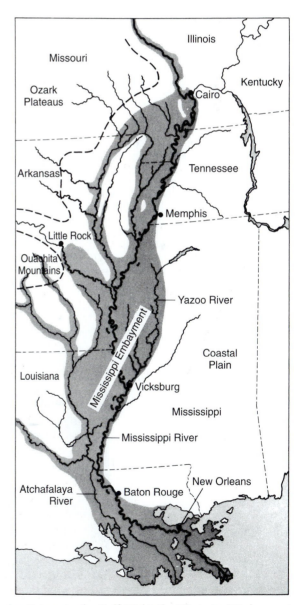

Fig. 2.10 Left, the Mississippi coursing its way to the Gulf. Right, the Mississippi Embayment and connecting lowlands. Notice the paralleling tributaries such as the Yazoo River.

Mississippi are diverted southward by the great river so that they parallel the Mississippi in their lower courses. Several, including the Black, Yazoo, and Ouachita rivers, follow the outer edge of the Embayment for more than 100 miles before joining the Mississippi. The result is a valley unique among North American rivers in that it is made up of a system of several large river channels occupying a single valley. During large floods, the Mississippi often turns floodwater back into the tributaries, which are actually slightly lower than the Mississippi itself where they parallel the main channel.

Inner Coastal Plain

On the **Inner Coastal Plain**, the land is generally higher and better drained than along the coast and in the Mississippi Embayment. There are two opposing trends to the topography here: *first*, systems of hills and valleys paralleling the coast that have formed along the outcropping edges of different rock formations. The lowlands are often marked by belts of deep, dark soils such as the Black Belt of Alabama and the uplands by sandy soils such as the Southern Pine Hills (Fig. 2.8). *Second*, intersecting these bedrock controlled land forms are ribbons of lowlands formed by the valleys of streams draining to the coast. Large lowlands are found along all the major river valleys, and most contain large areas of wetland and forests where they have not been drained and cleared for agriculture. West of the Mississippi River the inner border of the Coastal Plain is clearly marked by the southern fringe of the Interior Highlands and more abruptly by the Balcones Escarpment in Texas.

Flood problems

Flooding is frequent, widespread, and severe in all large river lowlands of the Coastal Plain. The damage it wreaks on property and life is enormous and has increased steadily throughout the twentieth century. The causes of this problem are manifold and include: (1) *high runoff* rates under the rainy climate of the South; (2) frequent tropical *storms and hurricanes* in coastal areas, which force heavy runoff and streams to back up in the Outer Coastal Plain; (3) extensive and growing *residential development* in high-risk zones along river lowlands and coastal areas; and (4) *engineering (structural) changes* such as the construction of levees in river valleys, which can constrict flows and increase the levels of floodwaters in certain locations (Fig. 2.11).

Climate and soils

The Coastal Plain lies almost entirely within the *humid subtropical climatic zone*. *Precipitation* generally ranges from 40 to 60 inches a year, with most occurring in the summer from thunderstorms. All the Coastal Plain is subject to hurricanes, and each

Fig. 2.11   A constructed levee designed to help control flooding along the Mississippi River in southern Louisiana. The floodplain on the left is frequently below river level.

section of the Outer Coastal Plain can expect at least one highly destructive event per decade. *Soils* tend to vary from sandy to clayey to organic depending on the underlying bedrock, nearness to the coasts, and river valley location. Most soils tend to be heavily leached, especially in nonwetland locations, and are reddish in color from *Vegetation* iron oxide residues below the topsoil. *Vegetation* patterns tend to follow the patterns of topography and drainage, with pines and oaks generally on the higher ground and tupelo, gum, and bald cypress in the wet lowlands. Only in Texas is the vegetation significantly different with forest grading into parkland and then into desert shrubs from east to west.

In dealing with planning problems in the Coastal Plain, it is always necessary to differentiate between upland and lowland terrain. The differences are often subtle, *Landscape assessment* especially on the edges, but vegetation and soil patterns can be helpful as indicators. Next, it is helpful to delineate units of upland ground and define their relationship to drainage systems, including channels, wetlands, floodplains, and habitats. This should include an assessment of the relative stability of different types of land related to soils, karst formations, and flooding from various sources. Finally, it is important to define the essential processes that operate in various environments, particularly in the river lowlands and coastal areas, and to understand the risks posed by each—mainly floods, storm surges, and hurricanes.

2.6 THE INTERIOR PLAINS

The heart of the North American landscape is a broad region of rolling terrain called the **Interior Plains** (Fig. 2.2). This region is made up of two large provinces, the *Boundaries* *Central Lowlands* and the *Great Plains*, and one small province, the *St. Lawrence Lowlands*. Roughly defined, the region has the shape of a great triangle with its base along the Rocky Mountains and its apex to the east pointing up the St. Lawrence Lowland. The western and northeastern borders are sharply defined by the Rocky Mountain Front and Canadian Shield respectively. The southern border varies from a neat edge, such as along the Balcones Escarpment in Texas, to a more transitional boundary along the Appalachian plateaus.

All three provinces are underlain with *sedimentary rocks* composed principally *Geology* of limestone, sandstone, and shale. These formations are flat lying or subtly bent into broad arches and basins. Michigan, for example, is at the center of a basin, whereas central Tennessee and Kentucky are situated across an arch. Bedrock exposures appear only in escarpments along the edges of basins and arches, in the walls of large stream valleys, and in areas of karst topography. But these exposures are small compared to area covered by surface deposits.

The surface deposits of the region are diverse in origin, composition, and thickness. In the Central Lowlands, north of the valleys of the Ohio and Missouri rivers, *Surface deposits* the deposits are mainly *glacial* and vary from sandy to clayey to mixed materials. In Illinois, Iowa, northern Missouri, Kansas, and Nebraska deposits of rich wind-blown silt, called *loess*, cover the glacial deposits and are the dominant surface material (Fig. 2.12). Much of the agricultural wealth of the Central Lowlands and Great Plains can be attributed to the fertile loess-based soils. Other deposits of the Interior Plains include sand dunes in northwestern Nebraska, southwestern Saskatchewan, and around the Great Lakes, clayey lake beds near the Great Lakes and Lake Winnipeg, and river deposits (alluvium) along most stream valleys throughout the region. It is these deposits, mainly glacial drift, loess, and alluvium, that provide most of the parent materials for the fertile agriculture soils of the Interior Plains.

The Interior Plains are drained by three major watersheds: (1) the *Mississippi* and *Drainage and climate* its tributaries upstream from the head of the Mississippi Embayment; (2) the *St.*

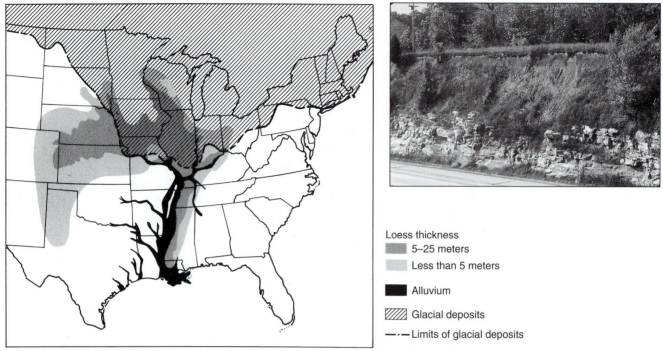

Loess thickness

▨ 5–25 meters

▨ Less than 5 meters

■ Alluvium

▨ Glacial deposits

—·— Limits of glacial deposits

Fig. 2.12 The surface deposits of the Interior Plains. Glacial drift is found north of the Missouri and Ohio rivers. Loess covers the drift in the central Midwest and extends over the adjacent section of the Great Plains. The photograph shows loess deposits (on top of the bedrock) exposed in a road cut in Missouri.

Lawrence, which drains the Great Lakes Basin; and (3) the *Nelson, Churchill,* and *Mackenzie* drainage basins, which drain the Canadian Plains and small portions of North Dakota and Minnesota. Water is generally abundant in the Central Lowlands, especially the Great Lakes area, but it declines westward into the Great Plains as mean annual precipitation falls and evaporation rates rise. From the Ohio River to the Rocky Mountain Front the mean annual precipitation drops from more than 40 inches to about 20 inches. At the 100th meridian the climate changes to *semiarid,* which dominates the Great Plains from central Texas to southern Alberta and Saskatchewan.

Great Plains Surface water is locally plentiful in the many stream valleys that cross the Great Plains. Streams such as the Saskatchewan, Yellowstone, Platte, Arkansas and many similar ones that follow the eastward slope of the plains, rise and fall with seasonal runoff from the Rockies. In some valleys such as the Missouri and Platte, systems of dams have been constructed for flood control and agricultural water supply. In addition, groundwater is abundant throughout much of the Great Plains; one aquifer, the *Ogalalla,* which is one of the largest in the world, stretches from South Dakota to northern Texas. Not surprisingly, agriculture in the Great Plains is overwhelmingly dependent on irrigation and is growing more so each decade.

Climate and vegetation There is also a marked *north–south climatic gradient* in the Great Plains related to temperature and evapotranspiration rates. At the southern end of the Great Plains, in the area of the Llano Estacado and Edwards Plateau, lake and reservoir evaporation rates exceed 5 feet a year and the climate approaches the dryness of true desert. Northward, evaporation declines with temperature and about 300 miles north of the U.S.–Canada border, scattered trees (mainly aspens) appear on upland surfaces. A little farther north, the plains are covered by boreal forest. Here the climate is subarctic, and although annual precipitation is no greater than farther south, evapotranspiration rates are much lower and the moisture balance is favorable. At the very northern end of the Great Plains Province, more than 2500 miles from the Edwards Plateau of Texas, the boreal forest gives way to tundra and we enter the Arctic Coastal Plain (Fig. 2.2).

Soils   Soils in the Central Lowlands are extremely diverse, especially in the upper Mississippi Valley, the Great Lakes region, and the St. Lawrence Lowland, owing to the extraordinary mix of glacial deposits left during the last glaciation 10,000 to 20,000 years ago. Environmental surveys for planning must recognize this diversity even at the local scale, and also that in some deposits soil materials change substantially with depth. Beyond the local variations related to different deposits (parent materials), soils also show a broad, regional trend related to the east–west bioclimatic gradient. Figure 2.13 illustrates these regional trends with the 100th meridian marking the change from humid region soils (mainly *Alfisols* and *Spodosols*) to dry region soils represented by *Mollisols* (or *Chernozems* in Canada). Mollisols are dark, organic rich soils that form under grass covers in the semiarid climatic zone (see Appendix A).

Landscape diversity   In the northern half of the Central Lowlands, lakes and wetlands are abundant and soils tend to be sandy in the newer glacial terrain. This province is second only to the Canadian Shield for its huge population of inland lakes. The landscape is generally diverse with mixed conifer–hardwood forests among the lakes and wetlands, but westward this diversity gives way to a more uniform landscape, the *Prairies*, and farther west, to the *Great Plains* (Fig. 2.13).

In the prairies and plains, the terrain tends to fall into two physiographic classes: river lowlands (for example, the valleys and floodplains of the Illinois, Mississippi, Missouri, and Iowa rivers), and broad, level, or gently rolling uplands between the

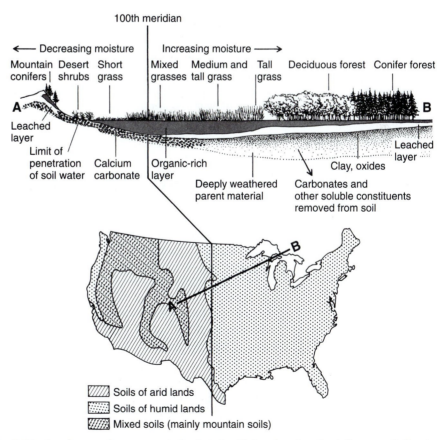

Fig. 2.13 Landscape change across the Interior Plains showing the influence of climate and vegetation on soil.

Vegetation river valleys, which make up the bulk of the landscape. The valleys support the principal tree covers, typically great corridors of willows, cottonwoods, sycamore, and other riparian species. The uplands originally supported *parkland* (mixed trees and tall grasses) that reached as far east as western Indiana, and *prairies* (tall, medium, and short grasses) that extended to the Rocky Mountains. Little if any of the original prairie remains; virtually all has been plowed and/or grazed and either destroyed or degraded. Today the region supports a massive agricultural economy dependent mainly on corn and wheat. The rural landscape that once contained great habitat corridors of grassland between linear corridors of woodland, now tends to be highly fragmented by farms, roads, and settlements. Both the Great Plains and the Central Lowlands are among the richest agricultural regions in the world, but in the past 50 years the Central Lowlands has also become a region of massive urbanization.

Urban development Superimposed over a large part of this North American heartland are major urban/industrial corridors, such as the one linking Toronto, Buffalo, Cleveland, Toledo, and Detroit. The growth of these corridors in the past several decades has resulted in many serious environmental problems including: (1) conflicts between urban and agricultural land uses; (2) misuses and eradication of floodplains, wetlands, and shorelands; and (3) widespread air and water pollution including threatened groundwater resources from buried wastes. The Central Lowlands and the St. Lawrence Lowlands are major sources of acid rain; and water pollution from both point (concentrated outfalls) and nonpoint (geographically diffused) sources are critical in both urban and agricultural areas.

Planning concerns Of a major concern to environmental planning and management in the Interior Plains are the problems arising at the interface between the expanding urban corridors and the rural landscape. As in the Eastern Seaboard, few planning problems in the Central and St. Lawrence Lowlands involve primary landscapes (virgin or near virgin lands); rather, most involve landscapes that have been subjected to several previous land uses, usually some combination of lumbering, early farming, modern farming, residential, and/or industrial development. Almost invariably, these used landscapes have suffered environmental damage: wetland drainage, stream channelization, habitat loss and fragmentation, and burial of wastes. Therefore, modern environmental planning in these provinces increasingly involves landscape repair and restoration as a part of community and economic development projects.

2.7 THE ROCKY MOUNTAIN REGION

The western border of the Great Plains is formed by the **Rocky Mountain Front**, one of the most distinct physiographic borders in North America. From the western Great Plains (at an elevation of 3500 to 5500 feet) the terrain abruptly rises 3000 to 6000 feet, and we enter the rugged Rocky Mountain Region. This region stretches about 2000 miles from central New Mexico to near the northern border of British Columbia. It is made up of four provinces: the *Southern, Middle,* and *Northern provinces* in the United States and the *Canadian Province* in Canada (Fig. 2.2). The Canadian Rockies form a long, narrow belt 50 to 100 miles wide that extends 500 miles northward from the Canadian–U.S. border. The American provinces, by contrast, are wider and far more irregular geographically.

Geology The Rocky Mountains are considered to be relatively young mountains because they are still active—though many of the rocks are of ages comparable to those in the Appalachians. Geologically, they are most diverse in the American provinces, where volcanic, metamorphic, and sedimentary rocks are all very prominent. The highest terrain in the Rockies, which is around 13,000 to 14,000 feet (4000 to 4300 meters) elevation, is found in Colorado, Wyoming, and along the British

Fig. 2.14 A view of the Rocky Mountain Trench looking southeastward. This great valley contains the upper reaches of both the Fraser and Columbia rivers.

Columbia–Alberta border. The largest areas of low terrain are the *Wyoming Basin* (5000–7500 feet elevation), which is underlain by sedimentary rocks and lies between the Northern and Southern Rocky Mountain provinces, and the *Rocky Mountain Trench*, a remarkably long, narrow valley stretching along the western edge of the Canadian Province in Southeastern British Columbia (Fig. 2.14).

Drainage   Americans have traditionally recognized the Rocky Mountains as the "continental divide"—the high ground that separates drainage between the Pacific and Atlantic watersheds. To the Gulf–Atlantic via the Mississippi go the Missouri, Platte, Arkansas, and many other large rivers; to the south goes the Rio Grande; to the Pacific via the Colorado and Columbia rivers goes the Okanagan, Snake, Green, Gunnison, and of course the heads of the Colorado and Columbia themselves. By contrast, in the northern part of the Canadian Province the Rockies do not form the continental divide. The Peace and Mackenzie Rivers, which drain into the Arctic Ocean, both have headwaters in the Rocky Mountain Trench west of the Rockies in British Columbia.

Owing to their diverse geology and rugged, high topography, the Rocky Mountains are highly varied in soils, climate, and biogeography. The American Rockies lie generally in an arid/semiarid climatic zone and most of the basins between ranges are relatively dry and dominated by prairie grasses and/or desert *Mountain climate* shrubs. But moisture conditions improve with elevation, not only because precipita-
and vegetation tion tends to increase higher up, but also because evapotranspiration rates decline at cooler temperatures. Around 7500 feet elevation in the Southern Rockies and 4000 to 5000 feet elevation in the Northern and Canadian Rockies, trees appear and forests of pine, fir, and aspens extend upslope several thousand feet until they are limited by cold temperatures. The elevation of the *tree line* varies from 11,000 feet or more in New Mexico to only 5000 to 6000 feet in the northern Canadian Rockies where the mountain forests merge into the boreal forest system. At the timber line, trees give way to *alpine meadow* consisting of grasses, wildflowers, and dwarf (shrub-sized) trees. For most mountains, alpine meadow marks the uppermost bioclimatic zone, but for mountains above 12,000 feet elevation, there is also a zone of permanent snowfields and/or glaciers (Fig. 2.15).

Besides the vertical zonation of climate, bioclimatic conditions are often quite different from range to range depending on location and orientation to prevailing winds and the sun. West slopes usually receive greater precipitation in their upper reaches than east-facing slopes because of the prevailing westerly airflow; and south-facing slopes are measurably drier owing to greater solar heating from the southerly

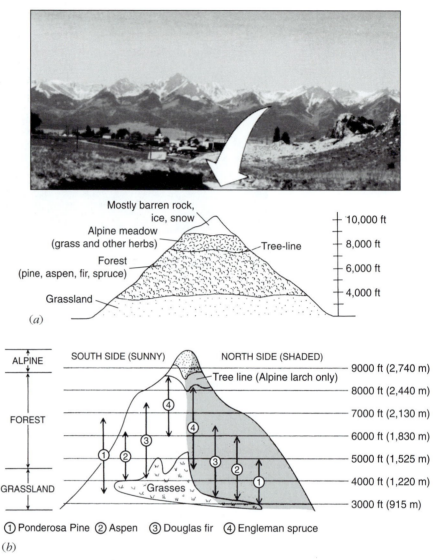

Fig. 2.15 (*a*) A simple model of the vertical zonation of vegetation in the Northern Rocky Mountains, a response to climate change with elevation. (*b*) the influence of sunny and shaded slope on vegetation zones.

exposures. Thus, vegetation zones often occur at different elevations from west to east as well as from south to north on individual mountains (Fig. 2.15*b*).

Water supply   The Rocky Mountains are the primary source of water for most of the American West and the Great Plains. The mountains induce relatively high precipitation rates (30–50 inches annually in many areas) much of which accumulates as snow that is released to streams in the spring and summer. West of the Rockies competition for a limited water supply is especially keen and growing more so with the expansion of agriculture and urban development. The water of the Colorado River System, for example, is partitioned by law between the Rocky Mountain states of Colorado, Wyoming, Utah, and New Mexico and three states, California, Nevada, and Arizona, that border on the lower Colorado River. Eight major dams have been constructed on the Colorado system, and their reservoirs provide huge amounts of water not only to agriculture and cities but for evaporation into the dry desert air (Fig. 2.16). By the time the Colorado reaches the Mexican border, little or no water remains.

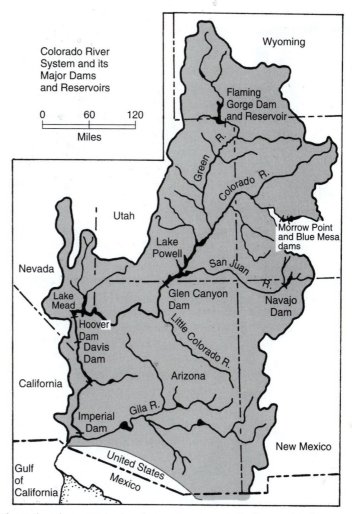

Fig. 2.16 The Colorado watershed and stream system, including major dams and reservoirs.

Land use   Settlement in the Rocky Mountains is light; there are no large cities within the region. (Denver, Salt Lake City, and Calgary lie on the borders of the region.) Most land is publicly owned. In both Canada and the United States, extensive tracts have been set aside as national parks, national monuments, and national forests (Fig. 2.17). Increasingly, landscape planning and management activity here and farther west are concerned with competition among mining interests, ranchers, preservationists, and others over the use of government lands. National forests and other federal lands are open to grazing and mining; both are a source of contention in many areas of the West because of their impacts on rangeland ecology and stream water quality.

2.8 THE INTERMONTANE REGION

Between the southern province of the Rockies and Pacific Mountain Region to the west lies an elevated region of plateaus and widely spaced mountain ranges. There are two major plateaus in the **Intermontane Region**: the *Colorado*, which lies at elevations around 6500 feet and is composed of sedimentary rocks, and the *Columbia*, which lies at elevations around 5000 feet and is composed of basaltic rock. The remainder of the region, about half its total area, is the *Basin and Range*

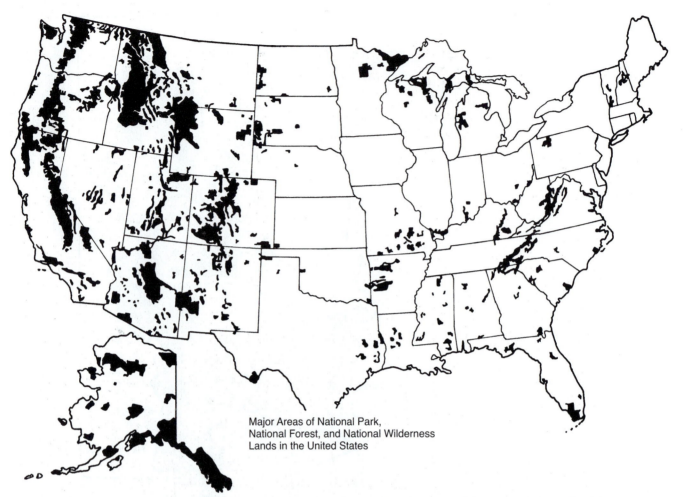

Major Areas of National Park,
National Forest, and National Wilderness
Lands in the United States

Fig. 2.17   The distribution of national parks, national forests, and national wilderness areas in the United States. In Canada, most of the country is held in public lands.

Province, which stretches from the Columbia Plateau on the north to Rio Grande Valley on the southeast (Fig. 2.18).

Landforms   The **Basin and Range** (also called the Great Basin) is characterized by disconnected, north–south trending mountain ranges formed by faulting and tilting of large blocks of diverse rock. The ranges are 50 to 75 miles long, 10 to 25 miles wide, and generally 7000 to 10,000 feet elevation. The basins between the ranges are larger in area and filled with thick deposits of sediment (thousands of feet deep) eroded from the adjacent mountains. The basin floors lie at elevations of 3000 to 5000 feet except in the southern part of the province where they are generally much lower (Fig. 2.18).

Climate and soils   The Basin and Range Province is severely arid (with less than 10 inches annual precipitation) except for the upper reaches of higher ranges where there is enough rain and snowfall to produce runoff. Some of the runoff delivered to the mountain slopes in winter and spring reaches the basin floors where it is trapped and evaporates, forming dry lake beds called *playas*. The runoff also brings sediment to the lower mountain slopes where, as flows decline, it is deposited forming *alluvial fans*, one of the distinctive landforms of the region. Soils on the valley floors belong to the *Aridisol* order: in and around the playas they are heavily saline. On the alluvial fans, the soils are far less salty, compositionally varied, and generally belong to the *Entisol* soil order (see Fig. 5.5).

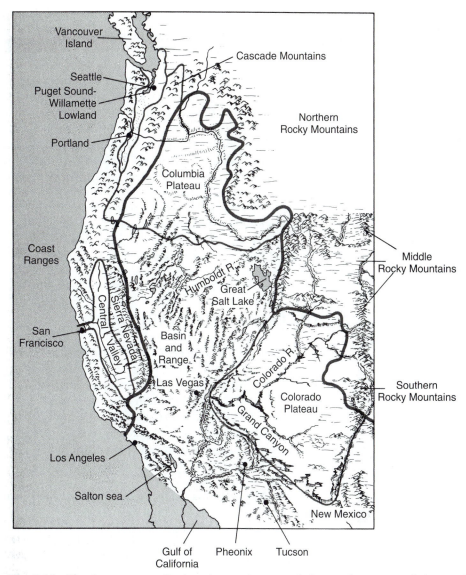

Fig. 2.18 The Intermontane Region, its provinces, and the southern part of the Pacific Mountain System.

Drainage

The northern half of the Basin and Range is the only area of *closed drainage* in the United States and Canada, that is, with no outlet to the sea. The largest stream in the area, the Humboldt River, rises in northeastern Nevada and ends in playas on the western side of the state. Most streams, however, are short, ending in local basins only miles from their headwaters. In the southern half of the Basin and Range, by contrast, drainage is *open* and dominated by the Colorado River and its lower tributaries (Fig. 2.16). The lower Colorado Valley is one of the driest areas of North America where reservoir evaporation (such as from Lake Mead) exceeds 100 inches a year.

Land use

In the past several decades, population has grown tremendously in the southern part of the Intermontane Region, principally in Arizona and southern Nevada. This has resulted in both urban and agricultural development, which has placed great demands on limited water supplies in this arid region. Groundwater aquifers are being rapidly depleted in some areas, and Arizona is being forced to turn to aqueduct systems fed by the Colorado to transport water to cities and farmlands. Los Angeles also uses the Colorado for part of its water supply.

Canyon lands

The **Colorado Plateau**, which lies between the Rocky Mountain and Basin and Range Provinces, is one of the truly celebrated physiographic provinces in the United States. It is a land of canyons, colorful landscapes, a rich Native American history, and home of the Grand Canyon, America's premier landform. The Colorado Plateau is a platform of sedimentary rocks with large areas of volcanic rocks and extensive surface deposits. The general elevation of the plateau is 5000 to 6000 feet but some mountain ranges within it, such as the Henry Mountains, reach 11,000 feet.

The Colorado River cuts through the heart of the plateau where it and its tributaries have carved a great system of canyons (Fig. 2.18). At its deepest cut—the Grand Canyon—4000 to 5000 feet of rock are exposed in the canyon walls, revealing a spectacular geologic cross section. The picture of geology and landforms in the province is enhanced by the arid climate (average precipitation is less than 15 inches a year) which limits vegetation to desert shrubs (mainly creosote bush and mesquite) and bunch grasses over most of the plateau. Trees grow at higher elevations with pine forest beginning around 7500 feet elevation.

2.9 THE PACIFIC MOUNTAIN REGION

Along the Pacific margin of North America is a complex region composed of many, large mountain ranges (Fig. 2.2). This region, known as the **Pacific Mountain System**, contains just about every geologic structure and mountain type imaginable and, by virtue of its location on or near major fault zones on the western edge of the North American tectonic plate, it is the most geologically active part of the continent. Earthquakes are commonplace throughout the region, and volcanism is active in the Cascade Range (which produced the Mount St. Helens eruptions of 1980 and 2004), in the Alaskan Peninsula, and in the Aleutian Islands, the most volcanically active area of North America.

Mountain ranges and types

The region is made up of eight provinces that together form a belt 200 to 400 miles wide extending from the Mexican border to near the Russian border at the end of the Aleutian Islands. Some of the more prominent mountain ranges include the *Coast Mountains* of British Columbia and Alaska, the *Sierra Nevada* of California, the *Cascades* of Washington and Oregon, the *Coast Ranges* of California and Oregon, and the *Alaskan Range* of southern Alaska. The Coast Mountains and the Sierra Nevada are mostly great masses of resistant granitic rock, whereas the Cascades are basically huge piles of weak (erodable) volcanic material. The Coast Ranges of California and Oregon are mainly deformed sedimentary rocks aligned into a series of ridges and valleys paralleling the coast. The Alaska Range, in the northernmost part of the region, is a complex body of mountains built by folding, faulting, and volcanic activity and is the highest range in North America with many mountains exceeding 15,000 feet in elevation. Southwest of the Alaska Range are the Alaska Peninsula and the Aleutian Islands, which comprise a chain of volcanic mountains.

Coastal landforms

Unlike the Atlantic Coast, the **Pacific Coast** lacks a coastal plain. In most places the mountains or their foothills run to the sea where wave erosion has carved a rugged and picturesque shoreline. Owing to ancient sea level changes and geologic uplift of the land, much of the coastline is terraced, that is, sculpted into steplike formations. In addition, the coastline is frequently intersected by streams that have cut narrow canyons down to the shore. Sandy beaches with sand dunes are found in the bays and near stream mouths. Accessibility to the coastline is difficult, and development is risky because of slope instability, the limited area suitable for building, and the often fragile character of the ecological environment.

Seismic activity

Earthquakes are a threat throughout the entire Pacific Mountain Region. The region is laced with fault lines, and many have an eventful history of activity. The most active earthquake zones are found in California and Alaska. California is clearly the most hazardous of the two, especially south of the San Francisco Bay area, not only because of the prominent fault systems there (such as the San Andreas), but also because of the massive urban development lying on or near active fault zones. California's urban/suburban population within this area approaches 20 million people. Recent earthquakes of modest magnitudes in Los Angeles and San Francisco portend the destructiveness of truly large earthquakes in these areas.

Major lowlands

Two major lowlands are found in the Pacific Mountain Region. The larger of the two is the *Central Valley of California*, which lies between the Sierra Nevada and the Coast Ranges (Fig. 2.18). The other is the *Puget Sound – Willamette Valley – Georgia Basin lowland*, which lies between the Coast Ranges and the Cascades in Oregon, Washington, and British Columbia. These lowlands are floored with deep deposits of sediment washed down from the surrounding mountain ranges over millions of years. And the Willamette is the center of drainage systems that trend north–south along the axes of the lowland. The three largest cities of the Pacific northwest—Portland, Seattle, and Vancouver—are located in the Puget Sound–Willamette lowland.

Biogeography

The climate, vegetation, and soils of the Pacific Mountain Region are as diverse as its geology and landforms. In Alaska and Canada, heavy precipitation in the Coast Mountains—the annual average exceeds 100 inches—nourishes large glaciers at elevations above 6500 feet and great conifer forests of hemlock, spruce, and cedar at lower elevations. The conifer forests extend down the coast into Washington and Oregon where the dominant tree is Douglas fir. Farther down the coast, in northern California, stands of huge redwoods become the dominant coastal forests. Precipitation declines sharply farther down the California coast, and between San Francisco and Los Angeles, where annual precipitation falls below 20 inches, the redwoods give way to a scrubby forest, called *chaparral*. Still farther south, in extreme southern California and the Baja California of Mexico, the chaparral gives way to bunch grass and shrub desert (Fig. 2.19).

Agriculture and forestry

In California, the combination of subtropical climatic conditions and water supplies from mountain streams has produced one of the most diverse and productive *agricultural regions* in the world. Grains, vegetables, grapes, and fruits are grown extensively in the southern two-thirds of the state, mostly with the aid of irrigation. The forest industry traditionally flourishes in the area between San Francisco and the Alaska Panhandle. In this area, as in the Rocky Mountains, public-owned forest and parks occupy large tracts of land in both Canada and the United States. In the United States, only about 10 percent of the original (old growth) forest remains and a contest is being waged among loggers, environmentalists, and the federal government for control of these forests. Elsewhere in the Pacific Region, south of Canada, lumbering is carried out in areas of second growth forest.

Environmental problems

The Pacific Mountain Region abounds with environmental planning and management problems. The scenic landscapes and pleasant climates are attractive to settlement, and population is growing in most areas. California's population now exceeds 33 million, more people than all of Canada. Growing population and expanding land use coupled with a diverse and active environment produce virtually every type of environmental problem known today and at a wide variety of geographic scales. California heads the list with serious air pollution, water supply, land stability, and habitat loss problems and contains more species on the federal list of rare and endangered species than any other state in the union. But the Pacific Northwest, including the Vancouver area in British Columbia, is rapidly developing its own list that includes forest management, urban sprawl, coastal development, and a host of pollution problems.

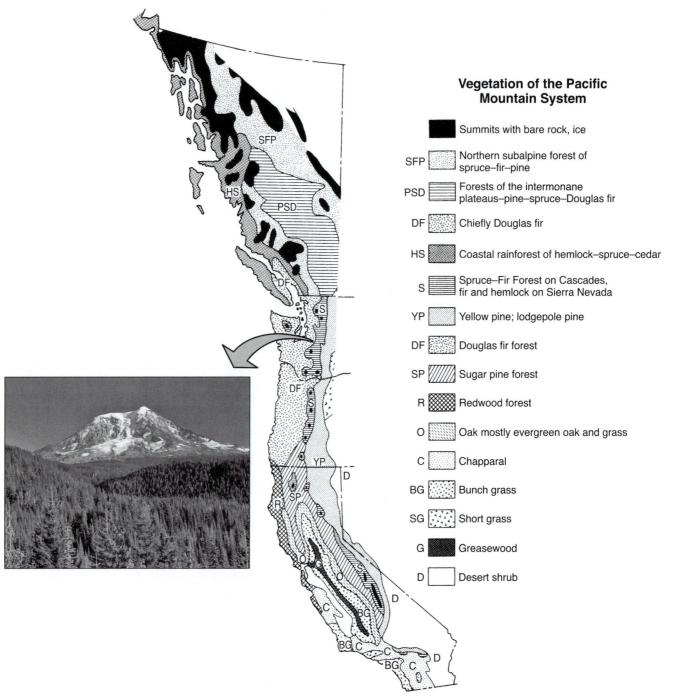

Fig. 2.19   The vegetation of the Pacific Mountain System from British Columbia to the Mexico border. Photograph shows a typical scene in the Cascade Mountains.

Among the region's most alarming problems is the dramatic decline of salmon. Over the past 20 years, most watersheds have lost a large share of returning populations of coho and some have declined to the point where they no longer have viable breeding stock. The cause of the decline is most certainly tied to multiple factors. At the top of the list are watershed changes—including degraded water quality and barriers to fish migration—and overfishing, but climatic change and increased predation may also be involved.

2.10 THE YUKON AND COASTAL ARCTIC REGION

Yukon Basin

South of Alaska's North Slope lies the Brooks Range (and its eastern limb in Canada, the British Mountains), a low, east–west trending mountain range. South of the Brooks Range and occupying the large interior of Alaska and the adjacent portion of the Yukon Territory is the **Yukon Basin** (Fig. 2.2). The Yukon River drains this large basin, flowing westward into the Bering Sea. Fairbanks, the principal city of central Alaska, is located near the center of the Yukon Basin. *Permafrost* is found over most of the Yukon Basin, but its coverage is discontinuous and its thickness highly variable. The landscape of the Yukon Basin is dominated by boreal forests in the south, but northward the tree cover grows patchy and gives way to tundra. The treeless tundra stretches in a broad belt along the entire Arctic Coast.

Arctic Coastal Plain

The Arctic Ocean is fringed by a coastal plain similar in topography to the Coastal Plain of southern United States. The **Arctic Coastal Plain**, however, is narrower and more desolate, being extremely cold and locked in by sea ice most of the year. The North Slope of Alaska, now famous for its oil reserves, is part of the Arctic Coastal Plain, as is the Mackenzie River Delta (just east of the Canada–Alaskan border) and the plain that fringes the northern islands of Canada. Virtually the entire Arctic Coastal Plain is underlain by permafrost, which in some areas extends offshore under the shallow waters of the Arctic Ocean.

The last frontier

Climate, permafrost, and ecology are the principal planning considerations of this region. The growing season, which is less than 60 days, prohibits agriculture, and permafrost limits development over much of the region because it leads to infrastructural damage. As one of the last great wilderness reserves on the continent, the tundra is given highest priority by environmentalists for long-term protection of its ecosystems. Conflict over economic development proposals and programs is destined to continue for decades.

2.11 SELECTED REFERENCES FOR FURTHER READING

Atwood, W. W. *The Physiographic Provinces of North America.* Boston: Ginn, 1940.

Bird, J. B. *The Natural Landscape of Canada.* Hoboken, NJ: Wiley, 1972.

Birdsall, S. S., and Flovin, J. W. *Regional Landscapes of the United States and Canada.* Hoboken, NJ: Wiley, 1985.

Bowman, Isaiah. *Forest Physiography.* Hoboken, NJ: Wiley, 1909.

Chapman, L. J., and Putnam, D. F. *The Physiography of Southern Ontario* (3rd ed.). Ontario: Ministry of Natural Resources, 1984.

Conzen, M. P. (ed). *The Making of the American Landscape.* Boston: Unwin Hyman, 1990.

Fenneman, N. M., and Johnson, D. W. *Physical Divisions of the United States* (U.S. Geological Survey Map). Washington, DC U.S. Government Printing Office. 1946.

Hunt, C. B. *Natural Regions of the United States and Canada.* San Francisco: Freeman, 1974.

King, P. B. *The Evolution of North America.* Princeton, NJ: Princeton University Press, 1959.

Leighly, John (ed.). *Land and Life: A Selection from the Writings of Carl Ortwin Sauer.* Berkeley: University of California Press, 1963.

Paterson, J. N. *North America: A Geography of the United States and Canada.* New York: Oxford University Press, 1989.

Pirkle, E. C., and Yoho, W. H. *Natural Landscapes of the United States* (4th ed.). Dubuque, IA: Kendall, 1985.

Thornbury, W. D. *Regional Geomorphology of the United States.* Hoboken, NJ: Wiley, 1965.

2.10 THE YUKON AND COASTAL ARCTIC REGION

Yukon Basin

South of Alaska's North Slope lies the Brooks Range (and its eastern limb in Canada, the British Mountains), a low, east–west trending mountain range. South of the Brooks Range and occupying the large interior of Alaska and the adjacent portion of the Yukon Territory is the **Yukon Basin** (Fig. 2.2). The Yukon River drains this large basin, flowing westward into the Bering Sea. Fairbanks, the principal city of central Alaska, is located near the center of the Yukon Basin. *Permafrost* is found over most of the Yukon Basin, but its coverage is discontinuous and its thickness highly variable. The landscape of the Yukon Basin is dominated by boreal forests in the south, but northward the tree cover grows patchy and gives way to tundra. The treeless tundra stretches in a broad belt along the entire Arctic Coast.

Arctic Coastal Plain

The Arctic Ocean is fringed by a coastal plain similar in topography to the Coastal Plain of southern United States. The **Arctic Coastal Plain**, however, is narrower and more desolate, being extremely cold and locked in by sea ice most of the year. The North Slope of Alaska, now famous for its oil reserves, is part of the Arctic Coastal Plain, as is the Mackenzie River Delta (just east of the Canada–Alaskan border) and the plain that fringes the northern islands of Canada. Virtually the entire Arctic Coastal Plain is underlain by permafrost, which in some areas extends offshore under the shallow waters of the Arctic Ocean.

The last frontier

Climate, permafrost, and ecology are the principal planning considerations of this region. The growing season, which is less than 60 days, prohibits agriculture, and permafrost limits development over much of the region because it leads to infrastructural damage. As one of the last great wilderness reserves on the continent, the tundra is given highest priority by environmentalists for long-term protection of its ecosystems. Conflict over economic development proposals and programs is destined to continue for decades.

2.11 SELECTED REFERENCES FOR FURTHER READING

Atwood, W. W. *The Physiographic Provinces of North America.* Boston: Ginn, 1940.

Bird, J. B. *The Natural Landscape of Canada.* Hoboken, NJ: Wiley, 1972.

Birdsall, S. S., and Flovin, J. W. *Regional Landscapes of the United States and Canada.* Hoboken, NJ: Wiley, 1985.

Bowman, Isaiah. *Forest Physiography.* Hoboken, NJ: Wiley, 1909.

Chapman, L. J., and Putnum, D. F. *The Physiography of Southern Ontario* (3rd ed.). Ontario: Ministry of Natural Resources, 1984.

Conzen, M. P. (ed). *The Making of the American Landscape.* Boston: Unwin Hyman, 1990.

Fenneman, N. M., and Johnson, D. W. *Physical Divisions of the United States* (U.S. Geological Survey Map). Washington, DC U.S. Government Printing Office. 1946.

Hunt, C. B. *Natural Regions of the United States and Canada.* San Francisco: Freeman, 1974.

King, P. B. *The Evolution of North America.* Princeton, NJ: Princeton University Press, 1959.

Leighly, John (ed.). *Land and Life: A Selection from the Writings of Carl Ortwin Sauer.* Berkeley: University of California Press, 1963.

Paterson, J. N. *North America: A Geography of the United States and Canada.* New York: Oxford University Press, 1989.

Pirkle, E. C., and Yoho, W. H. *Natural Landscapes of the United States* (4th ed.). Dubuque, IA: Kendall, 1985.

Thornbury, W. D. *Regional Geomorphology of the United States.* Hoboken, NJ: Wiley, 1965.

3

LANDSCAPE FORM AND FUNCTION IN PLANNING

3.1 INTRODUCTION

It is no news that we are in serious trouble with the landscape. In North America, there are few places where we have been able to achieve a lasting balance among land use activities, facilities, and environment. Part of the explanation for this state of affairs has to do with the direction of modern urban life. In the past generation or two, people have generally lost touch with the land and in turn with most traditional knowledge about the way the landscape works. Planning for the use and care of the local landscape is no longer part of personal and family tradition and responsibility. To a large extent it has been relegated to second and third parties and has been transformed into bureaucratic and business processes made up of inventories, checklists, permits, and contracts which ask for little understanding of the landscape's true character.

The route to finding the true character of the landscape lies in understanding the way the land functions, changes, and interacts with the life it supports. One of the most fundamental points of understanding is that the landscape is more dynamic than it is static, with forms and features in a continuous state of change. Change is driven by systems of processes that include rainstorms, streamflow, fires, soil formation, plant growth, and yes, the human activities that we call land use. These processes shape the landscape now as they did in the past. It is increasingly apparent that successful landscape planning must do more than respond to mere shapes and features but must also respond to the processes themselves and to the systems that drive them.

3.2 ESSENTIAL PROCESSES OF THE LANDSCAPE

One of the prevailing misconceptions about the landscape is that it is made up of features, especially natural features, that date from times and events millions of years ago. This view is especially common for landforms and soils that are typically regarded as products of the geologic past. Therefore, any attempt to understand their origins and development requires special knowledge of ancient chronologies and events. For most landscapes, however, this view is probably not valid.

Landscape of the here and now   By and large, the forms and features we see in the local landscape are the products of the processes that presently operate there. This perspective is important, because it implies that it is possible to understand the landscape according to the workings of the environment in *the present era*, more or less. Using an analogy from medical science, we know that the body is a product of long-term biological evolution but we also know that the individual organs in the body have sizes, shapes, and compositions related to their present function. As with the organs in the human body, form and function in the landscape go hand in hand. It follows that as the clever physician is able to read basic functional (physiological) problems from changes in organ shape and composition, so the insightful student of landscape can read changes in the functional character of the land from observable changes in landforms, soil, drainage features, vegetation, and so on.

Formative processes   What are the essential or formative processes of the landscape? They include waves, wind, glaciers, and runoff; however, there is little question that running water heads the list. This includes all the various forms of runoff fed directly or indirectly by precipitation—namely, overland flow, streamflow, soil water, and groundwater. The work accomplished by runoff, measured by the total amount of material eroded from the land, exceeds that of all other formative processes manifold, even in dry environments. Therefore, we can safely conclude that the landforms we see in most landscapes are mainly water carved, water deposited, or influenced by water in some sig-

Fig. 3.1 Water-carved landscapes. Runoff is the most effective natural agent in shaping the landscape over most of the earth, especially in terms of landform, soil, and habitat formation.

Landscape differentiation

way (Fig. 3.1). Excluded, of course, are landforms in glacial environments and sand deserts, but many of these, too, are often profoundly influenced by running water.

The *differentiation of terrain* into various physiographic zones and habitats begins with weathering and the sculpting of landforms by runoff. As the landforms take shape, different moisture environments emerge such as wet valley floors, mesic (intermediate) hillslopes, and dry ridge tops. These in turn give rise to different plant habitats, and it is the combination of moisture conditions, vegetation, and surface sediment that yields soil. Thus, in our search for fundamental order in the landscape it is advisable to begin with landforms and drainage. In most instances the remainder of the landscape, that is, vegetation, soils, and habitats will fall into place once the essential system of landforms and runoff has been worked out (Fig. 3.2).

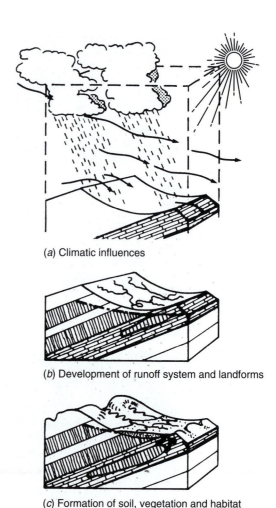

(*a*) Climatic influences

(*b*) Development of runoff system and landforms

(*c*) Formation of soil, vegetation and habitat

Fig. 3.2 A schematic diagram illustrating the development of landscape beginning with (*a*) climate, (*b*) runoff and landforms, and leading to (*c*) soil, vegetation, and habitat formation.

Moreover, this rule applies at essentially all spatial scales from that of microtopography to the macro forms of the great landform systems.

Form–function concept   If we agree that the landscape represents a basic *form–function* (or form-process) relationship, then we should be able to deduce a great deal about the processes that operate in it from the forms we can observe. This concept is a very important part of terrain analysis and environmental assessment because we rarely have the time and resources to undertake our own scientific investigations as part of planning projects. That is, we are usually unable to launch studies to generate first hand data leading to analysis of processes and their effects on the landscape. Therefore, we must learn to read the formative process of the land from the forms and features that can be observed and measured in the field and on maps and imagery.

The form–function concept also reveals that any actions taken as a part of land use planning and engineering that result in changed landscape forms, such as cut and fill grading, soil preparation, and vegetation alteration, must produce a change in the
The need for balance   way the landscape processes work. If certain balances exist on a slope among, for example, runoff and soils, slope inclination, and vegetation cover, then an alteration of one component without appropriate counter alterations in the other three will result in an imbalance. The slope may begin to erode, sediment may accumulate on it, or vegetation may decline. In any case, the problem of maintaining balance in the land-

scape in the face of physical changes brought on by land use demands that we view the land as a dynamic gameboard rather than a static stage setting. As a rule, our goal in landscape planning should be to guide change in such a way as to maintain the long-term performance of the critical processes and systems of the landscape—in other words, to guide change toward a sustainable landscape.

3.3 THE NATURE OF LANDSCAPE CHANGE

There is an old debate in science about the character of change in nature. Does nature change gradually over long spans of time, as traditional evolutionists thought, or does nature change in short bursts, as many earth scientists think? Both views are correct to some extent, but it appears that the second one is more appropriate for the larger features of the landscape such as stream channels, hillslopes, and land use.

Events and their force
The essential processes of the landscape work at highly uneven rates, rising and falling dramatically over time. Each rise or fall in, for example, streamflow or wind can be described as an *event*. Each event can be measured in terms of its magnitude, that is, its size such as the discharge of a stream or the velocity of wind in response to a storm. A huge streamflow that produces massive flooding is a high-magnitude event and it exerts much greater force on the environment than a small or modest event.

Very significant to our understanding of landscape change is the fact that the *force* exerted on the environment by an event *increases geometrically with its magnitude*. This means that to accurately interpret the potential for change (or work) by a force such as wind or running water we must understand, for example, that a three-fold-magnitude increase from, for instance, level 2 to level 6 represents an increase in force as great as 25-fold. This is termed an exponential relationship in which force increases as some power function of event magnitude.

Magnitude and frequency concept
If we examine the relationship between the *magnitude and frequency* of events in the landscape, we will find that almost regardless of the process involved (e.g., streamflow, wind storms, rainfalls, fires, earthquakes, snowfalls, oil spills, car accidents, or disease epidemics) the pattern is basically the same. There are large numbers of small events, much smaller numbers of medium-sized events, and very few large events. The truly giant events, which can render huge amounts of change, are very scarce indeed. As it turns out, the events that do the most work *in the long run* are not the giants (they are too infrequent) and not the high-frequency events (they are too small even when counted together), but fairly large events of moderately low frequency. In streams, for example, these events may be bankfull discharges and modest floodflows that occur once a year, every other year, or less.

Identifying formative processes
An inherent pitfall of landscape analysis for planning purposes is that we usually have no chance to observe or measure the events that really shape the landscape. For hydrologic and atmospheric processes, these are events, such as a very intensive thunderstorm, a floodflow, or a massive snowfall, which have a duration or life ranging from an hour or two to several days and a frequency of occurrence of once every year or two. The events we are apt to see on the typical field visit (usually a fairly nice day) are probably meaningless in terms of total effectiveness in shaping the landscape. Therefore, unless we are careful we run the risk of grossly misinterpreting how the site actually functions, how its features are shaped, and how things relate to each other. We may be led to infer that the processes we happen to observe are really the essential ones or that it was all shaped so long ago that the processes operating there in this era cannot possibly have anything to do with the site as we see it. Both conclusions would be mistaken.

Form–function relations
How do we avoid the pitfalls of misinterpretation of the landscape? The answer lies in understanding form-function relations. When we describe the forms and fea-

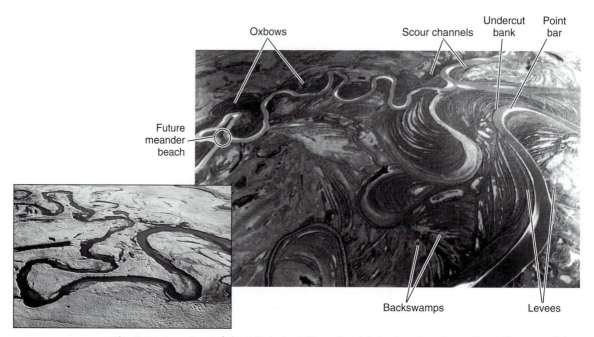

Fig. 3.3 A section of the Mississippi River floodplain showing the artifacts (forms and features) left in the landscape as a record of the river's formative processes. Inset shows channel formation in action where a stream is cutting into the bank along a series of bends.

tures of the landscape, we are actually observing the artifacts and fingerprints of the formative processes (Fig. 3.3). Through insightful field investigation we can begin to deduce which processes at which levels created, shaped, or affected different landscape features, and which processes and events are meaningful to both existing and future land uses and facilities.

Stream valleys and channels provide some of the best illustrations of this approach. Studies show that the stream channel is shaped principally by relatively large flows, mainly those that occur once or twice a year or once every two or three years. These may be floodflows or flows that fill the channel with modest spillover into adjacent low areas. Such flows are capable of scouring the channel bed, moving heavy sediment loads, eroding banks, and causing the channel to shift laterally as is illustrated in Figure 3.3. Bank vegetation may be undercut, and where flows overtop the banks, waterborne debris such as dead leaves is often stranded on shrubs and trees, marking the flood water level. Taken together, these features serve as markers of flow depths and extent as well as indicators of the distribution of energy and the work accomplished by running water. They reveal that stream channels and valleys are highly dynamic environments and that the formative events are both destructive and constructive—important information for environmental planning and management.

3.4 THE CONCEPT OF CONDITIONAL STABILITY

Whether or not a landscape is stable under the stress of the various forces applied to it depends not only on the strength of those forces, but also on the resisting strength of the landscape. Resisting strength is provided by forces that hold the landscape together, that is, keep soil from washing away, slopes from falling down, and trees from toppling over. Among the resisting forces—which include gravity, chemical cementing agents, and vegetation—living plants are the most effective in holding the soil in place.

Critical balance

In most natural landscapes, a state of balance exists between the driving forces, represented by water, wind, human, and other processes, and the resisting forces. Only when this balance is broken—usually because a powerful event exceeds the strength of the resisting force—is the landscape prone to massive change such as wholesale soil erosion and slope failure. Most landscapes, however, are resistant to breakdown from all but the strongest events. In some places, however, stability is maintained by an extremely delicate balance which is *conditional* on a special ingredient in the environment. That ingredient, such as a mat of soil-binding roots on an oversteepened slope, functions as the stabilizing kingpin in the landscape. If the kingpin is weakened or released, the landscape can literally fall apart under the stress of even modest events.

Conditional stability

Recognizing such conditional situations as critical to landscape planning because it is necessary to guide the change brought on by land use without pulling a kingpin and triggering a chain of damaging vents. Thus, in evaluating a site for a planning project, it is important to identify features that may be pivotal to the overall stability of the landscape. These may not be the most apparent features in terms of size or coverage. Among the conditional, or *metastable*, features commonly noted are steep, tree-covered slopes, vegetated sand dunes, stream banks made up of erodable sediment, certain wetlands, and groundwater seepage zones.

By way of example, let us illustrate the concept of conditional stability using a forest-covered sand slope in a coastal setting. In an unvegetated state, sand slopes (such as a wave-eroded bank or a dune face) can be inclined at an angle no greater than 33 degrees. When a cover of woody plants is added to the slope, especially trees, the angle can be as much as 45 to 50 degrees because the roots and stems lock the sand into a stable slope form. Such oversteepening of wooded slopes is a common occurrence in coastal environments where sand is added to an existing slope by wind or runoff. Alteration or removal of vegetation for roads, trails, structures, or lumbering can initiate slope failure, erosion, and a host of related problems on conditionally stable slopes (Fig. 3.4).

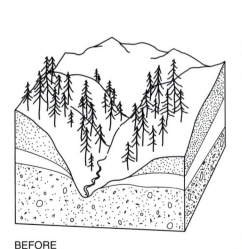

BEFORE

AFTER

Fig. 3.4 Slope failure following clear cutting of forested terrain in the American Northwest. These are not sand dune slopes, but they can exhibit the same tendency toward instability when they are deforested and logging roads are cut onto weak materials.

3.5 PERSPECTIVES ON SITE

Site as real estate

Invariably, projects and problems in environmental planning involve a parcel of space in the landscape called a site or a project area. It may range in size from less than 1 acre to thousands of acres. Its shape is usually some sort of rectilineal form, a product of mapping systems and the surveyor's coordinant lines. As a physical entity, the site has meaning mainly as a piece of real estate whose value is governed principally by its size and location. In land planning it is the envelope of space for which a use is sought or to which some land use has been assigned and within which a plan will be built.

Site as environment

From the standpoint of the environment and its functions, however, the site as conventionally defined has limited meaning and generally cannot be used to define the scope of environmental analysis for planning and design purposes. The reason why is that the spatial confines of site space usually have little to do with the workings of the environment, that is, with the processes and systems that shape and characterize the landscape of which the site is a part. Air, water, and organisms, for example, move in spaces and patterns that typically show little or no relationship to the space defined by a site. Therefore, as we pursue planning problems involving sites, we must deal with many different envelopes of environmental space as they ultimately relate to a prescribed piece of real estate.

Site in three dimensions

Although we commonly view the site as a two-dimensional plane defined by the surface of the ground and contiguous water features, in reality it is distinctly three dimensional. The third dimension, height and depth, extends the site upward into the atmosphere and downward into the ground. The relevance of atmospheric and subterranean phenomena to planning problems is generally secondary compared to surface phenomena. However, concern for groundwater contamination, air quality, and climatic change increasingly calls for serious consideration of these phenomena.

In addition, it is important to appreciate that the atmospheric and the subsurface realms are the source areas of forces that drive many of the surface processes. For example, groundwater is the principal source of stream discharge, and solar radiation is the primary source of surface heat. Alteration of these driving forces directly or indirectly affects terrestrial processes such as runoff, erosion, evaporation, and photosynthesis, which may in turn change the fundamental balance of the surface environment including slope stability, ground-level climate, wetland trends, streamflow regimes, as well as the conditions of facilities such as building foundations, roadbeds, and utility lines.

3.6 SPATIAL DIMENSIONS OF THE SITE

Site in dynamic space

Sites are traditionally described according to their forms and features and the spatial relations among them. However, sites can also be described according to their dynamics, that is, according to processes that shape their forms and features. In addition to running water, rainfall, wind, and animal movements, these processes are the various land use activities and their byproducts such as noise and air pollution. Each landscape process is part of a flow system and can usually be described in terms of direction, velocity, mass, and force.

If we stand on a site, no matter where it is located, you can imagine that you are at the intersection of several flow systems that occupy different levels or strata of space at, above, and below the surface. Each flow system originates somewhere, usually offsite, and goes somewhere else after it has crossed the site. As it moves through the site, the flow is inevitably changed in some way as it interacts with the various forms and features of the site (Fig. 3.5).

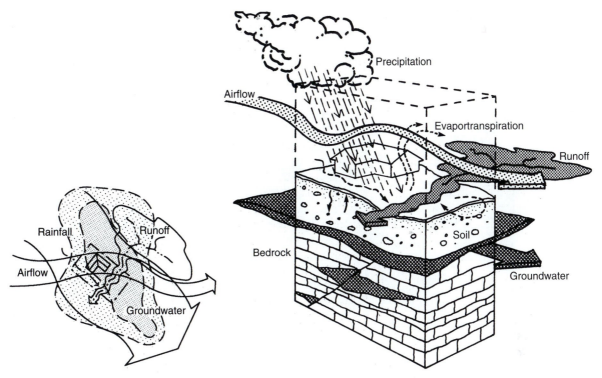

Fig. 3.5 Several important systems that intersect at a planning site: surface runoff, precipitation, evapotranspiration, groundwater, and air flow.

The upper tier
Air occupies the upper level or tier of the site. Airflow usually originates far beyond the site because most winds are driven by regional scale forces, namely, pressure differences among air masses. The movement of air over the surface can be described as a fluid layer called the *atmospheric boundary layer*. This layer, which measures about 1000 feet deep, is dragged over the earth's surface by the general motion of the larger atmosphere. As it slides over the surface, airflow in the lower levels is slowed substantially because of the frictional resistance imposed by the landscape. As a result, wind velocities at ground level are usually a small fraction of those just 25 to 50 feet above the surface (Fig. 3.6).

The change in wind velocity with height above the ground is described by a standard curve called the **normal wind velocity profile**. This curve reveals that the greatest change in velocity can be expected in the lower 100 feet of the boundary layer where most planning and design activities are focused. The exchange of heat, moisture, and pollutants with the overlying atmosphere takes place across this zone, and the faster the airflow the greater the rate of exchange. This explains the requirement that factory and power plant exhaust stacks be heightened to reach faster streamlines as a means of reducing local pollutant concentrations near the ground.

The middle tier
The middle tier of this model is the landscape per se, which extends from the upper limits of the vegetation canopies and tops of structures to the lower limits of root systems, buildings, and utility systems (Fig. 3.6). This is decidedly the most active layer of the landscape, with a host of flows taking on complex motions and exchanges. Although most movement is horizontal, some is decidedly vertical. Water, the dominant agent, is delivered vertically as precipitation then moves laterally as runoff. Runoff assumes a variety of flow forms as it moves over the surface, depending on slope, soil, and vegetation conditions. It may occur in thin sheets that move at velocities as low as 0.1 foot per second to streams that move at velocities up to 6 feet per second.

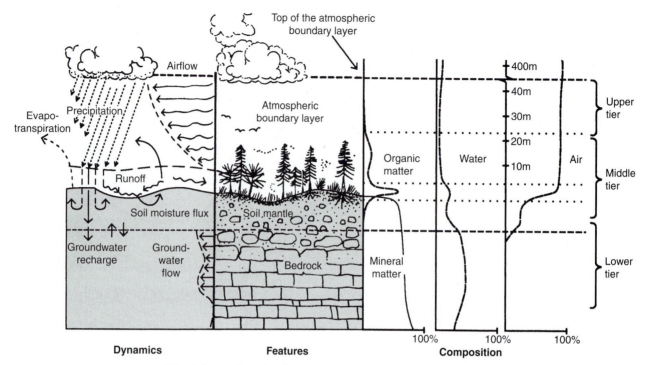

Fig. 3.6 A schematic illustration of the dynamics, features, and composition of the landscape in the upper, middle, and lower tiers of the landscape.

In site planning, runoff is one of the most serious considerations in terms of both water quantity and quality. Generally land development of any sort results in increased stormflows and decreased water quality. Since every site is part of a flow system, local loading of runoff is always passed on, and the problem tends to become cumulative downhill and downstream. Realization of this fact led to the first guidelines and rules for on-site management of stormwater in site planning to control flooding.

Where drainage is impeded by natural or other means, water may collect locally and initiate wetland formation. A dependable and abundant water supply encourages the development of special plant communities with distinctive floristic compositions and strong productivities. The organic matter added to the moist soil from plant remains further improves the water-holding capacity of the wetland, and advances its development and stability. This is one way in which the landscape becomes differentiated into zones of particular hydrologic, soil, and ecological character.

The lower tier   The lower tier of the site is comprised of the soil mantle and underlying bedrock (Fig. 3.6). Once considered relevant to planning only in terms of building foundations, utility lines, and water supply, today the subsurface environment must be given more detailed attention because of our heavy reliance on contaminant-free groundwater. Groundwater is the largest reservoir of fresh (liquid) water on the planet, but it is widely threatened by contamination from surface pollutants and buried waste. Unlike surface water, which is flushed from the land relatively quickly, groundwater moves so slowly that once contaminated it remains polluted for decades and even centuries. Groundwater movement in aquifers is limited to a fraction of a foot per day; therefore, exchange (flushing) times are typically decades or centuries. In addition, because of their large sizes, aquifers almost always extend well beyond the scale of individual sites (Fig. 3.5). In site planning, therefore, it is useful to know what part of an aquifer the site lies over. For example, does the site lie over the recharge (water-receiving) zone or over the output (water discharge) zone of the aquifer?

Below the soil mantle lies the zone of bedrock geology. In mountainous regions the bedrock protrudes through the soil mantle to the surface, and we usually view it as a sign of ground stability in land use planning. The bedrock can, however, be decidedly unstable in areas of active faults and cavernous limestone. Faults usually occur in swarms running along fault zones. Earthquakes can occur anywhere in the fault zone, and their destructiveness is governed by the magnitude of the energy release as well as the nearness of the earthquake to the surface and its proximity to urban type development. In the case of cavernous limestone, the concern is with surface collapses and associated groundwater flows. Not all limestone is cavernous, but where it is, such as in central Florida, collapse features and caverns should be located and taken into account in land use planning.

3.7 SOURCES OF ENVIRONMENTAL DATA FOR SITE PLANNING

Field investigation

Despite the technological advances in the acquisition and processing of environmental data, the analysis and evaluation of the landscape still rely heavily on field investigations. This is especially so for problems involving small to medium-sized sites (several acres to several hundred acres) for which the resolution of secondary data sources, such as satellite imagery, soil maps, and standard topographic maps, is not well suited. Although these sources are helpful in understanding the regional context, the internal character of the typical site falls between the cracks, as it were.

Secondary sources

For large problem areas we must, of course, rely mainly on secondary sources to gain a sense of the character of the landscape and its environmental setting. Inevitably, however, field observation is also necessary, but it must usually follow the examination of secondary sources and serve as a ground-truth exercise to test the validity of our initial ideas about the makeup and operation of the environment. Among the secondary sources, topographic contour maps are probably the most valuable.

Topographic maps

Topographic contour maps are published by the U.S. Geological Survey and in Canada by the Department of Energy, Mines, and Resources and various provincial mapping programs. The maps are published at a variety of scales and are available from both the government and private outlets. In the United States, coverage ranges from the entire coterminous United States on one sheet to coverage of a local area of about 55 square miles. The latter sheets, called 7.5-minute quadrangles (because they cover about 7.5 minutes of latitude) are the most useful for planning purposes. The comparable sheets in Canada are the 1:10,000 (1 cm to 100 m) topographic base maps.

The 7.5-minute quadrangles and their correlatives in Canada are excellent sources of information on drainage systems, topographic relief, and slopes and are helpful in locating land use features, water features, and wooded areas. The 7.5-minute maps are printed at a scale of 1:24,000 (about 0.4 mile to the inch) with a contour interval of 10 feet. Today most planning projects involving major facilities call for highly detailed topographic maps at much larger scales (often as large as 1 inch to 100 feet or 1 inch to 50 feet), and for these, aerial mapping companies must be specially contracted.

Soil maps

Soil maps in the United States are prepared by the U.S. Natural Resources Conservation Service (formerly the U.S. Soil Conservation Service) and published county by county in booklets called county soil reports. These map reports give the classification and description of soils to a depth of 4 to 5 feet as well as an indication of the representative slope of the ground over the area covered by each soil type. The soil type and slope are printed on the map in a letter code. Most boundary lines between soil types are highly generalized and should be scrutinized in the field for site planning problems. Unfortunately, the scale of the soil maps (usually at 1:20,000) is larger than that of the 7.5-minute U.S. Geological Survey topographic maps, prohibiting easy compilation of soil and topographic data on a single map. Chapter 5, "Soil, Land Use Suitability, and Waste Disposal," discusses soil maps in more detail.

Aerial photographs
Aerial photographs are available for virtually all areas of the United States and Canada. In the United States they were regularly produced by various governmental agencies including the U.S. Natural Resources Conservation Service, the U.S. Forest Service, and the U.S. Bureau of Land Management. Today a joint program called the National Aerial Photography Program (NAPP) works in behalf of these and other federal agencies, and the photographs (both old and new) are available from the EROS Data Center in Sioux Falls, South Dakota. Standard aerial photographs are available in 9-inch by 9-inch formats in black and white prints that are suitable for stereoscopic (three-dimensional) viewing. Individual photographs can be enlarged for planning purposes to any desired scale, and although they cannot be used as a source of precise locational information (because of inherent photographic distortions), aerial photographs are an excellent source of information on vegetation, land use, and water features.

Scanner and radar imagery
In addition to aerial photographs, a wide variety of nonphotographic imagery is available today. This imagery is provided mainly by *scanners* and *radars* mounted in satellites that relay data to earth receiving stations. Much of this imagery is produced for meteorological purposes and lacks the spatial resolution needed for most planning problems. However, several satellite systems show promise for environmental planning, including the *Landsat* satellites, which have been in operation since 1972, and the *SPOT* satellites which have been in operation since 1986. The newest among the Landsat satellites is a scanner system called *Thematic Mapper*, which provides resolution as fine as 30 by 30 meters in six bands. The high resolution SPOT satellite provides even finer resolution (10 by 10 meters) in one visible band. Landsat, SPOT, and other satellite imagery can be acquired from the EROS Data Center.

Special sources
Increasingly, special sources of data and information are available, especially for populous regions. Many states and counties contract their own aerial photographic surveys on a regular basis with imagery in black and white, color, and infrared formats. Information on water resources, wetlands, and other resources is available for selected areas as a result of research projects, environmental impact reports, planning projects, and the efforts of local watershed and streamkeeper groups.

The U.S. Geological Survey is active in every state. Besides topographic maps, the Survey also publishes a wide range of other maps, as well as data and reports for various local and regional problems and resources. These include earthquake hazard maps, stream discharge records, groundwater surveys, maps of geological formations, and regional- and state-based reports on the nation's water resources. The same generally holds for the U.S. Environmental Protection Agency (EPA) and the National Oceanic and Atmospheric Administration (NOAA), each of which produces a variety of maps, reports, and data on different topics for different regions and communities. The body of special information sources is huge and growing rapidly; space prohibits us from describing them here. Indeed, it is probably fair to say that it is impossible for any individual to keep track of this mass, especially for areas beyond your own theater of operation. Therefore, in the face of a planning problem in any location, it is advisable to move quickly to local clearinghouses such as planning commissions, environmental agencies, and universities to find the special sources.

Nongovernmental organizations (NGOs) are also becoming valuable sources of environmental data and information. NGOs have been around a long time, but until the past decade or two they concentrated almost exclusively on government lobbying activity. Now, however, many are also sponsoring projects that generate useful environmental data. Those that operate at a global scale are of limited value in site planning, but those that operate locally are often surprisingly helpful in site planning projects. It is not uncommon to find hundreds of organizations in a state or province involved, for example, in stream and watershed monitoring and restoration, wetland protection, and parkland acquisition and management.

3.8 THEORETICAL PERSPECTIVES ON LANDSCAPE

We should not conclude our overview discussion on landscape without mentioning some conceptual statements. Over the past century or so, several major concepts or models of landscape change and development have been proposed. Broadly speaking, these concepts constitute theoretical statements, and whereas no one concept is avidly followed today, each has contributed to our understanding of landscape and its formation. To some extent, these concepts represent the framing paradigms of landscape studies and as such lend perspective to landscape planning.

Evolution theory

If we ignore early theological and geological debate on the origins of the earth, the first major theoretical statements on landscape emerged in the late 1800s at a time when scientific thought was strongly influenced by evolution theory. In its simplest form, the central idea in evolution theory was change over time. The manner of this change was often conceived as a developmental process in which organisms, or whatever phenomena were being examined, tended to change sequentially or stagewise through time. In the late 1800s, for example, Russian soil scientists proposed that soils, like organisms, evolve toward a mature state in response to their environment, particularly the bioclimatic environment. A little later, the American geographer W. M. Davis articulated a unified model of landscape evolution based on stream erosion, which was to become a major theme in twentieth century educational circles.

Geographic cycle

Davis called his concept the **geographic cycle,** and in it he proposed that landscapes evolve through a series of developmental stages as streams deepen and widen their valleys. He envisioned three main stages of development, which he named *youth, maturity,* and *old age.* In the youthful stage the landscape is rugged and valleys are V-shaped as streams cut into the land. In maturity, valleys have begun to widen and take on U-shapes and in old age valleys are very wide with no intervening uplands (Fig. 3.7).

Davis's interests were in landforms, but it is apparent that the geographic cycle also applies to the other components of landscape. From youth to old age, the landscape becomes increasingly fluvial, or stream controlled, evolving into a low relief landscape dominated by broad floodplains, deep soil mantles, and great corridors of riparian vegetation. And indeed we find these landscapes in abundance across the earth, but they tend to be associated more with geographic position in watersheds than with a stage of evolutionary development of a land mass. Likewise, mature and youthful landscapes are also plentiful but they too appear less stage-controlled and more related to position in drainage systems. Generally, terrain is more youth-like in the upper reaches of watersheds and more old age-like in the lower reaches. The Mississippi system generally follows this pattern. Davis gave the term *peneplain* to the broad lowland of old age and he proposed that over geologic time peneplains were eventually uplifted and the cycle of erosion and landscape evolution renewed.

Base level and grade

Although the geographic cycle is not widely supported in modern scientific circles, several of the concepts on which it is based are still embraced. One of these is *base level,* the lowest elevation to which a stream can downcut into the land. For rivers terminating in the sea, base level is set by sea level. Inland, the base level may be set by a lake, wetland, another stream, or a reservoir. Reservoirs set artificial base levels that disrupt the normal relationship of a stream to the landscape and interrupt the flow of sediment. A related concept is *grade* or *graded profile,* which is the up-sloping longitudinal profile of a river. From mouth to headwaters, streams gradually increase in elevation, thus forming a long concave profile (Fig. 3.8). The graded pro-

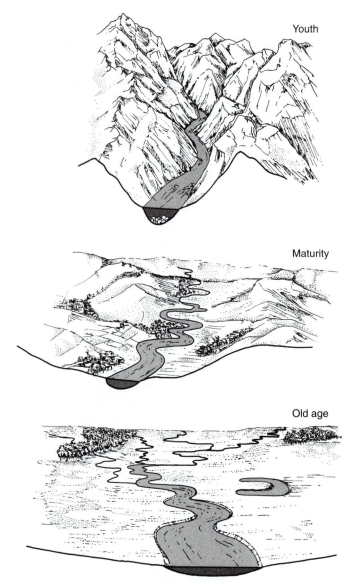

Fig. 3.7 The three main stages of landscape development in the geographic cycle: youth, maturity, and old age.

file reflects an equilibrium condition in a stream system representing a balance among channel slope (gradient), rate of flow, and the ability to move sediment. In other words as the stream develops a graded profile, it moves toward an equilibrium or near equilibrium condition.

Dynamic equilibrium concept The observation that stream systems trend toward equilibrium prompted another theoretical concept about river-sculptured landscapes based on the energy systems theory. This concept, called **dynamic equilibrium**, was advanced around 1960. In sharp contrast to the evolutionary perspective of the geographic cycle, the dynamic equilibrium concept argued that stream systems (and their landscapes) function as energy systems and as such are always trending toward a steady state. Instead of winding down to lower and lower energy levels (toward entropy) with the advancing stages of the geographic cycle, according to this concept the stream's energy system may increase or decrease with changes in the factors controlling its energy: climate (more precipita-

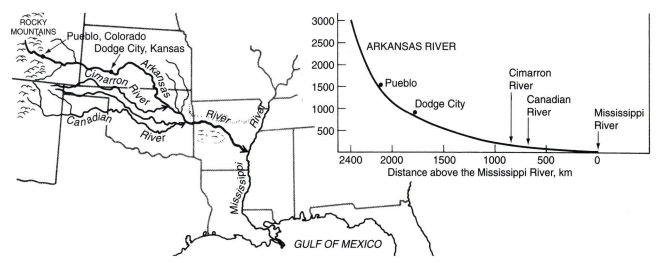

Fig. 3.8 The elevation profile of the Arkansas River illustrating the concept of the graded profile. The Arkansas' base level is the Mississippi River whereas the Mississippi's is the ocean (Gulf of Mexico).

tion means greater potential energy), runoff (more runoff means greater kinetic energy), and land elevation (higher elevation means greater potential energy).

Landscape change within watersheds need not be progressive and toward flatter, lower ground (with deep soil mantles), but can move in any direction (e.g., valleys may be downcut, filled, or maintained) depending on changes in the driving forces within the system. We see such trends in today's landscape; for example, at the local scale increased channel erosion and valley deepening in response to greater stormwater runoff with land clearing and urban development. With global warming in the twenty-first century, sea level rises are projected that will raise the base levels of streams draining to the ocean and in turn reduce potential energy in the lower parts of streams. The implications for landscape change include increased flooding and sedimentation in coastal lowlands.

Neither the geographic cycle nor the dynamic equilibrium concept addressed the influence of earth's varied climatic conditions on landscape development. W. M. Davis himself recognized this apparent shortcoming, and he and others wrote about an "arid land cycle" that addressed landform development peculiar to deserts. Later an attempt was made to build a broadly based concept incorporating more or less the full range of climate types and the geomorphic processes and landscape types associated with each. This concept was called **morphogenetic regions**, and it examined landscape as the product of different climatic regimes rather than only stream erosional systems. Much of the emphasis was on surface processes such as weathering and it underscored the differences, for example, between the wet tropics with strong chemical weathering and periglacial environments dominated by permafrost conditions and related processes. Outside of mountainous areas, where climates are arranged vertically, morphogenetic regions follow the broad zonal patterns of global climates. But climates change, so some zones exhibit features of two or more climates, and of course, the borders between climatic zones in most places are broadly transitional. Thus application of the morphogenetic regions concept must recognize past climatic environments, such as the glacial conditions of midlatitude zones only 10,000 years ago, as well as the inexact nature of the climatic borders (Table 3.1).

Morphogenetic region concept

Table 3.1   Morphogenetic Regions and Their Characteristics

Region	Present Climate	Past Climates	Active Processes[a]	Landforms
Zone of glaciers	Glacial (cold; wet)	Glacial	Glaciation	Glacial
Zone of pronounced valley formation	Polar, tundra (cool; wet, dry)	Glacial, polar, tundra	Cryogenic processes Stream erosion Mechanical weathering (Glaciation)	Box valleys Patterned ground Glacial forms
Extratropical zone of valley formation	Continental (cool, temperate; wet, dry)	Polar, tundra, continental	Stream erosion (Ground frost processes) (Glaciation)	Valleys
Subtropical zone of pediment and valley formation	Subtropical (warm; wet, dry)	Continental, subtropical	Pediment formation (Stream action)	Planation surfaces and valleys
Tropical zone of planation surface formation	Tropical (hot; wet, wet–dry)	Subtropical, tropical	Planation[b] Chemical weathering	Planation surfaces and laterite

[a] And former processes.
[b] Planation: processes leading to a low plain-like surface.
Source: adapted from Büdel, J., 1963.

Community-succession concept

Several notable concepts from ecology and plant geography also qualify as landscape theory. These deal mainly with the spatial and compositional dynamics of biotic communities and/or ecosystems, but can be extended to the larger landscape in many respects as well. The first of these, the **community-succession concept**, belongs with the evolution-based concepts of landscape change. This concept is based on the observation that communities of organisms have the capacity to stabilize the ground and progressively change the surface environment (especially soil, moisture, and ground-level climate), making it suitable for other communities. The succession model usually begins with a denuded (barren) surface, such as ground newly exposed by a retreating glacier, a harsh fire, or intensive wind erosion, onto which a community of resilient organisms, called *pioneers*, becomes established. The plants are usually small and hardy (e.g., grasses, mosses, and lichens) and they prepare the site, as it were, for another community, which is larger and more complex than the first. Each community represents a *successional stage* and each changes the landscape, making it more favorable for other biological communities, until eventually a relatively enduring community, called a *climax community*, is established. This community represents a state of bioequilibrium within the larger climatic environment.

Disturbance theory

Disturbance theory is in most respects a counterstatement to the community-succession concept. Rather than plants and related organisms acting as the principal change agents within a relatively passive or nurturing environment, disturbance theory holds that external (mainly abiotic) forces exert a shaping or controlling influence on the makeup and distribution of plant and animal communities in the landscape. The forces include fire, floods, drought, diseases, volcanic eruptions, hurricanes, land use, and pollution that operate at different magnitudes and frequencies around and within ecosystems. The organisms that make up ecosystems survive or perish depending on whether the impact of these forces, acting individually or collectively, exceeds their tolerance limits. When these forces are strong, such as with a hurricane, volcanic explosion, or construction activity, ecosystems in the affected area may be damaged, reduced, or destroyed. Disturbance theory argues that the earth surface is characterized by a more-or-less continuous string of such events operating at different magnitudes and frequencies, and that ecosystems, including soils, drainage, and other elements of the landscape are in a state of continuous

Fig. 3.9 On the left a photograph of the Pisgah Forest in New England, a 300-year-old climax forest destroyed (right) in one day by a hurricane in 1938.

adjustment to these forces. In other words, the patterns of ecosystems we see on the earth's surface are in a constant state of flux, and that progressive change ending in a climax state, as the succession concept argues, is too simple an explanation for this process. Thus, the disturbance concept views the environment's behavior as more irregular or chaotic than the succession concept does (Fig. 3.9).

Much of what we see as evidence of succession or disturbance depends on our perspective of time. If we look at the segments of time in the intervals between powerful events, succession appears to be a reasonable model. These intervals are relatively quiescent periods when growth and infilling can take place. However, if we take a longer view of time, then ecosystem changes look more like fluctuations in response to environmental perturbations and a particular spatial trend may or may not be apparent.

Human disturbance

No review of landscape concepts could ignore the human factor. When we consider the actions of humans, the time scale is especially critical. Set into geologic time (millions of years), the human epic on earth is a single event or disturbance of global proportions, the end of which we have yet to see. It is one of the major events of earth's recent geologic history. Set into historical time (several thousand years) however, the human epic can be broken down into thousands of events in different places with different effects. Examples include the land degradation in the Mediterranean Basin in ancient Greek times; industrialization and urban development in Europe in the eighteenth and nineteenth centuries; plowing of the North American prairies in the nineteenth and twentieth centuries; and the destruction of large areas of tropical rainforest in the twentieth century.

In many instances, some ecological recovery or succession takes place after major events; however, it is rarely complete. The Mediterranean lands can no longer support the forests and grass covers that existed before Christ. The lands in the Great Plains ravaged by erosion during the North American Dust Bowl of the 1930s once again support grasses, but the cover is weaker with fewer species (Fig. 3.10). Overall, the trend in historical times has been toward increased magnitude and frequency of disturbance by humans as our numbers, consumption of resources, and occupancy of the planet's surface have expanded. The periods or windows of time available for recovery, in turn, are shorter, leading to a lower order of quality in landscapes with fewer species, lower productivity, and often less resilience to future disturbances.

Fig. 3.10 Sources of landscape disturbance: (a) human in the Dust Bowl of the 1930s, which degrades prairie ecosystems, reduced topsoil, and lowered agricultural potential; and (b) natural in the devastation from the 1991 Mt. Pinatubo volcanic eruption in the Philippines.

Chaos theory   Some observers argue that the global environment has become more chaotic in the past several decades, and indeed **chaos theory** is receiving serious attention today. Among other things, they cite human-induced environmental change such as increased runoff and flooding related to urban development, increased ocean storminess related to global warming, and advancing land degradation related to desertification. Whether or not this is so is uncertain, but we are sure that today more people occupy disturbance-prone environments than in the past. Marginal environments that were traditionally beyond the bounds of significant agriculture and settlement, such as steep mountain slopes, deserts, stormy coastlines, and flood-prone river valleys, are clearly subject to more erratic changes and disturbances than nonmarginal environments. As population growth and economic development drive land uses farther into marginal environments, our susceptibility to disturbance rises. Natural disasters seem more common and more destructive (Fig. 3.10b). Not surprisingly, as these experiences mount, nature appears to many societies as less knowable, less predictable, and more chaotic. Chaos theory may emerge as the leading landscape concept of the twenty-first century.

3.9 SELECTED REFERENCES FOR FURTHER READING

Brunsden, D., and Thornes, J. B. "Landscape Sensitivity to Change." *Transactions of the Institute of British Geographers* 4, 1979, pp. 463–484.

Clements, F. E. *Plant Succession: An Analysis of the Development of Vegetation.* Washington, DC: Carnegie Institution No. 242, 1916.

Davis, W. M. *Geographical Essays* (D.W. Johnson, ed.). New York: Dover Publications, 1954.

Hack, J. T. "Interpretation of Erosional Topography in Humid Temperate Regions." *American Journal of Science,* Bradley Vol. 258-A, 1960, pp. 80–97.

Marsh, W. M., and Dozier, J. "Magnitude and Frequency Applied to the Landscape." In *Landscape: An Introduction to Physical Geography.* New York: Wiley, 1981.

Mitchell, C. W. *Terrain Evaluation* (2nd ed.). New York: Longman and Wiley, 1991.

Nikioroff, C. C. "Reappraisal of Soil." *Science,* 3383, 1959, pp. 186–196.

Peltier, L. C. "The Geographic Cycle in Periglacial Regions as It Is Related to Climatic Geomorphology." *Annals of the Association of American Geographers* 40, 1950, pp. 214–236.

Raup, H. M. "Vegetational Adjustment to the Instability of the Site." *Proceedings and Papers of the Sixth Technical Meeting.* Edinburgh: International Union for the Conservation of Nature and Natural Resources, 1959.

Selby, M. J. *Earth's Changing Surface: An Introduction to Geomorphology.* Oxford: Clarendon Press, 1985.

Wendell, Berry. *The Unsettling of America: Culture and Agriculture.* New York: Avon, 1978.

Wolman, M. G., and Gerson, R. "Relative Time Scales and Effectiveness of Climate in Watershed Geomorphology." *Earth Surface Processes* 3, 1978, pp. 189–208.

Wolman, M. G., and Miller, J. P. "Magnitude and Frequency of Forces in Geomorphic Processes." *Journal of Geology* 58, 1960, pp. 54–74.

4

TOPOGRAPHY, SLOPES, AND LAND USE PLANNING

4.1 INTRODUCTION

Given the choice of a place to live, most of us will choose hilly ground over flat ground. This is not surprising, for to most people, hilly terrain is more attractive because it has greater variations in vegetation, ground conditions, and water features, to say nothing of the opportunities it affords for vistas and privacy in siting houses. Our success with establishing and nurturing land uses on hillslopes, however, is not equal to our love for them. In fact for many land uses, slopes are decidedly inferior places to build. Many communities recognize this and require information on slopes to help guide development decisions.

Let us highlight a few slope–land use relations. Level or gently sloping sites are usually necessary for industrial and commercial buildings. Cropland is generally limited to slopes of less than 10 degrees (18 percent), because of the performance and safety restrictions posed by the operation of tractors and field machinery. In the era of horse and oxen power, slopes of 15 degrees (26 percent) or steeper could be used for cultivated crops. The influence of slopes and topography on the alignments of modern roads depends on the class of the road; the higher the class, the lower the maximum grades allowable. Interstate class expressways (divided, limited access, four or six lanes) are designed for high-speed, uninterrupted movement and are limited to grades of 4 percent, that is, 4 feet of rise per 100 feet of distance. On city streets, where speed limits are 20 to 30 mph, grades may be as steep as 10 percent, whereas driveways may be as steep as 15 percent.

Besides influencing land use, slopes may also have a pronounced influence on various biophysical aspects of the landscape. High on the list is the influence on runoff and ground stability. Runoff typically flows faster with greater erosive power on large, steep slopes. Likewise, steeply sloping ground is usually the first to fail in slides and mudflows when the land is saturated with massive amounts of rainfall. Slopes influence microclimate inasmuch as mountain sides with southern exposures are warmer and drier with fewer and smaller trees than their north-facing counterparts. The resultant differences in soil formation, runoff, and habitat are often striking. In short, slopes influence so many important aspects of landscape and land use that they have become one of the top two or three environmental criteria in regulating development at the community level. Not surprisingly, the slope map is probably the most widely used tool for evaluating the environmental suitability of development proposals by planning agencies in towns and counties.

4.2 SLOPE PROBLEMS

Misuse of slopes

The need to consider topography in planning is an outgrowth of the widespread realization not only that land uses have slope limitations but also that slopes have been misused in modern land development. The misuse arises from two types of practices: (1) the placement of structures and facilities on slopes that are already unstable or potentially unstable; and (2) the disturbance of stable slopes resulting in failure, accelerated erosion, and/or ecological deterioration of the slope environment.

The first type can result from inadequate survey and analysis of slopes in terrain that has a history of slope instability. More infrequently, however, it probably results from inadequate planning controls (for example, zoning and environmental ordinances) on development. In some instances, admittedly, surveys reveal no evidence of instability, and failure of a slope catches inhabitants completely unawares.

Disturbance of slope environments is unquestionably the most common source of slope problems in North America. Three types of disturbances stand out:

Fig. 4.1 Slope failure and erosion resulting from alteration of drainage, vegetation, and soil associated with residential development.

Causes of disturbance

■ *Mechanical cut and fill* in which slopes are reshaped by heavy equipment. This often involves steepening and straightening, resulting in a loss of the equilibrium associated with natural conditions; in Canada and the United States this is best exhibited in mining areas and along major highways.

■ *Deforestation* in hilly terrain by lumbering operations, agriculture, and urbanization. This not only results in a weakened slope because of the reduced stabilizing effect of vegetation, but also increases stress from runoff and groundwater because discharge rates are increased (see Fig. 3.4).

■ *Drainage alteration* related to improper siting and construction of buildings and other facilities, leading to an upset in the slope equilibrium and/or increased erosion from runoff (Fig. 4.1).

4.3 BUILDING SLOPE MAPS FOR LAND USE PLANNING

Years ago the configuration of the terrain could be measured only by field surveying. In its simplest form, this involved projecting a level line into the terrain from a point of known elevation and then measuring the distances above and below the line to various points in the terrain. Once elevation points were known and registered geographically to establish their locations, a contour map could be constructed.

Topographic contour map

Contour maps are comprised of lines, called *contours*, connecting points of equal elevation. In modern mapping programs, such as the one practiced by the U.S. Geological Survey, the contours are drawn from specially prepared sets of aerial photographs. These photographs and the optical apparatus used to view them enable the mapper to see an enlarged, three-dimensional image of the terrain. Based on this image, the mapper is able to trace a line, the contour, onto the terrain at a prescribed elevation. The contour elevation is calibrated on the basis of survey markers, called *bench marks*, placed on the land by field survey crews before the aerial photographs were flown.

Calculating percent slope

To determine the inclination of a slope from a topographic contour map, we must know the scale of the map and the elevation change from one contour to the next, called the *contour interval*. With these, the change in elevation over distance can be measured, and in turn a percentage can be calculated:

$$\text{percent slope} = \frac{\text{change in elevation}}{\text{distance}} \times 100$$

This is one of the two conventional expressions for slope, the other being degrees. Conversions can be made for degrees or percentages with the aid of the diagram in Figure 4.2. Degrees are commonly used in engineering calculations of slope stability, whereas percentages are the standard units used in planning for slope classifications.

Mapping slope

To avoid costly damage to the environment or to structures and utilities, it is necessary to make the proper match between land uses and slopes. In most instances this is simply a matter of assigning to the terrain uses that would (1) not require modification of slopes to achieve satisfactory performance, and (2) not themselves be endangered by the slope environment and its processes. Generally speaking, topographic contour maps alone do not provide information in a form suitable for most

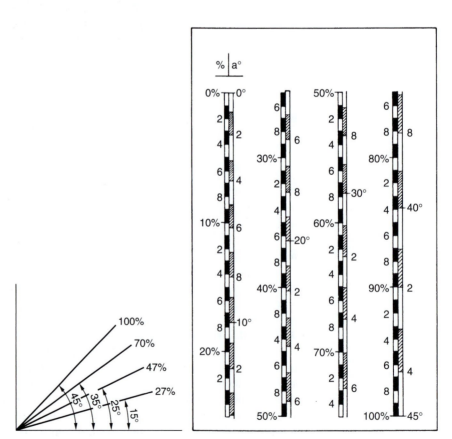

Fig. 4.2 Degree equivalence of percent slope up to 100 percent.

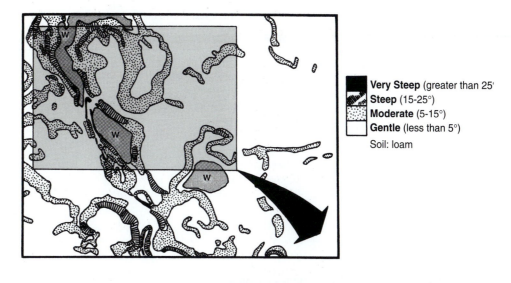

Very Steep (greater than 25°)
Steep (15-25°)
Moderate (5-15°)
Gentle (less than 5°)
Soil: loam

Presently active with erosion and/or failure underway.

Highly susceptible to failure and erosion should forest be removed.

Erosion imminent under present use.

Least susceptible to failure and erosion under agricultural, residential and related uses.

Natural waters presently influenced by sedimentation from eroded slopes.

Natural waters highly susceptible to sedimentation should nearly slopes be activated.

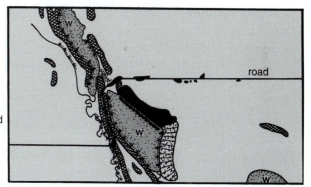

Fig. 4.3 Slope classification for the purpose of identifying slopes prone to erosion and failure. The area around the water features is abandoned farmland.

planning problems. The contour map must instead be translated into a map made up of slope classes tailored to planning problems. The utility of such slope maps is a function of (1) the criteria used to establish the slope classes, and (2) the scale at which the mapping is undertaken.

Map scale   The scale of mapping and the level of detail that are obtainable are strictly limited by the scale and contour interval of the base map. In areas where maps of two or three different scales are available, the scale chosen should be the one that best suits the scale of the problem for which it is intended, such as site plan review, master planning, or highway planning.

Setting slope classes   The criteria used to set the slope classes are dependent foremost on the problems and questions for which the map will be employed. The area mapped in Figure 4.3 is hilly and subject to heavy off-road vehicle traffic. The problem is how to manage this traffic with the least environmental damage such as soil erosion and lake sedimentation. Information on slope, both steepness and condition, is essential to development of a management plan. For areas under the pressure of suburban development the question is: Which land uses and facilities are appropriate and inappropriate for which slopes? One set of criteria would be the maximum and minimum slope limits of the various community activities. Another would be the natural limitations and conditions of the slopes themselves, which are taken up in the next section. With respect to land use activities, we would want to know the optimum slopes for parking lots, house sites, residential streets, playgrounds and lawns, and so on (Table 4.1).

Table 4.1   Slope Requirements for Various Land Uses

Land Use	Maximum	Minimum	Optimum
House sites	20–25%	0%	2%
Playgrounds	2–3%	0.05%	1%
Public stairs	50%	—	25%
Lawns (mowed)	25%	—	2–3%
Septic drainfields	15%[a]	0%	0.05%
Paved surfaces			
Parking lots	3%	0.05%	1%
Sidewalks	10%	0%	1%
Streets and roads		—	1%
20 mph	12%		
30	10%		
40	8%		
50	7%		
60	5%		
70	4%		
Industrial sites			
Factory sites	3–4%	0%	2%
Lay down storage	3%	0.05%	1%
Parking	3%	0.05%	1%

* Special drainfield designs are required at slopes above 10 to 12 percent.

In addition to the selection of slope classes and the appropriate base map, the preparation of a slope map also involves:

Building slope maps

1. *Definition of the minimum size mapping unit.* This is the smallest area of land that will be mapped, and it is usually fixed according to the base map scale, the contour interval, and the scale of the land uses involved. For 7.5-minute U.S. Geological Survey quadrangles (1:24,000), units should not be set much smaller than 10 acres, or 660 feet square.

2. *Construction of a graduated scale* on the edge of a sheet of paper, representing the spacing of the contours for each slope class. For example, on the 7.5-minute quadrangle, where 1 inch represents 2000 feet and the contour interval is 10 feet, a 10 percent slope would be marked by a contour every 1/20 inch.

3. *Placement of the scale on the map* in a position perpendicular to the contours to delineate the areas in the various slope classes (Fig. 4.4).

4. *Coding or symbolizing*, each of the areas delineated according to some cartographic scheme.

4.4 INTERPRETING SLOPE STEEPNESS AND FORM

In addition to inclination, we must often know a number of other things about slopes in order to make meaningful interpretations for land use planning. A first consideration is slope composition or lithology, that is, the soil and rock material that comprise the slope. For any earth material, there is maximum angle, called the **angle of repose**, at which it can be safely inclined and beyond which it will fail. The angle of repose varies widely for different materials, from 90 degrees in strong bedrock to less than 10 degrees in some loose, unconsolidated materials. Moreover, in unconsolidated material it may vary substantially with changes in water content, vegetative cover, and the internal structure of the particle mass. This is especially so with clayey materials. A poorly compact-

Angle of repose

Slope composition

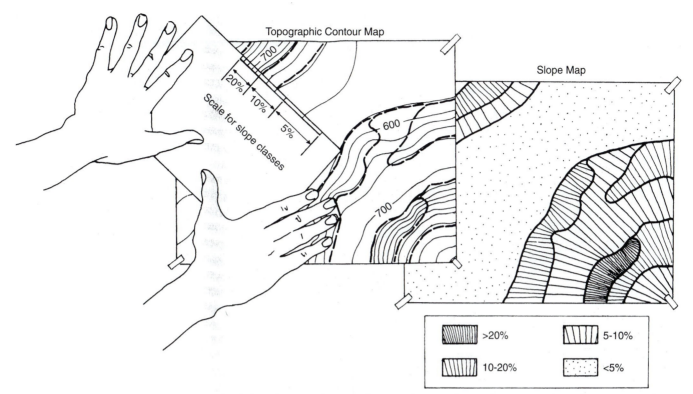

Fig. 4.4 Schematic diagram showing the use of a graduated scale for mapping slopes from a contour map. The lower map shows the results.

ed mass of saturated clay may give way at angles as low as 5 degrees (8 percent), whereas the same mass of clay with high compaction and much lower water content may be able to sustain angles greater than 45 degrees (100 percent). In most surface materials however, angles of repose are difficult, if not impossible, to define with much accuracy because influencing factors such as moisture content are so variable.

In surface deposits of more homogeneous composition, however, the angle of repose can be defined with more confidence. This is especially so for coarse materials, such as sand, pebbles, cobbles, boulders, and bedrock itself, which are less apt to vary with changes in water content and compaction. Representative angles of repose for some of these materials and others are given in Figure 4.5. Beyond the angles shown, these materials are susceptible to failure in which the slope ruptures and slides, slumps, falls, or topples, or in the case of saturated materials of clay, silt, and/or loamy composition, flows downslope.

Influence of Vegetation   The influence of vegetation on slopes is highly variable depending on the type of vegetation, the cover density, and the type of soil. Vegetation with extensive root systems undoubtedly imparts added stability to slopes composed of clay, silt, sand, and gravel, but the influence is limited mainly to the surface layer where the bulk of the roots are concentrated. For very coarse materials such as cobbles, boulders, and bedrock blocks, the influence of vegetation is typically not significant unless large trees buttress loose rock. On sandy slopes such as sand dunes, the presence of a forest cover can increase the inclination by 10 to 15 degrees above to 33 degree repose angle producing a metastable slope condition. Such slopes are sometimes marked by a distinctive convex profile.

Slope form   In addition to the overall angle, form or shape can also be an important factor in slope analysis. Form is expressed graphically in terms of a *slope profile*, which is

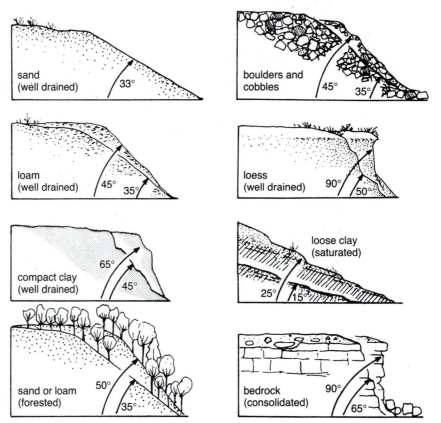

Fig. 4.5 Angles of repose for various types of slope materials. Angles are given in degrees.

basically a silhouette of a slope drawn to known proportions with distance on the horizontal axis and elevation on the vertical axis. The vertical axis is often exaggerated to ease construction and accentuate topographic details (Fig. 4.6).

Five basic slope forms are detectable on contour maps: straight, S-shape, concave, convex, and irregular (Fig. 4.7). These forms often provide clues about a slope's makeup, past behavior, and potential stability. For problems of land use planning and landscape management, it is necessary to understand the relationship between slope form and various geologic, soil, hydrologic, and vegetative conditions in order to identify typical and atypical slope shapes and conditions in different areas. In areas where slopes are comprised of unconsolidated materials (soils and various types of loose deposits), and bedrock is not a controlling factor, slope form is often the product of the interplay and balance among vegetation, soil composition, and runoff processes. Superimposed on this model, then, are the effects of undercutting by rivers, excavations for roads and buildings, changes in drainage, erosion associated with deforestation, and slope failures.

Common slope forms   *Smooth S-shapes* usually indicate long-term slope stability and a state of equilibrium among stabilizing and destabilizing slope forces. Such slopes rarely exceed 40 degrees inclination and are usually secured against heavy erosion by a substantial plant cover. Steep S-shaped slopes are commonly associated with hilly terrain, mountain foothills, and coastal areas where there is a permanent forest cover. *Concavities* in otherwise straight or S-shaped slopes are often signs of former failures, such as slides or slumps, and may indicate past disturbance from logging operations, road building, pipeline construction, and/or destabilizing subsurface conditions such as groundwater seepage.

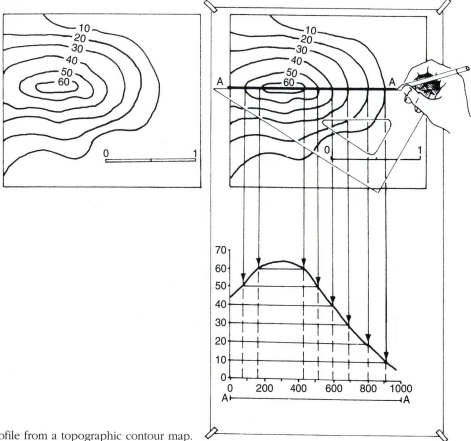

Fig. 4.6 Construction of a slope profile from a topographic contour map.

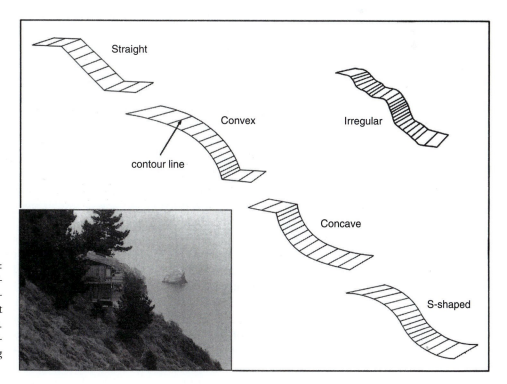

Fig. 4.7 Five basic slope forms: straight, convex, concave, *S*-shaped, and irregular with corresponding contour spacing as it appears on a topographic map. Inset photograph shows a dangerously steep convex slope leading down to the ocean.

(*a*)   (*b*)

Fig. 4.8   (*a*) Long, convex slopes on the Central California coast. The toe is cut back by wave erosion and the middle is held in place by bedrock. (*b*) Scars left by failures and erosion on mountain slopes stand out from the surrounding landscape (Fig. 4.8b).

Convex slopes may indicate the presence of resistant bedrock in the midslope, loading of the upper and middle slope by deposits (natural or human), and/or retreat of the footslope due, for example, to wave erosion or a road cut (Fig. 4.8*a*). Irregular slopes may reflect a variety of conditions. Where bedrock is a factor, variations in the resistance of adjacent strata to weathering and erosion may produce alternating gentle and steep slope segments. In other instances a rock unit or deposit within a slope is unstable and behaves erratically with changing moisture contents, giving rise to a chaotic slope profile.

4.5   ASSESSING SLOPES FOR STABILITY

Many criteria should be taken into account in assessing the susceptibility of slopes to failure. At the top of the list are slope angle, composition, history of instability. Steep slopes composed of rock formations, surface deposits, and/or soil types with an estab-

Indicators of instability   lished *record of instability* stand a greater chance of failure when subjected to land use activity because construction processes, loss of vegetation, and changes in drainage typically lower the stability threshold (Fig. 4.9). As planners in parts of California have shown, simply mapping unstable formations and their position in steep slopes can provide highly valuable information in land use planning. These formations are often marked by scallops in the slope profile and/or by visible failure scars imprinted into vegetation and soil covers (Fig. 4.8*b*). In time they are modified by runoff and erosion, but even ancient scars are often detectable on the basis of drainage and topographic irregularities and subtle variations in the types and patterns of vegetation.

Plant cover is another important criterion inasmuch as devegetated slopes show a much greater tendency to fail than fully vegetated ones. Studies of slopes in the

Conditions of instability   American West indicate that deforested slopes and slopes cut by logging roads in some areas fail more frequently under the stress of heavy precipitation than do fully

Fig. 4.9 The La Conchita landslide near Santa Barbara, California, March 1995. This slope had all the markings of a high risk setting, including steep inclination (>60%), evidence of former failure, susceptibility to instability under heavy rainfall, concentrated stormwater runoff from the upper slope, and land use disturbance (note the road cut near midslope). In 2005, after heavy rains, it failed again killing 10 people.

forested, undisturbed ones. From the air, scars caused by slope failure or severe erosion can often be identified on the basis of breaks in the plant cover. Fresh scars are marked by light tones on aerial photographs, and older scars by different plant species and forms (Fig. 4.8*b*). Although the stabilizing effects of vegetation are often not appreciated, there is no doubt about the overall effectiveness of a vigorous, dense plant cover, especially in reducing shallow failures and surface erosion. In addition to the strength added by roots, vegetation has many indirect effects on slope stability, including a remarkable capacity to dewater wet soils through transpiration. Accordingly, irregularities in an otherwise continuous and healthy plant cover are often clues to slope problems or potential problems.

Slope *undercutting* and *earthquake* activity are also significant. Active erosion at the foot of a slope by waves, rivers, or human excavation produces steeper inclinations and less confining pressure on the lower slope, thereby increasing the failure potential (Fig. 4.8*a*). Road cuts are particularly important on steep hillsides not only because the slope is physically weakened by the cut, but the cut may also intersect groundwater activating seepage and runoff that destabilize surface materials. When earthquakes jar rock and soil material, interparticle bonds may be weakened and the material's resistance to failure reduced. Some of the worst disasters from slope failure have been triggered by earthquakes, especially where material such as snow, rock debris, and sediments is perched on high slopes that are already near the failure threshold. .

Drainage conditions   *Drainage* is a critical consideration in assessing slope stability. Though often difficult to evaluate, runoff processes, soil water, and groundwater can have a pro-

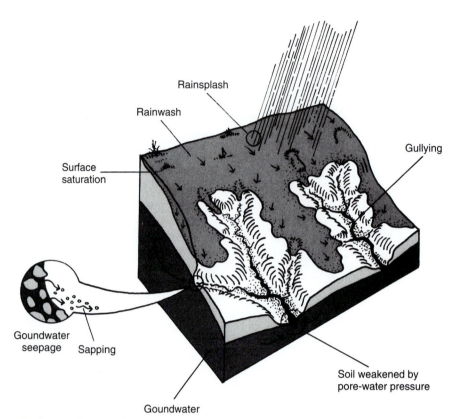

Fig. 4.10   Schematic diagram showing the influence of drainage on slope stability and erosion. The slope is least stable where gullies cut deeply and groundwater seeps from the ground.

nounced influence on slopes (see the crest slope in Fig. 4.9). Although runoff processes such as rainsplash and rainwash are usually ineffective in weakening the slope surface, they can saturate the soil and transform it into a muddy slurry that is prone to shallow mudflows. As runoff increases downslope and concentrates in channels, its erosive power increases immensely. This can result in gullying, which weakens the slope as it cuts through the soil and into the subsoil (Fig. 4.10). At depths of only 3 to 6 feet, gullies and channels can intercept groundwater, thus allowing it to seep out and saturate the soil material, which transforms it into a loose plastic or liquefied mass that is prone to erosion and mudflows.

In addition, groundwater seepage can produce undermining of slopes by processes known as *sapping and piping*, and since groundwater flow is driven by interparticle hydrostatic pressure, called *pore-water pressure*, this force may weaken the skeletal strength of materials within a slope causing failure. Pore-water pressure can be a significant cause of slope failure around reservoirs where the water table (and hydrostatic pressure) have been raised under reservoir side slopes. Pore-water pressure can also weaken ground where irrigation water or water from leaking sewer and water pipes has increased seepage along a slope. Both pore-water pressure and deep soil saturation leading to muddy soil masses were at the heart of the devastating slope failures in the Pacific Northwest in the winter of 1996–1997.

And finally, there is the *land use factor* to consider. Where slopes are cut by roads, mining operations, utility lines, or building foundations resulting in alteration of slope form, drainage and vegetative cover, the overall stability may be upset. Alterations in drainage may be especially significant. As we have noted, the raising of reservoirs is a well-established cause of slope failures. Earthen dams are known to fail by the same mechanism, namely, seepage and pore-water pressure. Drainage

Table 4.2 Slope Stability Checklist

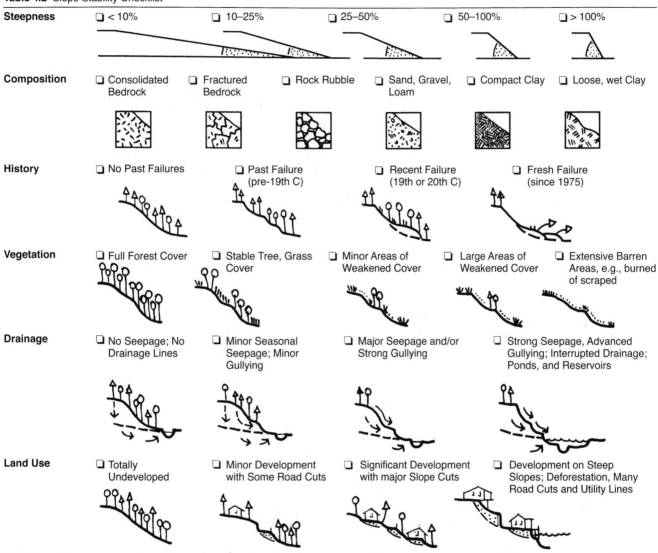

Steepness	❑ < 10%	❑ 10–25%	❑ 25–50%	❑ 50–100%	❑ > 100%
Composition	❑ Consolidated Bedrock	❑ Fractured Bedrock	❑ Rock Rubble	❑ Sand, Gravel, Loam	❑ Compact Clay	❑ Loose, wet Clay

History: ❑ No Past Failures — ❑ Past Failure (pre-19th C) — ❑ Recent Failure (19th or 20th C) — ❑ Fresh Failure (since 1975)

Vegetation: ❑ Full Forest Cover — ❑ Stable Tree, Grass Cover — ❑ Minor Areas of Weakened Cover — ❑ Large Areas of Weakened Cover — ❑ Extensive Barren Areas, e.g., burned of scraped

Drainage: ❑ No Seepage; No Drainage Lines — ❑ Minor Seasonal Seepage; Minor Gullying — ❑ Major Seepage and/or Strong Gullying — ❑ Strong Seepage, Advanced Gullying; Interrupted Drainage; Ponds, and Reservoirs

Land Use: ❑ Totally Undeveloped — ❑ Minor Development with Some Road Cuts — ❑ Significant Development with major Slope Cuts — ❑ Development on Steep Slopes; Deforestation, Many Road Cuts and Utility Lines

changes around roads, pipelines, and bridges can also cause increased seepage and redirected streamflows can weaken slopes and cause serious failures.

Synthesizing slope factors — Unfortunately, the means to integrate analytically all these variables have not been developed, and for large areas field and laboratory testing are not economically feasible. Engineering models for calculating stability are generally not reliable either, because most natural slopes (as opposed to slopes constructed in fill material) are too complex in terms of composition and drainage to fit the equations. In addition, the engineering formulas are site/slope specific and do not lend themselves to mapping programs. Therefore, evaluations of slope stability for purposes of land use planning and environmental management must be based on some sort of systematic review using the criteria cited previously; history of stability, composition, steepness, and the like, and a basic checklist of these criteria is provided in Table 4.2. Whether all these criteria can be used in mapping depends on the availability and reliability of data and one's abilities to synthesize and interpret them.

Data sources — Generally, topographic contour maps, aerial photographs, geologic maps, seismic maps, and soil maps are used as data sources, but they require interpretation and adaptation because of differences in scale, resolution, and units of measurement. When the desired criteria can be employed, the map overlay technique can be used to delineate areas of different levels of slope stability—GIS techniques are certainly helpful in such tasks. Finally, it should be noted that slope stability is always the product of a combination of conditions, some transient, some permanent, some visible, some not. Not surprisingly failure events are difficult to forecast. Some California slopes with no record of instability fall apart, as it were, after vegetation loss to fires is coupled with ground saturation from prolonged, heavy winter rains; other slopes that are steeper, wetter, and more denuded hold their own.

4.6 APPLICATIONS TO COMMUNITY PLANNING

In the absence of effective planning controls on development and the base information necessary to help guide decision making, many American and Canadian communities in the past several decades have seen development creep over the terrain, enveloping sloping and flat ground alike. The main force driving this development is real estate economics rather than terrain, drainage, and related factors. Unlike development of an earlier era when the decision maker, developer, and resident were one and the same, in modern development these roles are often played by different and separate parties.

Buyer beware   As a result, in the modern, large-scale development project, the developer may make poor decisions on siting and upon sale of the property the matter is passed on to the buyer. When a problem eventually arises such as foundation failure or flooding, it becomes the responsibility of the owner/resident. Complicating matters further, the property may have a second or third owner by the time the problem actually surfaces. The social and financial costs for remedial action are typically high and must be borne by the current property owner, by the community, by some higher level of government, or by all three parties. This harsh lesson has taught communities that the tendency to push development beyond the terrain's capacity must be curbed by land use regulations and/or the education of developers and buyers.

As a result, environmental ordinances (or by-laws) based on slope have been enacted in many North American communities. The rationale behind local slope ordinances varies. In some communities, they are designed for growth control, in others for environmental protection, and in still others for hazard management. In any case, the implementation of slope-based environmental ordinances necessitates (1) building a slope map showing suitable and unsuitable terrain, and (2) establishing a procedure to review and evaluate development proposals and site plans.

Slope-related ordinances   Communities use several different approaches in designing slope-related ordinances. One approach is based on development density and average slope inclination within specified land areas or units. Terrain units (such as valley floors or foothills) comprising many individual but similar slopes are assigned a maximum allowable density of development according to the average slope inclination (Fig. 4.11). The percentage of ground to be left undisturbed as required by three California communities is shown in Table 4.3. In Chula Vista, for example, terrain with slopes averaging above 30 percent requires that for each acre of development 9 acres be left in an undisturbed state.

A second approach is based strictly on the inclination of individual slopes. Each slope over a specified area, usually a proposed development site, is mapped according to the procedure described earlier in the chapter. Four or five classes are usually used, and for each there is a maximum allowable density set by ordinance. In Austin, Texas, for example, residential density is limited to one unit per acre for slopes in the 0 to 15 percent class. For slopes in steeper classes, 15 to 25 percent and 25 to 35 percent, development is allowable only on approval of a formal request.

The third approach employs several slope characteristics in addition to inclination. For example, slopes comprised of unstable soils, conditionally stable forested

Fig. 4.11 An example of terrain units in an area of mountainous topography. Highest densities would be allowed in class I where slopes average 10 percent or less.

slopes, and scenically valued slopes are mapped in combination with slope inclination. Development plans are compared directly to the distribution of each factor. This approach is commonly employed in areas of complex terrain where regulations based on inclination alone would prove inadequate.

None of the methods used by community ordinances relates slopes to the fundamental terrain or geomorphic unit in which they develop and are nested, namely, the watershed. The patterns of slope stability and instability are not randomly distributed in the landscape, but tend to show spatial patterns related mainly to steep ground (rugged topography) and drainage conditions such as soil saturation, groundwater seepage, and rapid runoff. These topographic and drainage conditions occur mostly in the upper parts of watersheds, in valley heads, foothills, and on mountain slopes.

Ordinances need to protect and manage these areas (zones) not only to reduce land use and environmental damage, but to hold down servicing costs, protect water

Slopes in watershed management

Table 4.3 Undisturbed Area Requirements for Sloping Ground in Three California Communities

Percent Slope (avg.)	Chula Vista	Pacifica	Thousand Oaks
10	14%	32%	32.5%
15	31%	36%	40%
20	44%	45%	55%
25	62.5%	57%	70%
30	90%	72%	85%
35	90%	90%	100%
40	90%	100%	100%

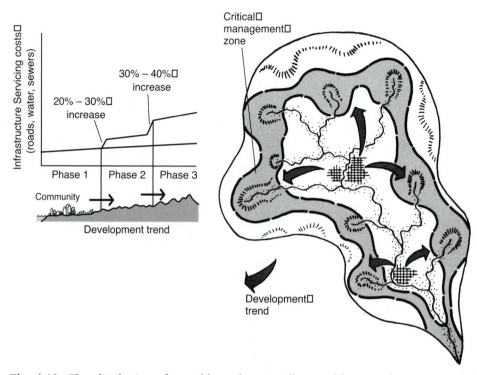

Fig. 4.12 The distribution of unstable and potentially unstable ground in a watershed (defined by the critical management zone). Development trends, shown by the arrows, are often toward this zone with residential building reaching higher and more rugged terrain where, as the graph shows, servicing costs are substantially greater.

supplies, manage flooding, and conserve open space (Fig. 4.12). All communities are situated in a watershed of some size and shape, and most are located at the receiving (downstream) end of the basin. Suburban and rural residential development, on the other hand, are pushing into the upper, more scenic and environmentally risky parts of watersheds. Unless communities recognize this trend and the need to formulate watershed-based approaches to land use planning that address landforms, slope stability, drainage, flooding, soils, forests, and so on, in an integrated framework, individual slope ordinances themselves will be of little value in the long run. We examine watersheds and their drainage systems later in Chapters 8 and 9.

4.7 WHY SLOPE MAPS MAY NOT WORK IN SOME AREAS

Slope and runoff rates

One of the chief rationales used by communities for limiting development on steep slopes is that facilities pose threats to the slope environment and related areas downhill. Among the commonly cited concerns is that development on slopes drives up runoff rates resulting in increased soil erosion, greater flooding and decreased water quality downslope. This assumes, of course, that the rate of hillslope runoff and related environmental damage rise with slope steepness. For a number of reasons, the relationship of slope inclination to runoff is often misunderstood. First, the total volume of rainfall on a slope actually decreases at steeper slope angles because the slope surface (catchment) area exposed to the sky is smaller for steeper slopes. For a vertical cliff, there is no catchment area. Second, the area above a slope is often much more significant than the slope itself as a source of runoff. Some gentle slopes may be more hazardous than nearby steep slopes because of large upslope contributions of runoff. Third, standard slope maps do not take slope runoff patterns into account, that is, whether runoff concentrates or diffuses as it spills over the slope face. The reason is that slope maps are only two-dimensional expressions. Three-dimensional expressions of slopes give us quite another view.

Slope form and runoff   As three-dimensional forms, most slopes are either concave or convex (they form either a nose or a hollow) and it is these forms that control the patterns of runoff. On a nose, runoff diffuses (spreads) downslope, whereas in a hollow, runoff concentrates downslope (see Fig. 9.10). Therefore noses, which are often steeper than hollows, tend to be dry, whereas hollows tend to be wet. Thus, the gentler slope form, the hollow, is in fact a more risky place for development both from the standpoint of facility damage and environmental degradation. The steeper slope, on the other hand, is drier with less upslope–downslope continuity of drainage. It follows that strict adherence to conventional slope maps (based on only steepness) could result in decisions opposite the ordinance's intent where the objective is limiting the impact of land use on slope drainage, water quality, and related factors such as habitat.

4.8 CASE STUDY

■ Slope as a Growth-Control Tool in Community Planning

W. M. Marsh

Years ago, when rural communities in the United States and Canada were setting up their systems of government, large areas of undeveloped land were often declared part of the community by virtue of the exaggerated alignment chosen for the town or village limits at the time of incorporation. These areas were often drawn into maps of the community with little or no attention to the nature of the terrain, and, in most cases, no realistic expectation on the part of the town fathers that the community itself would ever expand into them. Some communities even went so far as to plat these areas, that is, subdivide them on paper into residential lots, city blocks, streets, and neighborhoods, but most remained undeveloped.

In recent decades, many of these empty areas have begun to take on development. Because they were already platted, the communities often find that they have little control over where development takes place. Their only recourse is to formulate planning policy to help manage new development and keep the community in balance with its resource base.

Planning ordinances aimed at regulating new development can take various forms. One form is a growth-control regulation that simply sets limits on the amount, type, and rate of development. This can be accomplished, for example, by setting limits on land use density or by setting quotas on the approval of building permits. Another form is the environmental ordinance that sets some kind of environmental limitation on development. Some California communities use water supply as the limiting factor. More commonly, communities use one or more landscape features such as slope, soil, wetland, and threatened species to regulate development.

For the northern Michigan community shown here, platting was carried out in the late 1800s or early 1900s when steep slopes and other terrain obstacles were often ignored despite the fact that they were hard to get up, difficult to build on, and even more difficult to farm (see Map A). Today, however, building on these slopes is technically feasible and because of the vistas they provide over Lake Superior, they are highly attractive to prospective home builders. To protect this valuable part of the community environment, a slope ordinance (by-law) is clearly needed.

To be most effective, the ordinance should be based on rational criteria that justify the controls imposed on land users. In other words, the reasons for not building on slopes should appeal to the citizen's logic from both a personal and community standpoint. In this case, the slopes are composed on sandy soil held in place by hardwood forests, and changes in the balance among the forces acting on the slope face could easily lead to instability and extensive damage downslope. This can be demonstrated by documentation from scientific and planning literature reporting on problems in similar settings. In addition, residential development on these slopes would not be served by the city sewer systems and therefore would require on-site

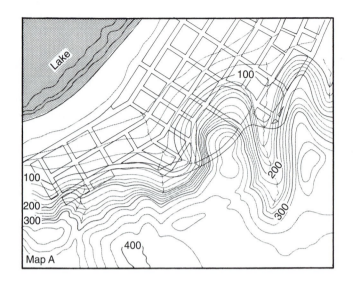

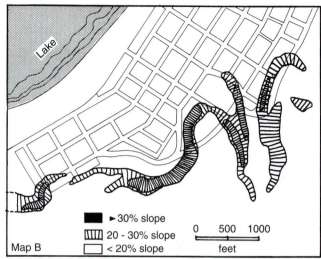

sewage disposal. Evidence is abundant and compelling on the severe limitation that steep slopes pose to on-site (septic) sewage systems. Based on these and other reasons, there is ample justification in this community for invoking slope as a growth-limiting factor for residential and other types of development and drafting a map (such as Map B) along with appropriate rationale to demarcate slopes that are off limits. ∎

4.9 SELECTED REFERENCES FOR FURTHER READING

Briggs, R. P., et al. "Landsliding in Allegheny County, Pennsylvania." *Geological Survey Circular* 728, 1975.

Carson, M. A., and Kirkby, M. J. *Hillslope Form and Process.* Cambridge, United Kingdom: Cambridge University Press, 1972.

City of Austin. *Comprehensive Watersheds Ordinance.* Austin, TX: 1986.

Cooke, R. U., and Doornkamp, J. C. *Geomorphology in Environmental Management* (2nd ed.). New York: Oxford University Press, 1990.

Crozier, M. J. "Field Assessment of Slope Instability." In *Slope Instability* (D. Brunsden and D. B. Prior, eds.). Hoboken, NJ: Wiley, 1984, pp. 103–142.

Dorward, Sherry. *Design for Mountain Communities: A Landscape and Architectural Guide.* Hoboken, NJ: Wiley, 1990.

Gordon, Steven I., and Klousner, Robert D., Jr. "Using Landslide Hazard Information in Planning: An Evaluation of Methods." *Journal of the American Planning Association* 52:4, 1986, pp. 431–442.

Keaton, J. R., and DeGraff, J. A. "Surface Observation and Geologic Mapping." In *Landslides: Investigation and Mitigation* (K. A. Turner and R. L. Schuster, eds.). Washington, DC: National Academy of Science Press, 1996, pp. 178–230.

Milne, R. J., and Moss, M. R. "Forest Dynamics and Geomorphic Processes on the Niagara Escarpment, Collingwood, Ontario." In *Landscape Ecology and Management.* Montreal: Polyscience Publications, 1988.

Schuster, Robert L., and Krizek, Raymond J. *Landslides: Analysis and Control.* Washington, DC: National Academy of Science Press, 1978.

Terzaghi, K. "The Mechanism of Landslides." In *Application of Geology to Engineering Practice* (S. Paige, ed.). Geological Society of America, Berkey Volume, 1950, pp. 83–123.

Turner, K. A., and Schuster, R. L. *Landslides: Investigation and Mitigation,* Special Report 247. Washington, D C: National Academy of Science Press, 1996.

5

SOIL, LAND USE SUITABILITY, AND WASTE DISPOSAL

5.1 INTRODUCTION

The relationship between soil composition and agriculture is apparent at practically any scale of observation, especially in areas of more traditional agriculture. The relationship between soil composition and other types of land uses, however, is often not so apparent. At least part of the reason for this in North America is that modern developers have given precious little attention to soils. Among the factors contributing to this state of affairs is a sense of complacency about soils created by a popular impression that modern engineering technology can overcome problems of the soil. Granted, the technology does exist to build practically any sort of structure in or on any environment. However, the added costs for construction and environmental mitigation in areas of weak, unstable and/or poorly drained soil are often prohibitive. Increasingly, decision makers are looking to soil surveys to help guide in site selection and site planning for residential, industrial, and other forms of development that involve surface and subsurface structures.

An additional consideration is the heightened concern over solid and hazardous waste disposal in the soil. Finding new disposal sites is a perennial problem requiring, among other things, careful evaluation of soil, topography, and drainage. It has also come to light that most urban areas contain scores of disposal sites and waste dumps hidden in the soil that pose not only an environmental threat but also a serious liability to development. If buried waste is encountered in constructing facilities, such as roads, buildings, and utilities, the site may have to be abandoned and/or the waste excavated and properly disposed somewhere else. For many landowners, the costs of mitigation are proving to be insurmountable; thus the demand for predevelopment soil surveys in many areas.

Barring special needs such as buried waste or unstable slopes, soils nevertheless tend to be treated less seriously by planners than other major environmental factors. Too often data and information from agricultural soil surveys, such as those conducted by the U.S. Natural Resources Conservation Service, are taken at face value and transposed to site maps without much regard for their scale and other limitations. In other instances, soils are purposely left to the engineers to be investigated as a part of facility design and construction; these investigations are usually restricted to building sites and subsoil materials. Neither approach is acceptable in environmental planning, although both agricultural soil surveys and engineering investigations are important sources of soil data for our purposes.

5.2 SOIL COMPOSITION

Several features, or *properties*, are used to describe soil for problems involving land development. Of these, texture and composition are generally the most meaningful; from them we can make inferences about bearing capacity, internal drainage, erodibility, and slope stability.

Composition

Composition refers to the materials that make up a soil. Basically, there are just four compositional constituents: mineral particles, organic matter, water, and air. *Mineral particles* comprise 50 to 80 percent of the volume of most soils and form the all-important skeletal structure of the soil. This structure, built of particles lodged against each other, enables the soil to support its own weight as well as that of internal matter such as water and the overlying landscape, including buildings. Sand and gravel particles generally provide for the greatest stability and, if packed solidly against one another, will usually yield a relatively high bearing capacity. *Bearing capacity* refers to a soil's resistance to penetration from a weighted object such as a building foundation. As a whole, clays tend to have lower bearing capacities than

Bearing capacity

Table 5.1 Bearing Capacity Values for Rock and Soil Materials

Class	Material		*Allowable Bearing Value tons per square foot*
1		Massive crystalline bedrock, e.g., granite, gneiss	100
2	(rock)	Metamorphosed rock, e.g., schist, slate	40
3		Sedimentary rocks, e.g., shale, sandstone	15
4		Well-compacted gravels and sands	10
5		Compact gravel, sand/gravel mixtures	6
6		Loose gravel, compact coarse sand	4
7		Loose coarse sand; loose sand/gravel mixtures, compact fine sand, wet coarse sand	3
8	(soil materials)	Loose fine sand, wet fine sand	2
9		Stiff clay (dry)	4
10		Medium-stiff clay	2
11		Soft clay	1
12		Fill, organic material, or silt	(fixed by field tests)

Source: Code Manual. New York State Building Code Commission.

sands and gravel. Loosely packed, wet masses of clay tend to compress and slip laterally under the stress of loading (Table 5.1).

The role of organic matter The quantity of organic matter varies radically in soils, but it is extremely important for both negative and positive reasons. Organic particles usually provide weak skeletal structures with very poor bearing capacities. Organic matter tends to compress and settle differentially under roadbeds and foundations, and when dewatered, it may suffer substantial volume losses as well as, decomposition, wind erosion, and other effects. Deep organic soils, which may reach thicknesses of 20 feet or more, of course pose the most serious limitations to facility development and land use in general.

On the positive side, organic matter is vital to the fertility and hydrology of soil. Indeed, the loss of topsoil from agricultural lands by runoff and wind erosion is viewed as one of the most serious environmental issues of our time. The role of organic matter in the terrestrial water balance is underscored by the water storage function of topsoil and organic deposits in wetlands. Topsoil takes up significant amounts of precipitation and therefore helps reduce runoff rates. Organic deposits often serve as moisture reservoirs for wetland vegetation as well as points of entry for groundwater recharge.

5.3 SOIL TEXTURE

The mineral particles found in soil range enormously in size from microscopic clay particles to large boulders. The most abundant particles, however, are sand, silt, and clay, and they are the focus of examination in studies of soil texture (Fig. 5.1).

Definition of texture **Texture** is the term used to describe the composite sizes of particles in a soil sample, (usually several representative handfuls). To measure soil texture, the sand, silt, and clay particles are sorted out and weighed. The weights of each size class are then expressed as a percentage of the sample weight.

Since it is unlikely that all the particles in a soil will be clay or sand or silt, additional terms are needed for describing various mixtures. Soil scientists use 12 basic terms for texture, at the center of which is the class *loam*, an intermediate mixture of sand, silt, and clay. According to the United States Natural Resource Conservation Service, the loam class is comprised of 40 percent sand, 40 percent silt, and 20 per-

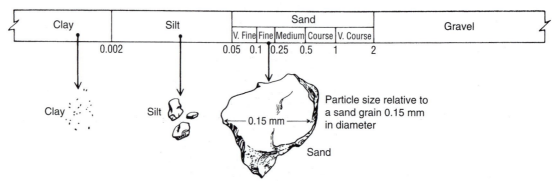

Fig. 5.1   A standard scale of soil particle sizes; the illustrations give a visual comparison of clay and silt with a sand particle 0.15 mm in diameter.

cent clay. Given a slightly heavier concentration of sand, for instance 50 percent, with 10 percent clay and 40 percent silt, the soil is called *sandy loam*. In agronomy, the textural names and related percentages are given in the form of a triangular graph (Fig. 5.2). If we know the percentage by weight of the particle sizes in a sample, we can use this graph to determine the appropriate soil name.

Field hand test   In the field, soil texture can be estimated by using the **field hand test**. The procedure involves extracting a handful of soil and squeezing it into three basic shapes: (1) *cast*, a lump formed by squeezing a sample in a clenched fist; (2) *thread*, a pencil shape formed by rolling soil between the palms, and (3) *ribbon*, a flattish shape formed by squeezing a small sample between the thumb and index finger (Fig. 5.3). The behavioral characteristics of the soil when molded into each of these shapes, if they can be formed at all, provides the basis for a general textural classification. The sample should be damp to perform the test properly.

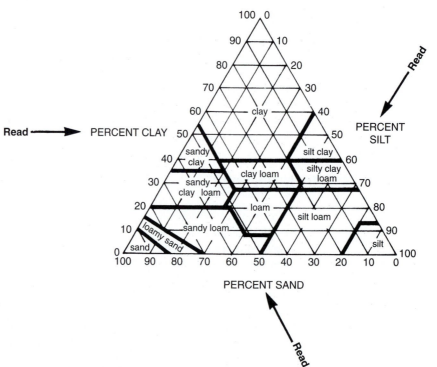

Fig. 5.2   Soil textural triangle, including the major subdivisions, used by the U.S. Department of Agriculture.

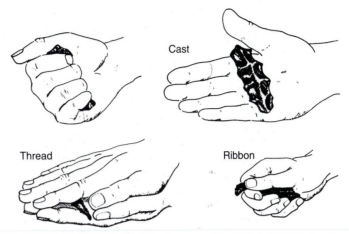

Fig. 5.3 The basic shapes used in the field hand test. See Table 5.2 for evaluation criteria in determining soil textural type.

Behavior of the soil in the hand test is determined by the amount of clay in the sample. Clay particles are highly cohesive, and, when dampened, they will behave as a plastic. Therefore, the higher the clay content in a sample, the more refined and durable the shapes into which it can be molded. Table 5.2 gives the behavioral traits of five common soil textures.

Sieving   Another method of determining soil texture involves the use of devices called *sediment sieves*, screens that have been built with a specified mesh size. When the soil is filtered through a group of sieves, each with a different mesh size, the particles become sorted into corresponding size categories. Each category can be weighed and the weight expressed as a percentage of the total sample weight, in order to make a textural determination.

Although sieves work well for silt, sand, and larger particles, they are not appropriate for clay particles. Clay is too small to sieve accurately; therefore, in clayey soils the fine particles are measured on the basis of their settling velocity when suspend-

Table 5.2 Behavioral Characteristics of the Basic Shapes Used in the Field Hand Test and the Soil Type Represented by Each

Field Test (Shape)	Soil Type				
	Sandy Loam	*Silty Loam*	*Loam*	*Clay Loam*	*Clay*
Soil cast	Cast bears careful handling without breaking	Cohesionless silty loam bears careful handling without breaking; better-graded silty loam casts may be handled freely without breaking	Cast may be handled freely without breaking	Cast bears much handling without breaking	Cast can be molded to various shapes without breaking
Soil thread	Thick, crumbly, easily broken	Thick, soft, easily broken	Can be pointed as fine as pencil lead that is easily broken	Strong thread can be rolled to a pinpoint	Strong, plastic thread that can be rolled to a pinpoint
Soil ribbon	Will not form ribbon	Will not form ribbon	Forms short, thick ribbon that breaks under its own weight	Forms thin ribbon that breaks under its own weight	Long, thin flexible ribbon that does not break under its own weight

ed in water. Since clays settle so slowly, they are easily segregated from sand and silt. The water containing the clay can be drawn off and evaporated, leaving a residue of clay particles, which can be weighed.

5.4 SOIL MOISTURE AND DRAINAGE

Forms of soil water

The water content of soil varies with particle sizes, local drainage and topography, and climate. Most water in the soil occupies the spaces between the particles; only in organic soils and certain clays do the particles themselves actually absorb measurable amounts of water. Two principal forms of water occur in both mineral and organic soils: capillary and gravity. **Capillary water** is a form of molecular water, so-called because it is held in the soil by the force of cohesion among water molecules. Under this force, water molecules are mobile and can move from moist spots to dry spots in the soil. In the summer, most capillary water transfer is upward toward the soil surface as water is lost in evaporation and transpiration.

Water table

Gravity water is liquid water that moves in response to the gravitational force. Its movement in the soil is preponderantly downward, and it tends to accumulate in the subsoil and underlying bedrock to form groundwater. Groundwater completely fills the interparticle spaces: thus below the water table the soil is largely devoid of air.

Drainage processes

References to *drainage* in soil reports usually refer to gravity water and a soil's ability to transfer this water downward. Three terms are used to describe this process: (1) *infiltration capacity*, which is the rate that water penetrates the soil surface (usually measured in cm or inches per hour); (2) *permeability*, which is the rate that water within the soil moves through a given volume of material (also measured in cm or inches per hour); and (3) *percolation*, which is the rate that water in a soil pit or pipe within the soil is taken up by the soil (used mainly in wastewater absorption tests and measured in inches per hour). Reference to poor drainage, for example, means that the soil is frequently or permanently saturated and may often have water standing on it. Terms such as "good drainage" and "well drained" mean that gravity water

Meaning of well-drained

is readily transmitted by the soil and that the soil is not conducive to prolonged periods of saturation. Soil saturation may be caused by the local accumulation of surface water (because of river flooding or runoff into a low spot, for example), or by a rise in the level of groundwater within the soil column (because of the raising of a reservoir or excessive application of irrigation water, for example), or because the particles in the soil are too small to transmit infiltration water (because of impervious layers within the soil or clayey soil composition).

5.5 SOIL, LANDFORMS, AND TOPOGRAPHY

Parent deposits

Nearly all soils are formed in deposits laid down by geomorphic processes of some sort; for example, wind, glacial meltwater, ocean waves, river floods, and landslides. In Canada and the United States, three types of surface deposits dominate the landscape: *glacial drift*, *alluvium*, and *aeolian* (loess and sand dunes) (see Fig. 2.12). These deposits form the **parent material** for most soils, and it is important to know not only which material is likely to occur in a particular location but what sort of spatial variability it is likely to have. In broadly uniform terrain such as the central Great Plains, there is little variability in soil materials; loess blankets virtually all surfaces except for stream valleys. In diverse terrain, however, it is not uncommon to find deposits of markedly different compositions over areas as small as 10 acres or so. For example, in river valleys, deposits of sandy gravels and organic matter are often complexly intermingled across floodplains; and in glaciated terrain, deposits may range

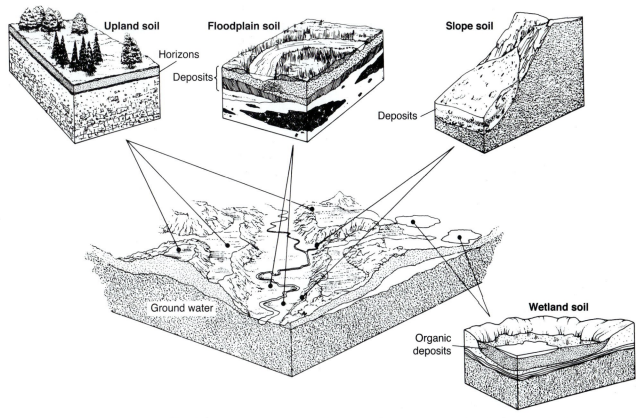

Fig. 5.4 Common sources and settings of deposits in which soils form: wetland, slope, flood-plain, and upland deposits.

from loose sands to dense clays over a distance of several hundred meters or less. Each deposit, however, tends to be associated with certain indicators: landforms, water features, and/or vegetation types (Fig. 5.4).

The solum   Once in place, the surface layer of all deposits is subject to alteration by soil forming processes associated with climate, vegetation, surface drainage, and land use. These processes in time produce a complex medium characterized by a sequence of zones or *horizons* of different physical, chemical, and biological compositions. This medium is usually 3 to 8 feet deep, and is known as the **solum** or "true soil". This is what soil scientists measure and map in national, regional, and county soil surveys. (Appendix A provides information on the principal soil classification systems used in the United States and Canada.) Below the solum, the bulk of most deposits, however, retains the basic properties imparted to them during their formation. This fact is especially significant in land development problems, because landfills, basements, roadbeds, slope cuts, and the like are built in these materials. The Unified Soil Classification System, which is widely used in civil engineering, applies to these materials (Table 5.3).

Landforms and soils   Geomorphologists have developed a basic taxonomy of the landforms associated with various types of deposits, making it possible to infer certain things about soil materials based on a knowledge of landforms. Any attempt to correlate landforms and soils, however, must recognize the limitations of geographic scale. Generally, the extremes of scale provide the least satisfactory results. At the scale of individual landform features, such as wetlands, sand dunes, or outwash plains, the correlation can

Table 5.3   Unified Soil-Classification System

Letter	Description	Criterion	Further Criteria
G	Gravel and gravelly soils (basically pebble size, larger than 2 mm diameter)	Texture	Based on uniformity of grain size and the presence of smaller materials such as clay and silt:
S	Sand and sandy soils	Texture	W Well graded (uniformly sized grains) and clean (absence of clays, silts, and organic debris) C Well graded with clay fraction, which binds soil together P Poorly graded, fairly clean
M	Very fine sand and silt (inorganic)	Texture Composition	Based on performance criteria of compressibility and plasticity:
C	Clays (inorganic)	Texture Composition	L Low to medium compressibility and low plasticity
O	Organic silts and clays	Texture Composition	H High compressibility and high plasticity
P	Peat	Composition	

be quite good for planning purposes. Appendix B lists a number of landform features along with the soil composition and drainage associated with each.

Toposequence   There are also predictable trends in soil makeup related to topographic gradients on individual landforms (Fig. 5.5). Such trends are called **toposequences**, and they can provide useful information for site planning. In the case of an alluvial fan, for example, texture tends to grade course to fine from the top to the toe. In addition, *alluvial fans* are structurally complex, being comprised of many layers of sediment, some of which may be saturated and unstable for building. Another type of deposit, *talus slopes*, is comprised of rock fragments that have fallen into place from an upper slope. In contrast to alluvial fans, particle size tends to increase downslope, and drainage is good at all levels. On vegetated hillslopes, toposequences are usually more subtle and limited largely to the surface layer. Topsoil, in particular, will vary with slope steepness, having the weakest development near midslope where the inclination is greatest and runoff removes most organic litter. Near the toe of the slope, topsoil thickens as runoff slows down and organic matter is deposited (Fig. 5.6).

5.6 APPLICATIONS TO LAND PLANNING

Published soil surveys   In both the United States and Canada, government agencies have conducted extensive surveys to map and classify soils. **Soil surveys** were originally designed to serve agriculture, but since 1960 or so they have been expanded to serve community land use and environmental interests as well. As we noted in Chapter 3, the Natural Resources Conservation Service, an agency of the U.S. Department of Agriculture, is responsible for soil surveys in the United States. Surveys are organized by county, and the results are published in a volume called a county soil survey that contains soil descriptions, various data and guidelines on soil uses and limitations, and maps of soil distributions. At the most fundamental level, the soil surveys provide data and information in three key areas: soil drainage, soil texture, and soil composition. They are, however, limited in three significant respects: (1) major metropolitan areas are

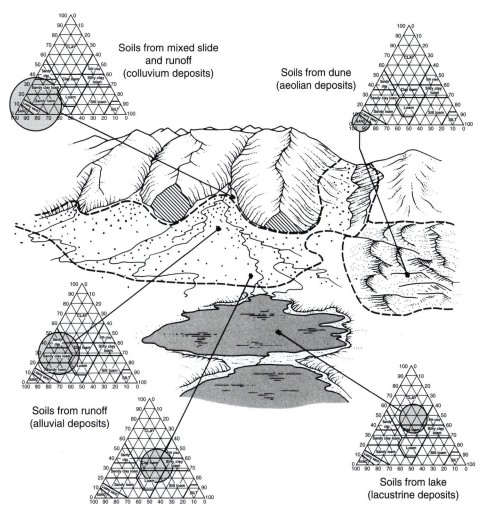

Fig. 5.5 The relationship between soils and landforms in mountainous terrain. Note the texture changes with distance into the valley.

omitted; (2) the map scale and accuracy are marginal for sites of less than 100 acres; and (3) the depth of analysis is limited to the upper 4 to 5 feet of soil. Furthermore, there are inherent accuracy problems with traditional soil surveys owing to a mapping process that is highly inferential in most areas.

Dealing with organic soils

In planning problems dealing with community development, including residential, industrial, commercial, and related land uses, soil composition is one of the first considerations. Organic soils are at the top of the list because they are highly compressible under the weight of structures, and they tend to decompose when drained. Decomposition and compression lead to subsidence and studies show that subsidence rates of 1–2 m over several decades are common where organic soils have been drained, farmed, or built over. Where the organic mass is shallow, it is possible to excavate and replace the soil with a better material, but this is often financially expensive, ecologically harmful, and may lead to drainage problems. Excavation and fill costs generally range from $5 and $10 per cubic yard. Also bear in mind that organic soils always form in areas where surface and subsurface water naturally collects; therefore, excavation will not eliminate these conditions; in fact, it may exacerbate them by releasing the confining pressure on inflowing water. From the ecological standpoint, we must recognize that organic soils are often indications of the

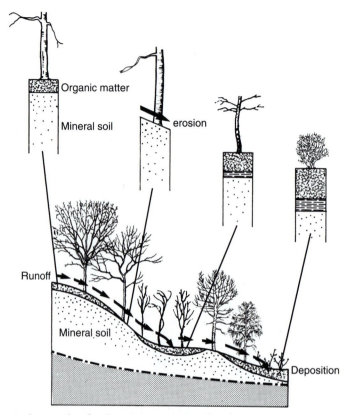

Fig. 5.6   A typical example of soil variation associated with a toposequence across a hillslope. Topsoil is thin on steep sections and thickens on the gentler parts of the slope where runoff is slower.

presence of wetlands that may support valued communities of plants and animals. On balance, then, organic soils should be avoided as development sites.

Suitability of mineral soils

For mineral soils, texture and drainage are the primary considerations. Coarse-textured soils, such as sand and sandy loam, are preferred for most types of development because bearing capacity and drainage are usually excellent (see Table 5.1). As a result, foundations are structurally stable and usually free of nuisance water. Clayey soils, on the other hand, often provide poor foundation drainage, whereas bearing capacity may or may not be suitable for buildings. In addition, certain types of clays are prone to shrinking and swelling with changes in soil moisture, creating stress on foundations and underground utility lines (Fig. 5.7). Therefore, development in clayey soils may be more expensive because special footings and foundation drainage may be necessary. Accordingly, site analysis often requires detailed field mapping followed by engineering tests to determine whether clayey soils may pose bearing capacity, drainage, and other problems.

Scale considerations

In preparing soil maps, the scale of analysis is often as important as the types of analysis. For regional-scale problems, Natural Resources Conservation Service (NRCS) maps are generally acceptable especially if they are used in conjunction with topographic maps, aerial photographs, and surface geology or landform maps. At the site scale, however, NRCS reports and maps are usually too general for most projects. In these cases, a finer grain of spatial detail is called for, which can be provided in part by refining the NRCS soil boundaries with the aid of a large, site-scale topographic map. This is based on the fact that many borders of soil types on NRCS maps follow topography, and at a site scale of detail, topographic trends and details are much more apparent than the soil scientist could detect in building the original map.

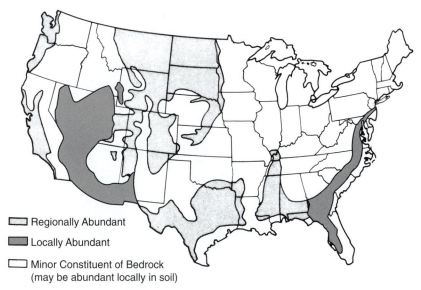

Fig. 5.7 The distribution of near-surface rock containing clay prone to swelling and shrinking. In soil, this clay (montmorillonite) is responsible for stress and cracking of foundations and utility lines.

Soil analysis for site planning

Therefore, where adjacent soils are clearly terrain-specific, the configuration of the border can be articulated much more accurately on large-scale site maps (Fig. 5.8).

Beyond NRCS considerations, or in areas not covered by NRCS maps, soil analysis for site planning should begin with two basic considerations. The *first* is the recent geomorphic history of the area so that we will have some idea of what types of materials to expect. This information can be obtained through a telephone call to a local university (geography, geology, or soil science department), the state geologist's office, or the county NRCS office. In any case, we need to know at the outset whether we are in an area of marine clay, sand dune deposits, glacial outwash, or other type material. The *second* consideration is local topography and drainage. Using topographic maps and field inspection, drainageways (stream valleys, swales, and floodplains) and areas of impeded drainage (wetlands, ponds, and related features) should be mapped. These are the zones where poorly drained and organic soils will most likely be found. The remaining area can be defined as slopes and upland surfaces,

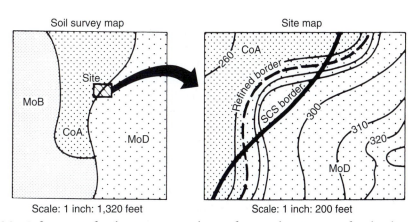

Fig. 5.8 Refinement of soil survey map polygons for site planning can often be done using topographic features such as slopes that are discernable only on large-scale contour maps such as the site map on the right.

Fig. 5.9 A backhoe excavating soil test pits and a photograph of a test pit revealing buried organic matter at a depth of 1.5 m (5 ft).

and, although soil texture may vary within these zones, drainage should be better and organic deposits less likely.

Field testing   Within this framework, a limited testing program can be carried out by digging holes at selected points along swales, floodplains, hillslopes, and ridgetops. This is best accomplished with a backhoe, which can quickly excavate to a depth of 2 meters or more, but a shovel or soil auger will suffice for shallower depths (Fig. 5.9). Once a test pit is excavated, the following questions can be answered:

■ Is the material below the surface mainly organic?

■ Does water rapidly seep into and eventually fill most of the pit?

■ Do the sides of the pit slump away rapidly with excavation and seepage?

■ Is buried waste (such as agricultural debris, industrial rubble, municipal garbage, or hazardous waste) present?

If the answer to any of these questions is yes, the soil should generally be classed as unsuitable for development. Further analysis may be required for engineering purposes, but for site planning, this level of investigation is adequate for laying out use zones, circulation systems, and defining building sites. *A note of caution:* What this does not tell you is what seasonal conditions may be like. For example, a soil that is dry and stable in summer may become saturated and soften and liquify in winter and spring. Therefore, a single sampling in one season may be misleading in locations subject to seasonal extremes in ground conditions.

When to bore   Soil borings are test holes used to measure the strata and composition of soil materials at depths greater than 10 feet. In most instances, borings are necessary only where facilities are to be built or where there is a suspected problem concerning groundwater or bedrock. The depth to which a boring is made depends on the building program requirements (for example, heavy buildings or light buildings), the types of materials encountered underground, and the depth to bedrock. If the soil mantle is relatively thin, say, less than 30 feet, and the bedrock stable, the footings for large buildings will usually be placed directly on the bedrock. In areas of unstable soils, such as poorly consolidated wet clays, or weak bedrock, such as cavernous limestone, special footings are usually required to provide the necessary stability.

5.7 PLANNING CONSIDERATIONS IN SOLID WASTE DISPOSAL

Types of waste

One of the most pressing land use problems in urbanized and industrial areas today is the disposal of various types of waste: municipal garbage, chemical residues from industry, rubble from mining and urban development, various forms of industrial and agricultural debris, and, most recently, nuclear residue from power plants and military manufacturing and development installations. This material is classed as either *solid waste* or *hazardous waste*, depending on whether or not it contains toxic materials. Solid waste is assorted materials variously described as trash, garbage, refuse, and litter. It is considered nonhazardous, that is, not harmful to humans and other organisms, and most (about 80 percent) is produced in rural areas by mining and agriculture. The remainder is *municipal solid waste* and *industrial solid waste*, which is disposed of in landfills in and around urban areas and is a major concern in developed countries.

Solid waste disposal

Landfills (or sanitary landfills) are managed disposal sites. They involve burial of waste in the ground and are the preferred method of solid waste disposal in the United States and Canada (Fig. 5.10). Most other disposal methods are either environmentally unacceptable or too expensive for most waste. Examples include burning, open pit dumping, and ocean dumping. Urban and industrial waste, though low in total output compared to agriculture and mining, present the most serious solid waste disposal problem for the planner. Several factors account for this, including the issue of groundwater contamination and public health, the growing concern over surreptitious hazardous waste within the solid waste, the conflict over aesthetic and real estate values, the limited availability of land around cities for disposal sites, and the high cost of garbage collection and hauling in urban areas. In addition, the production of urban and industrial waste has been rising in most parts of the United States and Canada.

Site selection criteria

The selection of a disposal site for urban and industrial solid waste is one of the most critical planning processes faced in suburban and rural areas today. Properly approached, it should be guided by three considerations: (1) *cost*, which is closely tied to land values and hauling distances; (2) *land use and environment* in the vicinity of the site and along hauling routes; and (3) *site conditions*, which are largely a function of soil and drainage. In both urban and industrial landfills, the chief site problem is containment of the liquids that emanate from the decomposing mass of waste.

These fluids are collectively called **leachate** and are composed of heavy concentrations of dissolved compounds that can contaminate local water supplies. Leachates are often chemically complex and vary in makeup according to the composition of the refuse. Moreover, the behavior of leachates in the hydrologic system, especially in groundwater, is poorly understood. Therefore, the general rule in landfill planning and management is to restrict leachate from contact with either surface or subsurface water. Instances of groundwater contamination by leachates typically result in the loss of potable water for many decades (Fig. 5.11).

Leachate control

Leachate containment may be accomplished in some cases by merely selecting the right site. An ideal landfill site should be excavated in soil that is effectively impervious, neither receives nor releases groundwater, and is not subject to contact with surface waters such as streams or wetlands. Dense (compact) clay soils and relatively high ground are the preferred site characteristics. In addition, the clay should not be interlayered with sand or gravel, not subject to cracking upon drying, and stable against mass movements such as landslides.

Fig. 5.10 A sanitary landfill operation. The mass of garbage (left) is mounded up and inter-laid with clayey soil. The final mass (right) is covered with a soil sheet several feet thick and then planted to stabilize it and improve its visual quality.

Where these conditions cannot be met, the site must be modified to achieve the satisfactory performance. If the soils are sandy and permeable, as was the case illustrated in Figure 5.11, a liner of clay or a synthetic substance such as vinyl must be installed to gain the necessary imperviousness. The same measure must be taken in stratified soils, wet soils, sloping sites, and sites near water features and wells. In many areas of the United States and Canada, however, the decision to modify a site is not left up to the landfill developer or operator because strict regulations often govern which sites can even be considered for landfills. Thus, inherently poor sites are often eliminated from serious consideration by virtue of planning policy.

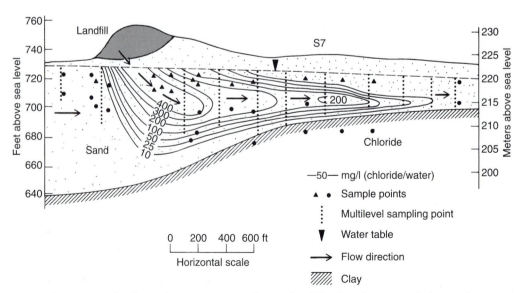

Fig. 5.11 A leachate plume emanating from a landfill. The plume traveled 2200 ft over 35 years, contaminating local groundwater to a depth of 40-80 ft.

Management planning

In addition to a formal site selection and preparation plan, a growing number of local and state/provincial governments are requiring the preparation of a management plan for the design and operation of landfills. In the case of sanitary landfills for municipal garbage, the planner is asked to address the following:

1. Compartmentalization of the landfill into cells or self-contained units of some sort.
2. Phasing of the operation; for example, excavation and filling of a limited area at any one time.
3. Landscaping, pest control, and protection of the site during operations.
4. Limiting the total thickness of refuse by interspersing layers of soil within the garbage.
5. Backfilling over the completed fill using a layer of soil with a provision for vent pipes to release gases, if necessary (Fig. 5.10).
6. Preparation of a plan for grading, landscaping, and for future use of the site.

5.8 HAZARDOUS WASTE REGULATION AND DISPOSAL

In 1976, the U.S. Congress enacted the **Resource Conservation and Recovery Act (RCRA)**, which provided a legal definition of hazardous wastes and established guidelines for managing, storing, and disposing of hazardous materials in an environmentally sound manner. The most widely known regulatory program for hazardous waste management in the United States is the **Superfund** program. It was created in 1980 as an act of Congress (under the formal title Comprehensive Environmental Response Compensation and Liability Act) for the purpose of cleaning up hazardous waste sites and responding to the uncontrolled release of hazardous materials into the environment. Central to Superfund is that it empowers the federal government to provide funding for the cleanup of severely contaminated hazardous waste sites.

Superfund status

At the end of the twentieth century, there were about 2000 sites with priority designations, which the EPA estimates will cost more than $30 billion to clean up. More than 10,000 waste sites are awaiting evaluation for possible priority status, and although most will not gain Superfund status, many will, so the list of Superfund sites is bound to grow. Since 1980 Congress has appropriated $13.5 billion to the program, but only about 10 percent of the money has actually been used for cleanup operations and only a handful of sites have been remediated. The lion's share of the funding has, unfortunately, gone to legal and administrative fees, principally for legal action to get responsible parties to help pay for cleanups. Because of the legal entanglement related to this polluter-pays provision, the Superfund program has become bogged down. In addition, there are serious criticisms of EPA's solutions to several of the sites already remediated. Understandably, the Superfund program is being reevaluated by both public and private groups, and it appears that significant changes will be recommended to Congress.

Polluter-pay

Most hazardous waste sites do not qualify for federal government assistance in remediation. In response to growing concern by property buyers about the liability of hidden wastes on a site, several states have enacted **polluter-pay laws**. These laws place the responsibility for site cleanup on past property owners under whose ownership the waste was buried. Such laws are proving to be effective because the cost of remediation is added to real estate, thereby returning pollution costs to the market system rather than passing them along to the state or federal government. Polluter-pay provisions are, not surprisingly, playing a major role in reusing (recycling) urban land, especially industrial **brownfields**, that is, land given up by old industrial operations.

Disposal versus treatment

Several strategies are available for managing hazardous waste, though none is without drawbacks. In general they fall into two classes: disposal and treatment. *Disposal* involves collecting, transporting, and storing waste with no processing or treatment. *Treatment* involves submitting the waste to some sort of processing to make it less harmful. Because of increasingly stringent health and safety requirements, as well as endless local land use and public opinion problems, land disposal has become a very costly undertaking. For this reason and others, the trend in hazardous waste management is toward on-site treatment, but off-site disposal is still the most common management method.

Secure landfill

The most widely used land disposal method is the **secure landfill**. The objective of this method is to confine the waste and prevent it from escaping the disposal site. The first step in planning a secure landfill is selection of an appropriate site, one that minimizes the risk of groundwater and surface water contamination. Normally this is a very lengthy and complex process involving a host of field tests, analyses, reviews, public hearings, and permit applications. Site selection is followed by landfill design, which must include: (1) a clay liner backed up (underlain) by a plastic liner; (2) a clay dike and clay cap; (3) a leachate collection and drainage system; and (4) monitoring wells around the site to check that leachate is not leaking into the soil and groundwater. It is argued that despite these measures, secure landfills can still leak. Therefore, it is extremely important that only the very best sites are used—namely, those with deep, dense-clay soils, little topographic relief (elevation variation), and water tables at great depth (Fig. 5.12).

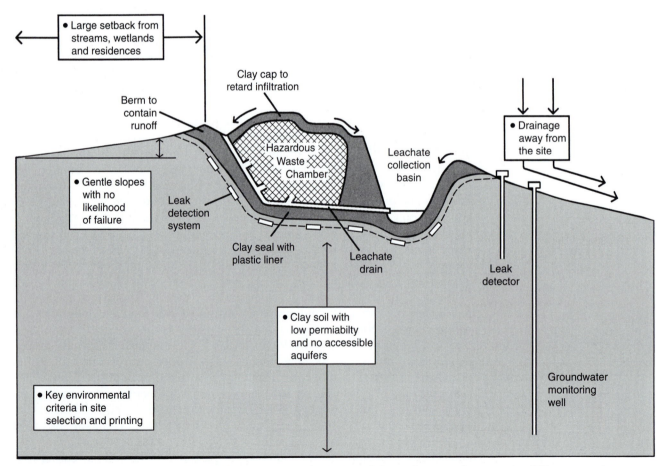

Fig. 5.12 Environmental criteria and design features of a secure landfill site according to current standards in the United States.

5.9 CASE STUDY   ■ **Mapping Soil at the Site Scale for Private Development**

W. M. Marsh

The aim in mapping soils for planning and architectural projects is clearly different than it is for agricultural purposes. Where the main question is the suitability of soil for facilities such as roads and buildings, rather than the suitability for crops, the scope of the study and parameters investigated must be defined differently. This requires a clear understanding of the development program: that is, what is to be built, how much floor space is called for, what ancillary facilities will be needed (parking and the like), what utility systems are necessary, what open space, habitat, and landscaping areas are scheduled, and so on. These elements of the program must be known to determine (1) the scales at which mapping must be conducted; (2) the soil features (parameters) that should be recorded; and (3) the depths to which soil must be examined.

Each building project requires its own soil survey. The results of this survey, along with the results of related studies such as drainage, slope, and traffic are used to formulate alternative design schemes for development of the site. In this particular project, which called for a large research/office facility in a campuslike setting

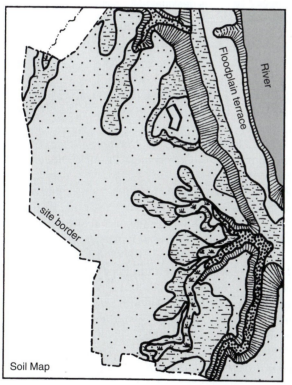

Soil Map

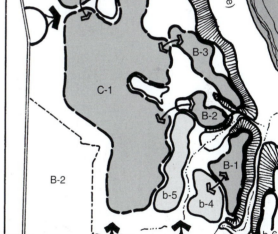

Land Units Map

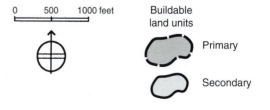

▢ **Upland Soil**: silty loam; well drained in most months but may be moist at depths of 1-2 feet; bearing capacity generally good, but foundation drainage may be needed.

▢ **Valley Soil**: diverse with heavy concentrations of large boulders; channel and slope deposits; prone to flooding and high water table.

▢ **Slope Soil**: dry surface; rapid runoff; prone to failure and erosion if disturbed and/or deforested.

▢ **Wetland Soil**: silty clay loam with appreciable organic fraction; serious drainage problems in most months; unsuitable for building.

0   500   1000 feet

Buildable land units

Primary

Secondary

▢ **Transition Soil**: silty loam; seasonally very wet; runoff collection zones; good potential for storm-water management facilities; poorly suited for building.

along the Mohawk River in New York, it was important to determine not only certain standard soil parameters, such as texture and drainage, but also the nature of the soil environment in terms of the processes and factors that shape soil formation, especially runoff processes. The resultant soil map (map A) was instrumental in defining buildable land units (map B), which in turn served as a framework for the formulation of alternative development plans.

The fact that soils are closely related to physiography, especially topography, drainage and vegetation, is critical to the logic of defining land units. Land units are spatial entities that possess a set of unifying physical traits such as upland topography, well-drained soil, and old field plant cover. Although the focus is on buildable land units, other types of land units are also significant in development planning. In particular, units featuring wetlands, slopes, and valley floors are important to other parts of development programs such as habitat conservation, landscaping, and stormwater management.

Unlike private development of the past, when buildings were all that mattered, much private development today strives to create comprehensive programs that relate to many types of land. The process of building alternative land use layouts leading to a master plan should aim to integrate buildable and nonbuildable land units into a workable whole. Following the selection of a preferred site plan, more detailed and spatially focused soil analyses should be conducted. This involves soil borings taken at specific locations designated for buildings and other facilities in the plan as well as testing soil fertility and moisture conditions for landscape and habitat restoration. ∎

5.10 SELECTED REFERENCES FOR FURTHER READING

Briggs, David. *Soils.* Boston: Butterworths, 1977.

Colonna, Robert A., and McLaren, Cynthia. *Decision-Makers Guide to Solid Waste Management.* Washington, DC: Government Printing Office, U.S. Environmental Protection Agency, 1974.

Davidson, Donald A. *Soils and Land Use Planning.* New York: Longman, 1980.

Eckholm, Erik P. *Losing Ground.* New York: W. W. Norton, 1976.

Gordon, S. I., and Gordon, G. E. "The Accuracy of Soil Survey Information for Urban Land Use Planning." *Journal of the American Planning Association* 47:3, 1981, pp. 301–312.

Hillel, Daniel. *Out of the Earth: Civilization and the Life of the Soil.* Berkeley: University of California Press, 1992.

Hills, Angus G., et al. *Developing a Better Environment: Ecological Land Use Planning in Ontario, A Study Methodology in the Development of Regional Plans.* Toronto: Ontario Economic Council, 1970.

Hopkins, Lewis D. "Methods of Generating Land Suitability Maps: A Comparative Evaluation." *Journal of the American Institute of Planners*, October 1977, pp. 388–400.

Metropolitan Area Planning Commission. *Halifax-Dartmouth Metro Area, Natural Land Capability.* Halifax, Nova Scotia: Nova Scotia Department of Development, 1973.

Pettry, D. E., and Coleman, C. S. "Two Decades of Urban Soil Interpretations in Fairfax County, Virginia." *Geoderma* 10, 1973, pp. 27–34.

Soil Society of America. *Soil Surveys and Land Use Planning.* Madison, WI: Soil Society of America, 1966.

6

SOILS AND WASTEWATER DISPOSAL SYSTEMS

6.1 INTRODUCTION

Until the twentieth century, soil served as the primary medium for the disposal of most human waste. Raw organic waste was deposited in the soil via pits or spread on the surface by farmers. In either case, natural biochemical processes broke the material down, and water dispersed the remains into the soil, removing most harmful ingredients in the process. But this practice often proved ineffective where large numbers of people were involved, because the soil became saturated with waste, exceeding the capacity of the biochemical processes to reduce the concentration of harmful ingredients to safe levels.

The results were occasionally disastrous to a city or town. Water supplies became contaminated and rats and flies flourished, leading to epidemics of dysentery, cholera, and typhoid fever. In response to these problems, efforts were made in the latter half of the 1800s to dispose of human waste in a safer manner. In cities, where the problems were most serious, sewer systems were introduced, allowing waste to be transported through underground pipes to a body of water, such as a river or lake, beyond the city. Chicago, for example, built its first sewer system in 1855. This practice gave rise to critical water pollution problems, principally in receiving streams, but it did remove waste from cities, and thus helped solve the immediate problem of human health. Outside the cities, people continued to use pit-style privies until well into the twentieth century. However, with the growth of suburban neighborhoods after 1930, the outdoor toilet proved unacceptable. In its place came the septic tank and drainfield for individual homes. Within the cities, large-scale treatment systems were introduced to help manage the water pollution problem created by sewer discharges.

In the United States the federal government spent huge amounts of money to upgrade and expand municipal treatment facilities in the 1970s. The principal aim was to reverse the degradation of the nation's receiving waters, the thousands of streams and lakes where poorly treated effluent was being dumped. In the 1980s, the federal government reduced funding for the development of large integrated or centralized treatment systems, encouraging communities to explore other options as well, including small-scale treatment systems such as the septic tank and drainfield systems, bioengineered wetlands, and small-scale mechanical systems. Here we examine onsite systems that rely mainly on soil to absorb and filter wastewater.

The septic drainfield systems used for individual homes are dependent on the soil environment for safe disposal of human waste. The system is designed to function underground and depends on the soil's capacity to take in and filter wastewater at a particular rate. Not all soils can do this; therefore, it is critical to distinguish suitable from unsuitable soils as part of environmental assessment for residential planning in areas not served by municipal sewer systems.

6.2 THE SOIL-ABSORPTION SYSTEM

The septic drainfield method of waste disposal is one version of a **soil-absorption system (SAS)**, so-called because it relies on the soil to absorb, filter, and disperse wastewater. The system is designed to keep contaminated water out of contact with the surface environment and to filter chemical and biological contaminants from the water before they reach groundwater, streams, or lakes. The contaminants of greatest concern are biological agents (pathogens), such as bacteria in the coliform group, which are hazardous to human health and nutrients like nitrogen and phosphorus that accelerate algae growth in aquatic systems.

SAS components   Most SAS systems have two components: (1) a holding or *septic tank* where solids settle out and bacteria decompose organic matter; and (2) a *drainfield* through which

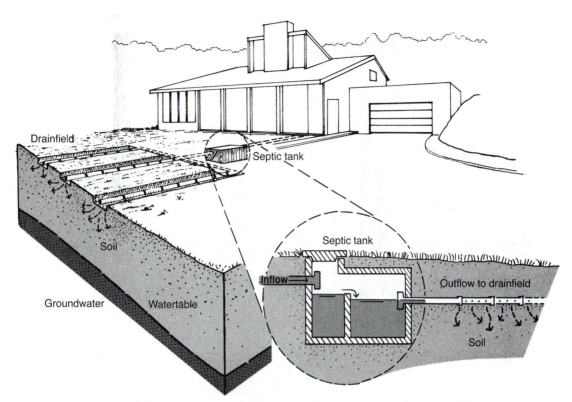

Fig. 6.1 The basic design for a standard soil absorption system. Solid waste accumulates in the septic tank while effluent water flows into the drainfield where it seeps into the soil.

wastewater is dispersed into the soil. The drainfield is made up of a network of perforated pipes or jointed tiles from which the fluid seeps into the soil (Fig. 6.1). Several different layout configurations are commonly employed. The size of the drainfield varies with the rate of wastewater input and the capacity of the soil to absorb it.

Soil permeability
    Critical to the operation of a soil-absorption system is the rate at which the soil can receive wastewater and diffuse it into the soil column. Soils with high permeabilities are clearly preferred over those with low permeabilities. **Permeability** is a measure of the amount of water that will pass through a soil sample per minute or hour. In the health sciences, permeability is measured by the **percolation rate**, the rate at which water is absorbed by soil through the sides of a test pit such as the one shown in Figure 6.2. The *"perc" test* is usually conducted *in situ*, meaning that it is conducted in the field rather than in the laboratory.

Controls on percolation
    The percolation rate of a soil is controlled by three factors: *soil texture, water content*, and *slope*. Fine-textured soils generally transmit water more slowly than coarse-textured soils, and thus have lower capacities for wastewater absorption. On the other hand, fine-textured soils are more effective in filtering chemical and bacterial contaminants from wastewater. The ideal soil balances the two attributes and is a textural mix of coarse particles (to transmit water) and fine particles (to act as an effective biochemical filter). In general, loams are the preferred texture, but they vary with compaction and moisture content.

    Soil moisture tends to reduce permeability in any soil. Soils with high water tables, for example, are limited for wastewater disposal because they are already saturated and additional water can cause the systems to back up and overflow. Soils with very low permeabilities, such as compact clays, can cause the same problem.

    In addition to soil drainage, slope is a consideration. Slope influences percolation inasmuch as it affects the inclination of the drainfield. If the slope of the ground

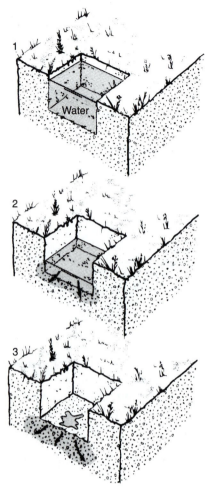

Fig. 6.2 The basic idea of the percolation test. A small pit is excavated, filled with water, allowed to drain, then refilled with water and allowed to drain again. The rate of fall in the water surface is the "perc" rate.

exceeds 3 to 4 feet per 100 feet, it is difficult to arrange the drainfield at an inclination gentle enough (2 to 4 inches per 10 feet is recommended) to prohibit wastewater from flowing rapidly down the drain pipes and concentrating at the lower end of the drainfield. And finally the thickness of the soil layer must be considered. Where the soil mantle is thin (less than 4 to 6 feet) and bedrock is correspondingly close to the surface, the downward percolation of wastewater may be retarded by the bedrock. This varies with the type and condition of the bedrock, but where it has limited permeability, wastewater may build up over it and cause system malfunction and/or surface contamination (Fig. 6.3).

6.3 ENVIRONMENTAL IMPACT AND SYSTEM DESIGN

Soil-absorption systems are a standard means of sewage disposal throughout the world. In the United States, as much as 25 percent of the population relies on these systems; in Canada the percentage may be somewhat higher, and in developing countries such as Mexico, fully 80 to 90 percent of the population uses some form of soil absorption for waste disposal, including pit-style privies. Not surprisingly, a

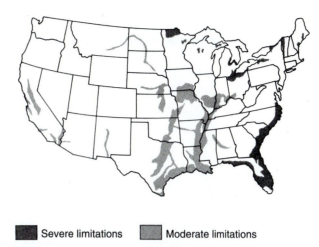

■ Severe limitations   ■ Moderate limitations

Fig. 6.3 The general distribution of soils with limitations to soil absorption systems in the coterminous United States. Locally, areas with limitations may be found just about anywhere in the United States and Canada though they are far more common in the areas highlighted.

high percentage of these systems do not function properly, often resulting in serious health problems and environmental damage.

SAS failure   The causes of system failure are usually tied to one or more of the following: *improper siting* with respect to soil texture, soil drainage, and slope; *improper design* of the drainfield; *overloading* (that is, overuse); *inadequate maintenance* of the septic tank and tile system; and *loss of soil-percolation capacity* due to clogging of interparticle spaces or groundwater saturation of the drainfield bed. Failure often results in the seepage of wastewater into the surface layer of the soil and onto the ground. Here humans are apt to come into contact with it which may lead to severe health consequences, including, in the extreme, such virulent diseases as cholera. Wastewater may enter groundwater and contaminate wells, and it may enter lakes and streams, thus contaminating water supplies, fostering the growth of algae, and degrading the overall quality of the aquatic environment.

Water pollution   Nutrient enrichment of lakes, ponds, and reservoirs with wastewater seepage is a leading cause of the deterioration of recreation waters in the United States and Canada. Where nitrogen and phosphorus, the nutrients of greatest concern, enter a lake, for example, the productivity of algae and other aquatic plants usually rises substantially, resulting in greater organic mass. This mass not only helps fill in the lake bottom, but as it decays the consuming bacteria use up available oxygen in the water. In time, the water's oxygen content declines, fish species change to less desirable species and the lake develops an overgrown (i.e., overfed) character that is often unsightly and smelly (Fig. 6.4). Chapter 11, "Water Quality, Runoff, and Land Use," addresses the problem of nutrient loading in more detail.

SAS sitting and design criteria   Avoidance of public health and water pollution problems stemming from soil-absorption systems begins with site analysis and soil evaluation. Soils with percolation rates of less than 1 inch per hour are considered unsuitable for standard soil-absorption systems. For soils with acceptable percolation rates, additional criteria must be applied: *slope, soil thickness* (depth to bedrock), and *seasonal high water table*. Given that these criteria are satisfied, the system can now be designed. The chief design element is drainfield size, and it is based on two factors: (1) the actual soil percolation rate, and (2) the loading rate, that is, the rate at which wastewater will be released to the soil. For residential structures, the loading rate is based on the number of bedrooms; each bedroom is proportional to two persons. The higher the

Fig. 6.4   An algae bloom resulting in part from seepage of nutrients into the lake from drainfields and other sources.

loading rate and the lower the percolation rate, the larger the drainfield required (Fig. 6.5).

Successful operation of the system necessitates regular maintenance, especially the removal of sludge from the septic tank, avoidance of overloading, and the lack of interference from high groundwater. The last-named may occur during unusually wet years or because of the raising of water in a lake, wetland, or a nearby reservoir, for example. On the average, the lifetime of a drainfield is 15 to 25 years, depending on local conditions. By this time the soil may be too wet to drain properly, because the water table has been raised after years of recharge from drainfield water, and/or the soil may have become clogged with minute particles. In addition, the buildup of chemicals such as phosphorus and nitrogen may become excessive, there-

Maintenance and life cycle

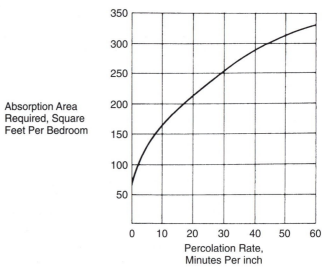

Fig. 6.5   Size of the absorption area required for soil-percolation rates ranging from 0 to 60 minutes per inch.

by reducing the filtering capacity of the soil. In any event, old drainfields should be abandoned and a new one constructed in a different location.

6.4 ASSESSING SOIL SUITABILITY FOR ONSITE DISPOSAL

Map and data sources

The suitability of an individual site for wastewater disposal is best determined by field inspection and percolation tests. For planning problems involving large areas, however, this is not always possible to carry out in detail, and it is necessary to resort to less expensive methods for an assessment of soil suitability. In most cases, the investigator must rely on existing maps and data, as is illustrated by the Nova Scotia case study at the end of the chapter. The most prevalent sources in the United States are the county soil reports produced by the Natural Resources Conservation Service; but as we illustrated in Chapter 5, NRCS maps are often too general in most areas for site-scale work, especially at the residential level. In Canada, comparable rough soil maps are produced by the provinces or by regional environmental agencies. Where NRCS studies are not available, or where their reliability is in question, the investigator must turn to other sources and synthesize existing maps and data into information meaningful to wastewater disposal problems.

The maps in Figure 6.6 are part of a set of maps produced by the U.S. Geological Survey for a land use capability study in Connecticut. Three of these maps (slope, seasonal high water table, and depth to bedrock) are directly applicable to wastewater disposal problems. However, the fourth map, called "unconsolidated materials," requires translation before it can be used in place of soil type or soil texture. The six classes of materials must be translated into their soil counterparts, and the soils, in turn, into their potential for wastewater disposal. For example:

Unconsolidated Material* on USGS Map	Approximate Soil Equivalent	Potential for Wastewater Disposal
Till	Loam	Good
Compacted till	Loam, perhaps clayey, that drains poorly	Fair/poor
Clay deposit	Clay; clayey loam	Poor
Sliderock deposit	Rock rubble	Poor
Swamp	Organic: muck, peat	Poor
Sand or sand and gravel deposits	Sand; sandy loam; gravel	Fair

* Also called surface deposits.

Map synthesis

With this classification in hand, we can proceed to combine the four maps and produce a suitability map for wastewater disposal. One procedure for this sort of task involves assigning a value or rank to each trait according to its relative importance in wastewater absorption. In this case, three ranks are possible for each of the four traits, and each can be assigned a numerical value (Table 6.1): good (3), fair (2), and poor (1) according to the previous translation. For any area or site, the relative suitability for wastewater disposal can be determined by simply summing the four values. The final step is to establish a numerical scale defining the limits of the various suitability classes. The classes should be expressed in qualitative terms, for example, as high, medium, or low, because as numerical values they are meaningless and often misleading in terms of their relative significance.

Modern soil surveys by the U.S. Natural Resources Conservation Service (NRCS) present a somewhat different arrangement. *First*, all data and information are orga-

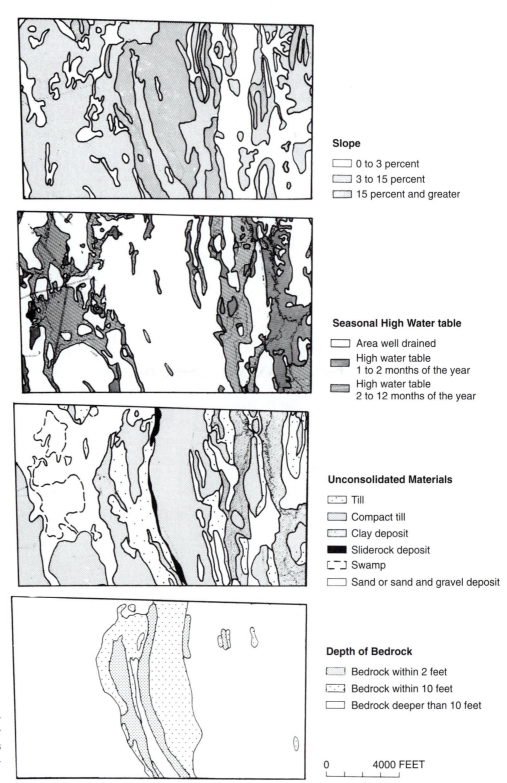

Slope

☐ 0 to 3 percent

☐ 3 to 15 percent

☐ 15 percent and greater

Seasonal High Water table

☐ Area well drained

☐ High water table
1 to 2 months of the year

☐ High water table
2 to 12 months of the year

Unconsolidated Materials

☐ Till

☐ Compact till

☐ Clay deposit

■ Sliderock deposit

☐ Swamp

☐ Sand or sand and gravel deposit

Depth of Bedrock

☐ Bedrock within 2 feet

☐ Bedrock within 10 feet

☐ Bedrock deeper than 10 feet

0   4000 FEET

Fig. 6.6 A series of maps compiled as part of a land capability study that can be used to assess on-site wastewater disposal suitability.

NRCS maps   nized and presented according to soil type. *Second*, in many areas the NRCS provides information on soil suitability for wastewater disposal by SAS. In that case, all one need do is to map the various NRCS classes—taking care, of course, to field check the NRCS distributions—and record the criteria and rationale for each. However, the

Table 6.1 Criteria and Numerical Values for SAS Suitability for Maps in Figure 6.6

		Trait			
		Material	*Depth to Bedrock*	*Seasonal Water table*	*Slope*
Rank	**3 (good)**	Till	> 10 ft	Well drained	0–3%
	2 (fair)	Sand, sand and gravel, compacted till	2–10 ft	1–2 mos. high water table	3–15%
	1 (poor)	Clay, sliderock, swamp	< 2 ft	2–12 mos. high water table	> 15%

criteria and rationale used are not always clear, or one may wish to add or delete certain criteria depending on the land use problem and/or local conditions. In such instances, the procedure would be largely the same as that outlined previously, except that the basic mapping unit is already established (Fig. 6.7) and the data base

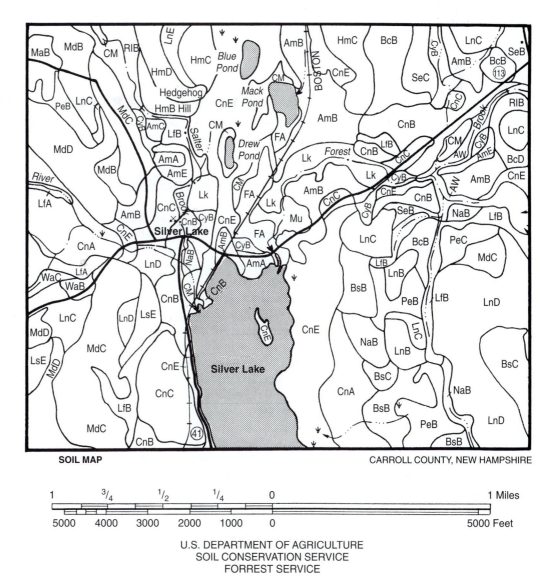

SOIL MAP   CARROLL COUNTY, NEW HAMPSHIRE

```
1        3/4       1/2       1/4       0                                    1 Miles

5000   4000   3000   2000   1000      0                                   5000 Feet
```

U.S. DEPARTMENT OF AGRICULTURE
SOIL CONSERVATION SERVICE
FORREST SERVICE

Fig. 6.7 An excerpt from a standard U.S. Natural Resources Conservation Service map. The letter codes denote different soil types and each code is defined in county soil reports.

Table 6.2   Procedure for Soil Suitability Classification for SAS

Step 1:   Select the traits for evaluation and build a matrix listing each soil type and each trait. For example:

| | | *TRAIT* | | | |
Soil	Texture	Depth to Bedrock	Depth to Seasonal Watertable	Permeability	Slope
Becket:					
Berkshire:					
Chocorua:					

Step 2:   Set up a ranking system such as that in Table 6.1 and complete the matrix by assigning a numerical value to each soil trait.

Step 3:   Total the ranking values for each soil type or series.

Step 4:   Based on these totals, classify the soils as high, medium, or low.

Step 5:   Go to the soils map and color the areas that correspond to each suitability class. For this you will need to devise a three-part color scheme and map legend.

Step 6:   Check the boundaries of the soil polygons (areas) against other map sources such as detail topographic maps and in the field and adjust accordingly (see Fig. 5.8).

may be different. County soil reports often include data on permeability, seasonal high water table, and depth to bedrock as well as soil texture and slope. Where data are available for these criteria, the steps listed in Table 6.2 can be used to build a soil suitability map for wastewater disposal.

6.5 ALTERNATIVES TO STANDARD SAS

Waterless system

It is necessary to mention some alternatives to the standard soil-absorption system. These are treatment systems designed to overcome various soil and topographic limitations at the site scale. The actual use of alternative systems varies from place to place, depending on local health and planning regulations. The simplest alternative is the waterless toilet, which eliminates the need for all or part of the drainfield in the standard residence. The **waterless system** is designed to concentrate solid waste and dispose of it in an environmentally safe manner; for example, as soil compost after it is free of disease agents. Many communities view this system very favorably because (1) it eliminates the need to handle and process large amounts of contaminated water, and (2) it greatly reduces domestic water use. Another system that reduces the need for a large drainfield involves separating toilet water from wastewater from sinks, washers, bathtubs, and so on. The toilet water is directed to the drainfield, whereas the other water, which is not laden with pathogens, can be reused to irrigate lawns and gardens.

Earth mound system

Where soils are the limitation, several other alternatives are possible. One involves replacing the upper 2 feet of soil with a fill medium of the desired percolation and absorption values. A related system, called **earth mound**, involves building a drainfield above ground level in a pile of soil medium that contains the release pipes. A pump is used to force the wastewater into the drainfield.

In areas where slopes are the chief limitation, drainfields can sometimes be laid out in several terraces. The drainfield is compartmentalized, and wastewater is distributed according to the size and capacity of each terrace compartment. Another method utilizes a pipe system to transfer wastewater to a suitable disposal site. This system usually incorporates a grinder to break up the solids, a storage tank, and a

Grinder/pump system pump to force the water to the disposal site. The **grinder/pump system** can be used to tie a number of houses into a single disposal system.

With the exception of the waterless toilet, construction and maintenance costs are usually greater for alternative systems than they are for the standard soil-absorption system. In addition, they require more attention in their design, and permitting by local regulatory agencies may be more difficult in many parts of the United States and Canada. In many places the waterless toilet, in particular, is not accepted for standard residential units. However, where water supply is an issue or where there is a concern about the contamination of groundwater and streams, the waterless toilet has gained serious attention.

6.6 SMALL-FLOW WETLAND SYSTEMS

Wetlands are also gaining serious attention in wastewater disposal planning. It is an established fact that wetlands can serve as effective treatment systems for impure water (see Case Study 11.8). They are known to perform several physical and biochemical functions that can reduce suspended sediment, harmful biological agents, and chemical contaminants. Among the wetland's water cleansing processes are:

Wetland processes
- ■ *entrapment* of coarse sediment as runoff slows down and plant stems, roots, and organic debris capture particles;
- ■ *filtration* of suspended sediment as water penetrates and percolates through the plant and soil mass;
- ■ *adsorption* (bonding) of nutrients, organic compounds, and biological agents by organic microparticles (colloids);
- ■ *plant uptake* of nutrients and other dissolved chemical ions;
- ■ *breakdown* of certain contaminants by microbial action within organic masses.

Both natural and constructed wetlands are being used today for disposal of residential sewage. Although viewed with skepticism by some public health agencies, wetland disposal systems are gaining increased acceptance in the United States and Canada as the list of successful experimental systems grows. At the top of the experimental list are small-scale, constructed wetland systems designed to serve individual *System components* houses or small groups of houses. These systems consist of a septic tank (with a pump system if necessary), wetland cell (with a substrate of saturated gravel supporting a cover of wetland plants), and an infiltration bed (to disperse water released from the wetland). The system is equipped with check valves to prevent reverse flows and the wetland cell is underlined with plastic to prevent water loss to the ground. The size of the wetland cell varies with climatic conditions and household size, but they typically cover an area of 800 to 1200 square feet. For an average-sized household producing 600 gallons of water per day, the effluent detention time in the wetland is two weeks. During this time suspended solids are filtered out and nutrients are consumed by bacteria and higher plants. Periodically, the wetland needs maintenance, including thinning of wetland plants and removal of debris and extraneous vegetation.

Community wetland systems Larger wetland systems have been used effectively in wastewater treatment in several communities. They too are often viewed with skepticism in both the public health and environmental engineering fields. Among other things, these professions tend to be more comfortable with community treatment systems they can control externally, that is, with mechanical systems. Although the design and operation of wetland treatment systems vary with local soil, drainage, and climatic conditions, most systems rely

Fig. 6.8 A small-scale community wastewater system using a series of wetland basins for effluent treatment.

on a series of wetland basins through which wastewater is dispersed after removal of solids in a settling basin (Fig. 6.8). The capacity of the wetlands to process effluent water varies seasonally with, among other things, water temperature and supply; therefore, systematic monitoring and day-to-day management are mandatory.

6.7 CASE STUDY

Planning for Wastewater Disposal Using Soil Maps, Nova Scotia

Michael D. Simmons

Although population growth in Nova Scotia has been slow, recent decades saw a relatively large number of housing starts as a result of the breakup of households, the availability of government housing assistance, and a general trend toward rural and suburban settlement. Many of these new homes were built on unserviced lots within commuting distance of urban centers. To help in planning where homes should and should not be built, the Province of Nova Scotia appointed a task force to determine soil suitability for onsite wastewater disposal.

The basic soil units and maps of the Nova Scotia soil survey served as the geographic base. The criteria for wastewater disposal were based on guidelines established by the Nova Scotia Department of Public Health:

- The percolation rate must be less than 30 minutes per inch; a rate of up to 60 minutes per inch may be acceptable with suitable modification.
- Depth to bedrock must be 4 feet or more below the disposal field; that is, about 6 feet below the surface.
- The groundwater table must also be at least 4 feet below the disposal field, and the drainfield site must not be in a marshy area or an area subject to flooding.
- Topography, or relative elevation within the lot, must be considered.

SOIL OF NOVA SCOTIA
LIMITATIONS FOR SEPTIC TANK ABSORPTION FIELDS[1]

Legend

Map Unit Symbol	Primary Limitation	Probability of Occurrence[2]	Secondary Limitation	Probability of Occurrence[2]	Acres ('000)	Hectares ('000)	% of Province
N	NONE TO MODERATE LIMITATIONS	85	NONE	·	709	287	5.5
B	BEDROCK	65	NONE	·	1016	411	7.9
BB	BEDROCK	90	NONE	·	6439	2606	49.4
P	PERCOLATION	90	NONE	·	420	170	3.3
PS	PERCOLATION	90	PERIODIC SATURATION	80	1322	535	10.3
PW	PERCOLATION	90	PROLONGED SATURATION	90	264	107	2.0
PB	PERCOLATION	90	BEDROCK	75	498	201	3.9
SB	PERIODIC SATURATION	90	BEDROCK	80	1073	434	8.3
WB	PROLONGED SATURATION	90	BEDROCK	80	388	157	3.1
R	RAPID PERCOLATION	80	NONE	·	163	66	1.2
WR	PROLONGED SATURATION	90	RAPID PERCOLATION	80	30	12	0.2
F	FLOODING	90	PERCOLATION (RAPID or SLOW)	80	171	69	1.4
O	OTHER UNSUITABLE AREAS [Swamps, peat bogs, salt marsh. coastal beach, eroded land]	95		·	444	179	3.5

[1]NOTE: Site modifications may be introduced which overcome these limitations. Techniques include the addition of fill, drainage to lower the water table and possibly deep tillage to increase the percolation rate. Detailed investigation of local site conditions are required to determine the feasibility and cost of overcoming the site limitations.
[1]PROBABILITY OF OCCURRENCE

In addition to the factors covered by the guidelines, three other limitations were considered:

■ Prolonged or periodic saturation of the surface soil layers associated with a perched watertable or flooding.

■ Excessively rapid percolation, permitting contamination of groundwater supplies.

■ Excessive slope of the land surface.

These criteria were compared with the soil survey data base, and six operational criteria were defined: (1) percolation rate too slow, (2) bedrock less than 6 ft from surface, (3) prolonged saturation, (4) seasonal saturation, (5) susceptibility to flooding, and (6) percolation rate too high. In addition to the soil survey, data from public health inspectors, well-drilling logs, as well as additional field tests were used in the mapping program.

In attempting to synthesize the data, inconsistencies were often uncovered between the soil descriptions included in the soil survey and observations based on field inspection. This was attributed to two factors: the scale of mapping and the time or season of field testing. Soil moisture varies seasonally; therefore, in periods of low moisture, soils may appear to perform satisfactorily if subjected only to a single percolation test. To identify soils with seasonal limitations, it was found that percolation tests should be performed only in May, early June, and November in most years.

The scope of mapping was a more difficult problem. Because the Province is covered by 14 soils survey reports at small map scales (1 inch to 1 mile, 1 inch to 2 miles, and 1 inch to one-third mile), the detail necessary for site-specific investigations was missing. For this reason, it was necessary to use a mapping technique based on the probability of occurrence of a limitation. This consisted of identifying the one or two major limitations of each soil type and then sampling that soil unit to determine the consistency of occurrence of those limitations. The overall consistency of occurrence ranged from 80 to 90 percent, although in some areas with a bedrock limitation, bedrock was identified in only 50 percent of the sites. A probability of 80 percent indicates that a given limitation could be expected on four out of five development sites selected at random within a given soil mapping unit.

Since mapping was based on the Nova Scotia soil survey, the largest scale maps that could be produced by this project were 1:50,000. These maps have been put to a variety of uses; *first*, as public information vehicles to illustrate the extent and distribution of problem soils; and *second*, in communities where there has been a reluctance to accept on-site test results, to provide corroborating evidence of problem soils. This has not only strengthened local planning regulatory processes, but has provided guidelines for municipal and regional development plans as well.

Michael D. Simmons is a resource planner with Maritime Resource Management Service, Amherst, Nova Scotia.

6.8 SELECTED REFERENCES FOR FURTHER READING

Clark, John W., et al. "Individual Household Septic-Tank Systems." In *Water Supply and Pollution Control* (3rd ed.). New York: IEP/Dun-Donnelley, 1977, pp. 611–621.

Cotteral, J. A., and Norris, D. P. "Septic-Tank Systems." *Proceedings American Society of Civil Engineers, Journal of Sanitary Engineering Division* 95, No. SA4, 1969, pp. 715–746.

Environmental Protection Agency. "Alternatives for Small Wastewater Treatment Systems." *EPA Technology Transfer Seminar Publication*, EPA-624/5-77-011, 1977.

Huddleston, J. H., and Olson, G. W. "Soil Survey Interpretation for Subsurface Sewage Disposal." *Soil Science* 104, 1967, pp. 401–409.

LaGro, J. A. "Designing with Nature: Unsewered Residential Development in Rural Wisconsin." *Landscape and Urban Planning*. 33, 1996, pp. 1–9.

Last, J. M. *Public Health and Human Ecology*. East Norwalk, CT: Appleton and Lange, 1987.

Public Health Service. "Manual of Septic Tank Practice." United States Public Health Service Publication No. 526. Washington, DC: U. S. Government Printing Office, 1957.

Wilhelm, S., et al. "Biogeochemical Evolution of Domestic Wastewater in Septic Systems: 1. Conceptual Model." *Ground Water*, 32, 1994, pp. 905–916.

World Health Organization. *The International Drinking Water Supply and Sanitation Decade Review of Regional and Global Data*. Geneva: WHO Offset Publication No. 92, 1986.

7

GROUNDWATER, LAND USE, AND AQUIFER PROTECTION

7.1 INTRODUCTION

Not many years ago, the only consideration given to groundwater in land use planning was as water supply. Groundwater is still an important source of water for residential, industrial, and agricultural land uses, and locating dependable supplies of usable water is still important in planning. Today, however, there is another problem to contend with: groundwater pollution. Contamination of groundwater is one of the most alarming of today's environmental problems, especially when we consider that groundwater is the single largest reservoir of fresh, liquid water on the planet.

Although groundwater is often pictured as a remote and complex part of the environment and traditionally is not the domain of planners, it is necessary to remind ourselves that virtually all groundwater begins and ends in the landscape. Therefore, changes in the surface environment, especially those involving land use activity can, and usually do, affect groundwater in some way. For land use planners and developers the problem is not only knowing what the polluting activities are, but also finding the proper location for them relative to the local groundwater systems. Indeed, the first line of defense in groundwater protection is judicious site selection for land uses with high impact potentiality.

The responsibility of the land use planner today goes even further, for it is also necessary to contend with hidden sources of soil and groundwater contamination. In most states and provinces, there are thousands of leaking underground storage tanks that are releasing oil, gasoline, and other petroleum products into the ground. There are also thousands of landfills containing hazardous waste in the United States and Canada. Most of these lie in and around metropolitan regions, and for a surprisingly large number of them, the locations are unknown. Increasingly, new landowners are uncovering storage tanks and hazardous landfill materials and are being forced to realize the expense and liability connected with cleanup and environmental restoration. Thus, to the list of considerations in environmental assessment, site selection, and site analysis, we must add that of unrecorded landfills, hazardous waste, and underground storage tanks.

7.2 GROUNDWATER SOURCES AND FEATURES

Groundwater begins with surface water seeping into the ground. Below the surface, the water moves along two paths: (1) some is taken up by the soil, and (2) some is

Gravity water   drawn by gravity to greater depths. The latter, called **gravity water**, eventually reaches a zone where all the open spaces (interparticle voids in soil materials and the cracks in bedrock) are filled with water. This zone is called the *zone of saturation* or the *groundwater zone.*

Water table   The upper surface of the groundwater zone is the **water table** (Fig. 7.1). In some materials the water table is actually a visible boundary line, but in many materials it is more a transition zone with an irregular configuration. Below the water table the groundwater zone may extend several miles into the earth, but at great depths the water content is very small. In the upper 5000 feet or so, in the zone of usable groundwater, the actual water content varies significantly with different soil and rock materials.

Porosity   The total amount of groundwater that can be held in any earth material is controlled by the material's porosity. **Porosity** is the total volume of void space in a material. It commonly varies from 10 to 40 percent in soils and near-surface bedrock (Fig. 7.2). In general, porosity decreases with depth, and at depths of several miles into the earth's crust, where the enormous pressure of the rock overburden closes out void spaces, it is typically less than 1 percent.

In most areas, the materials underground are arranged in layers, formations, or zones with different porosities and groundwater capacities. Those materials with espe-

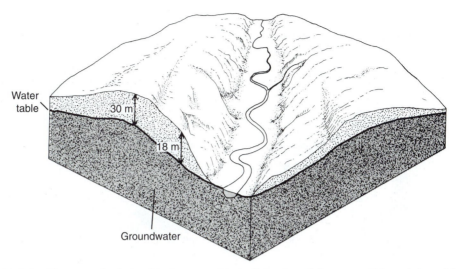

Fig. 7.1 The general relationship between the configuration of the water table and that of the overlying terrain. The variation in the elevation of the water table is usually less than that of the land surface, resulting in many intersections between groundwater and the surface of the land.

Aquifers cially large concentrations of usable groundwater are widely known as **aquifers**. Many different types of materials may form aquifers, but porous material with good permeability, such as beds of sand and fractured rock formations, are usually the best. An aquifer is evaluated or ranked according to (1) how much water can be pumped from

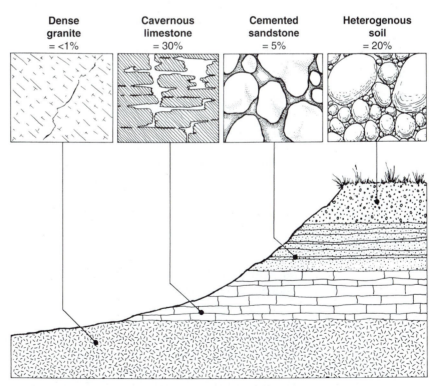

Fig. 7.2 Porosities of four materials near the surface. In all materials, porosity decreases with depth because the great pressure exerted by the soil and rock overburden tends to close the voids. Porosity can be as high as 45 percent in some near-surface bedrock, but it is rarely higher than 2 to 3 percent at depths exceeding 2000 m.

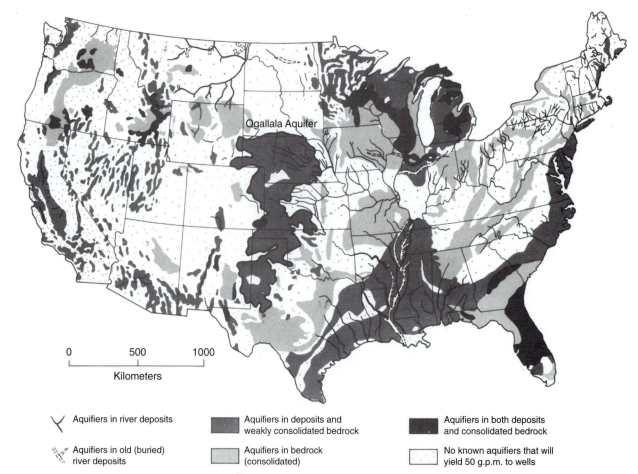

Fig. 7.3 Distribution of major aquifers in the coterminous United States. A major aquifer is defined as one composed of material capable of yielding 50 gallons per minute or more to an individual well of water generally not containing more than 2000 parts per million of dissolved solids.

it without causing an unacceptable decline in its overall water level; and (2) the quality of its water. With respect to water quality, highly mineralized water, such as saltwater, is not generally usable for agriculture and municipal purposes, and aquifers containing such water are usually not counted among an area's groundwater resources.

Aquifer types and materials   Aquifers are grouped into two classes depending on the type of material in which they form: *consolidated* (mainly bedrock) and *unconsolidated* (mainly surface deposits) (Fig. 7.3). In the central part of North America, where glacial deposits of 50-70 to 500-ft thickness lie over sedimentary bedrock, aquifers are found in both the bedrock and the surface deposits. Owing, however, to the extreme diversity of glacial deposits in many areas, aquifers in the unconsolidated class tend to vary greatly in size, depth, and water supply. Bedrock aquifers, on the other hand, tend to be more extensive—especially in areas of sedimentary rocks—often covering hundreds of square miles in area. The largest aquifer in North America is the Ogallala, which stretches more than 1000 miles from South Dakota into northern Texas, and is composed principally of sandstone (Fig. 7.3).

In the lowlands of large rivers such as the Illinois, Wabash, Platte, Arkansas, and hundreds of other streams, extensive shallow aquifers (at depths less than 500 feet)

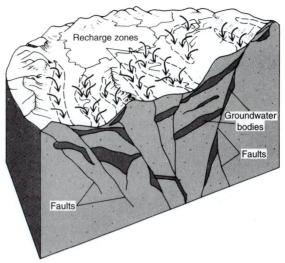

Fig. 7.4 The concept of a groundwater basin: different bodies of groundwater fed by one or more recharge areas within a common geological setting.

lie beneath the valley floors. These are *alluvial aquifers* (i.e., composed of buried stream deposits). They are recharged (i.e., replenished) by river water, and their water supplies fluctuate with the seasonal changes in streamflow. In addition, river aquifers are distinctive for their geographical distributions as they tend to form ribbons following the floors of river valleys and their tributaries.

In the American West, aquifers are found in both deposits and bedrock, but many dry areas contain only marginal aquifers with small yields, say, less than 50 gallons of water per minute. Locally, however, river aquifers fed by runoff from mountain ranges are very important sources of groundwater in otherwise dry mountain valleys. In general, the diverse mountainous terrain of western North America, with deep valley deposits, alluvial fans on mountain flanks, and fractured bedrock formations, produces a highly varied groundwater environment closely linked to geologic structures, mountain runoff, and related deposits (Fig. 7.3).

Groundwater basins   A group of aquifers linked together in a large flow system is called a **groundwater basin**. Groundwater basins are typically complex three-dimensional systems characterized by vertical and horizontal flows among the various groundwater bodies—both those that would qualify as aquifers and those that would not—and between groundwater bodies and the surface (Fig. 7.4). The spatial configuration of a groundwater basin is determined largely by regional geology, that is, by the extent and structure of the deposits and rock formations that house the groundwater bodies. Because these deposits and formations usually differ vastly in their size, composition, and shape, exactly how the various bodies of groundwater in a basin are linked together at different depths is seldom clear. This uncertainty is significant not only in planning for water supplies, but also in understanding the spread of contaminants from among aquifers.

7.3 THE GROUNDWATER FLOW SYSTEM

If we were to map the elevations of aquifers, we would find that essentially all bodies of groundwater are inclined (tilted) to some degree. This can be verified for most shallow aquifers by tracing the elevation of the water table across the landscape.

Hydraulic gradient

Generally, the water table rises and falls with broad changes in surface topography. In deeper aquifers groundwater commonly slopes with the dip of rock formations. The rate of change in elevation across an aquifer or segment of the water table is termed the **hydraulic gradient**, and it is calculated in the same fashion as a topographic slope, that is, change in elevation between two points divided by the distance between them.

Flow velocity

The flow of groundwater is driven by gravity along the hydraulic gradient. For a given material, the steeper the hydraulic gradient, the faster the rate of flow. To determine the flow velocity of groundwater, it is necessary to know not only the hydraulic gradient, but also the resistance imposed on the water by the material it is moving through. *Resistance* is a function of permeability that is determined by the size and interconnectedness of void spaces. For clays permeability is low; for sands and gravels, it is high. The basic formula for groundwater velocity, known as **Darcy's Law**, is

$$V = I \bullet k$$

where V is velocity, I is hydraulic gradient, and k is permeability.

Compared to the flow velocities of surface water, groundwater is extremely slow. Typical velocities for large aquifers are only 50 to 75 feet per year. The time it takes water to pass completely through an aquifer, which is called *residence time* or *exchange time*, measures in decades and centuries. In the case of a typical aquifer 3 or 4 miles in diameter, for example, exchange time could be as great as 250 to 350 years. This time, of course, may be much shorter for wells in the aquifer that withdraw the water after it has passed only part of the way through the system. In any case, groundwater systems are very slow and the time required to flush contaminants from polluted aquifers is exceedingly long by human standards.

Recharge

Recharge is the term given to gravity water supplied to a body of groundwater from surface sources, such as soil, wetlands, and lakes. Although some aquifers, especially shallow ones, receive recharge water from a broad (nonspecific) surface area, many aquifers are recharged from specific areas, called *recharge zones*. These zones may be places where (1) surface water accumulates such as in a wetland or a topographic depression, (2) where there is highly permeable soil or rock formation at the surface, or (3) where an aquifer is exposed at or near the surface (Fig. 7.5). In all cases, recharge zones are critical to aquifer management because they are the points of most ready access for contaminants from land use activity.

System balance

When recharge water enters the aquifer, its continued movement depends on the permeability of the material it encounters and the aquifer's hydraulic gradient. If an aquifer receives rapid recharge, but the material is of low permeability (i.e., high resistance), the new groundwater tends to build up. As a result, the hydraulic gradient increases, but as it does, so does flow velocity, according to Darcy's Law. Under such conditions the hydraulic gradient will continue to rise until the rate of lateral flow, or *transmission*, is equal to the rate of recharge. Where recharge is low and permeability is high, on the other hand, the hydraulic gradient will fall until the rates of transmission and recharge are equal.

Recharge rates

In general, shallow aquifers have a capacity to recharge much faster than deeper ones. Although average recharge rates decrease more or less progressively with depth, as a whole aquifers within 3000 feet of the surface require several hundred years (300-year average) to be completely renewed, whereas those at depths greater than 3000 ft require several thousand years (4600-year average) for complete renewal. This difference is related not only to the distance recharge water must travel, but also to the fact that percolating water moves slower in the smaller spaces found at great depths.

Although large quantities of groundwater remain in some aquifers for thousands of years, most water remains underground for periods ranging from several years to

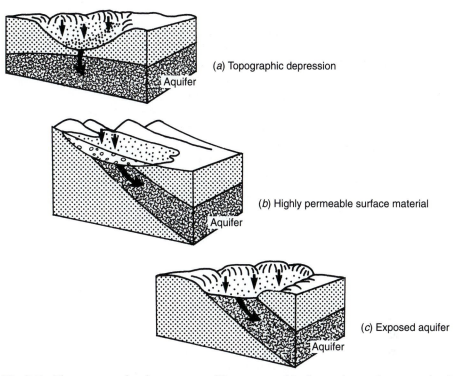

Fig. 7.5 Three types of recharge zones: (1) a topographic depression such as a wetland; (2) a highly permeable surface material such as an alluvial fan or talus slope; and (3) an exposed area of aquifer such as near an escarpment.

several centuries. Aside from pumping, groundwater is released to the surface mainly through (1) *capillary rise* into the soil from which it is evaporated or taken up by plants and released in transpiration; and (2) *discharge* or seepage into streams, lakes, and wetlands where it becomes part of the surface runoff system. Both modes of release involve huge amounts of water but both are nearly imperceptible in the landscape.

Discharge

Many different geologic and topographic conditions can produce seepage. In mountainous areas, for example, it can be traced to fault lines and outcrops of tilted rock formations (Fig. 7.6). In areas where bedrock is buried under deep deposits of soil, **seepage zones** are usually found along the foot of slopes. Steep slopes represent "breaks" in elevation that may be too abrupt to produce a corresponding elevation change in the water table. Under such conditions, the water table may intercept the surface, especially in humid regions where the water table is high. If the resultant seepage is modest, a spring is formed, but if it is strong, a lake, wetland, or stream may be formed. Seepage zones, also called discharge zones, are ecologically significant in any geographic setting because they provide dependable water supplies and temperate thermal conditions. In stream channels, groundwater seeps into the streambed providing an important long-term source of discharge, called baseflow.

Seepage zones

Which type of water feature develops from groundwater seepage depends on a host of conditions, including the rate of seepage, the size of the receiving depression, and the rate of water loss to runoff and evaporation. Most of the thousands of inland lakes of Minnesota, Wisconsin, Michigan, and Ontario, for example, are "seepage"-type lakes whose water levels fluctuate with the seasonal changes in the elevation of the water table. Most of these lakes are either connected to wetlands or them-

Seepage lakes

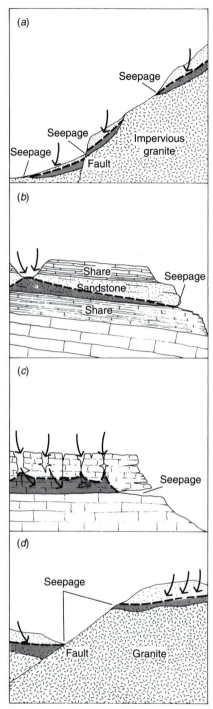

Fig. 7.6 Seepage zones associated with different physiographic settings: (a) unconsolidated sediments over impervious bedrock; (b) inclined sandstone over shale; (c) cavernous limestone over unweathered limestone; (d) unconsolidated materials broken by a fault.

selves evolve into wetlands as they are filled in with organic debris. Where, for example, groundwater is contaminated by nutrients from residential sewer systems, fertilizers, or stormwater, lakes undergo accelerated growth of algae and other aquatic plants, which hastens the filling process.

7.4 GROUNDWATER WITHDRAWAL AND AQUIFER IMPACT

Groundwater is used for drinking, agriculture, and industry in every part of North America. In rural areas, it provides more than 95 percent of the drinking water and about 70 percent of the water used in farming. Between 25 and 50 percent of the communities in the United States and Canada depend on groundwater for public water supplies.

Water supply   The best aquifers for water supply are those containing vast amounts of pure water that can be withdrawn without causing an unacceptable decline in the head elevation of the aquifer. Two conditions must exist if an aquifer is to yield a dependable supply of water over many years. First, the rate of withdrawal must not exceed the transmissibility of the aquifer. Otherwise, the water is pumped out faster than it can be supplied to the well, and the safe well yield will soon be exceeded. *Safe well yield* is defined as the maximum pumping rate that can be sustained by a well without lowering the water level below the pump intake. Second, the rate of total withdrawal should not outrun the aquifer's recharge rate, or else the water level in the aquifer will fall. When an aquifer declines significantly, the *safe aquifer yield* has usually been exceeded, and if the *overdraft* is sustained for many years, the aquifer will be depleted.

In arid and semiarid regions where groundwater is pumped for irrigation, the safe aquifer yield is typically exceeded and often greatly exceeded. In the American West, many aquifers took on huge quantities of water during and after the Ice Age several thousand years ago; today these reserves are being withdrawn at rates far exceeding present recharge rates. Unless pumping rates are adjusted to recharge rates, the water level will decline, wells will have to be extended deeper and deeper, and these aquifers will eventually be depleted.

Aquifer decline   In parts of Arizona, the water level in some aquifers is declining as much as 20 feet a year because of heavy pumping for agriculture and urban uses. In the Great Plains the huge Ogallala Aquifer is declining as much as 3 feet per year in some areas because of heavy irrigation withdrawals for agriculture (Fig. 7.3). With the current shifts in the U.S. population toward the south, not only are demands on groundwater reserves rising but the risk of contamination is rising as well. In Florida, where the population is growing by about 3 million people a decade and is overwhelmingly dependent on groundwater, a major problem is declining quality caused by stormwater pollutants transported through the sandy soils with recharge water (Fig. 7.7).

In and around urbanized areas (or agricultural areas with high concentrations of large wells), the groundwater level in aquifers may not only be greatly depressed by heavy pumping but also develop a very uneven upper surface. This happens where a high rate of pumping from an individual well is maintained for an extended period of time, causing the level of groundwater around the well to be drawn down by many meters or tens of meters. The groundwater surface takes on a funnel shape, as *Cone of depression*   on the water surface above an open drain in a bathtub, and is called a **cone of depression**.

As the cone of depression deepens with pumping, the hydraulic gradient increases, causing faster groundwater flow toward the well. If the rate of pumping is not highly variable, the cone usually stabilizes in time. However, if many wells are clustered together, as is often the case in urbanized areas, they may produce an overall lowering of the water table as the tops of neighboring cones intersect each other as they widen with drawdown. If the wells are of variable depths, this may result in a loss of water to shallow wells as the cones of big, deep wells are drawn beyond the shallower pumping depths.

Drawdown of groundwater over a large area can also lead to loss of volume in the groundwater-bearing materials. This may result in subsidence in the overlying ground,

Subsidence and
saltwater intrusion

as it has in Houston, Texas, for example, where much of the metropolitan area has subsided a meter or more with groundwater depletion. In addition, cones of depression can accelerate the migration of contaminated water because they increase hydraulic gradients and transmission velocities. In coastal areas, groundwater drawdown may also lead to the intrusion of salt groundwater into wells from the bottom up. **Saltwater intrusion** occurs when the mass (weight) of fresh groundwater is reduced by pumping and the underlying salt groundwater rises in response to the reduction in pressure. Once contaminated by saltwater, the wells are effectively lost as a water source.

7.5 SOURCES OF GROUNDWATER CONTAMINATION

The sources of groundwater contamination are very widespread in the modern landscape. They include all major land uses: industrial, residential, agricultural, and transportational. Planning for groundwater protection, therefore, is not limited to urban and industrial landfills, as is widely thought, but includes agricultural, mining, residential, highway, and railroad activities as well. The level of concern or precaution, however, is not everywhere equal because (1) groundwater susceptibility to contamination varies widely from place to place, (2) the contaminant loading rate varies with land use type and practices (such as in pesticide applications with different farming methods), and (3) the contaminants released to the environment vary in their harmfulness to humans and other organisms.

The following are the six major classes of groundwater contamination sources:

Contamination sources   **Landfills**   Buried wastes, both solid and hazardous, which discharge contaminated liquids called *leachate* (also see Section 5.7 in Chapter 5). The composition of leachate varies with composition of the landfill. For urban landfills made up of residential garbage, the leachate may contain organic compounds such as methane and benzene; for agricultural wastes, the leachate may be heavy in nutrients such as phosphorus and nitrogen as well as organic compounds; and for industrial wastes it is commonly heavy in trace elements, that is, metals such as lead, chromium, zinc, and iron, as well as a host of other contaminates including organic compounds, petroleum products, and radioactive materials.

Farmlands   Agricultural fertilizers and pesticides that are carried by soil water and gravity water to aquifers. Fertilizers are composed principally of nitrogen and phosphorus. Nitrogen, which is the most mobile of the two in soil and groundwater, commonly shows up in aquifers and poses a serious health problem in public water supplies. Pesticides are heavy in organic compounds, mostly synthetic compounds such as diazinon, fluorene, and benzene.

Urban stormwater   Runoff from developed areas, especially streets, parking lots, industrial and residential surfaces, is normally rich in a wide variety of contaminants. Most stormwater is discharged into streams, but a significant share of it goes directly into the soil (Fig. 7.7). Although concentrations of most contaminants are reduced by soil filtration, in coarse-grained soils an appreciable load may also be transported through to the groundwater zone. This includes metals (e.g., lead, zinc, and iron), organic compounds (mainly insecticides such as diazinon and malathion), petroleum residues, nitrates, and road salt.

Drainfields   Release of sewage effluent water into the soil through seepage beds. The sewage is produced by household drainfields or community drainfields and when fed to the ground, large amounts of nitrogen, sodium, and chlorinated organic compounds may be discharged into groundwater.

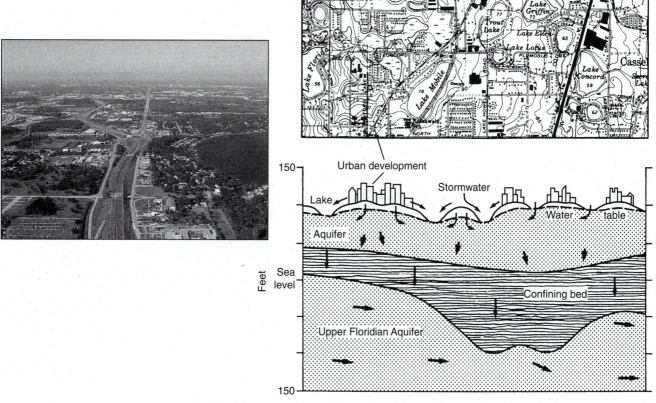

Fig. 7.7 The relationship of the water table aquifer to surface drainage in the urbanized area around Orlando, Florida which is shown in the map and photograph.

Mining Mineral extraction and related operations, such as refinement, storage, and waste disposal, yield a variety of contaminants to surface and groundwater. Some operations are especially harmful, such as certain gold mining operations that employ a leaching technique based on the application of cyanide to crushed rock. In most operations, such as coal, iron ore, and phosphate, contaminants are discharged in leachate from decomposing waste rock.

Spills and leakage Here the possibilities seem endless. Spills made up of petroleum products, various organic compounds, fertilizers, metals, and acids are most common along highways, railroads, and in and around industrial complexes. Leakage from underground storage tanks (of which there are more than 10 million in the United States and Canada), pipelines, and chemical stockpiles are also widespread. Residential land uses also contribute: paint, cleaning compounds, car oil, and gasoline, for example. Spills are typically point sources, and where they are associated with a known accident, remedial measures can be applied. However, underground leakages are usually hidden and may go undetected for years (Fig. 7.8).

7.6 APPLICATIONS TO LANDSCAPE PLANNING

Unlike surface water, which we can see, measure, and map with comparative ease, groundwater is far more elusive as an environmental planning problem. Among other things, it occupies complex three-dimensional space, with different aquifers and

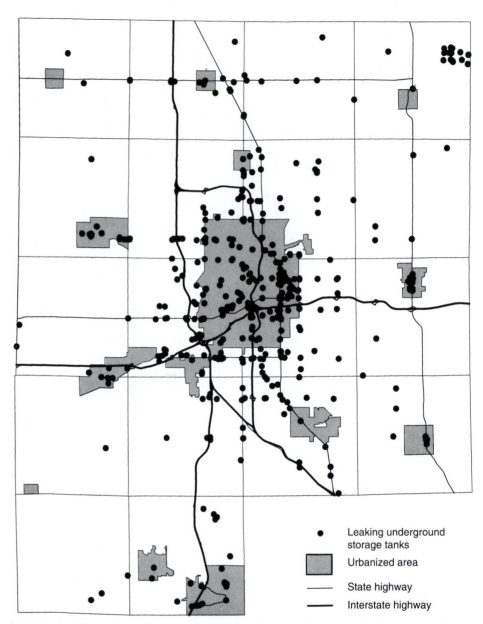

Fig. 7.8   Leaking underground storage-tank sites that have been recorded in Genesee County, Michigan, a typical midwestern urban area about 75 miles from Detroit. Many more leaking tank sites are unrecorded.

processes operating at different levels. Not surprisingly, it is difficult to determine just how a land use at some location will relate to this vast underground environment. Thus we approach landscape planning for groundwater protection more or less as a probability problem, that is, by estimating the *likelihood* of a land use's impact on groundwater given different locations, layouts, densities, and management arrangements.

In the case of proposals for new land uses, planning for groundwater protection *Production sources* begins with an understanding of the potential for contaminant production. Among the land uses of special concern are *industrial facilities*, including manufacturing installations, fuel and chemical storage facilities, railroad yards, and energy plants;

urban complexes, including highway systems, landfills, utility lines, sewage treatment plants, and automotive repair facilities; *agricultural operations*, including cropland, feedlots, chemical storage facilities, and processing plants. Land uses of less concern are single-family residential, institutions such as schools and churches, commercial facilities (excluding those with large parking lots), parks, and open space.

Site evaluation   Once the potential for contaminant production is established, the next step is to assess proposed sites for groundwater susceptibility to pollution. This entails learning about aquifer depths and linkage to the surface as well as significance as drinking water sources. Information on water use and well location is usually obtainable from local or state health departments. Most states require that a log be registered with the health department for residential wells that identifies the materials penetrated and the finished depth of the well. In some cases, the results of a water quality test may also be on record.

A key question in projects involving land uses with high contaminant production potential is the location of recharge zones. Generally, recharge zones are to be avoided, especially those that feed shallow aquifers (Fig. 7.9). Another important question is the permeability of surface materials because permeability controls the rate at which contaminated water and leachate from spills infiltrates the ground.

The shallowest of aquifers is the water table aquifer. It may lie only several meters underground, making it especially prone to contamination. Although it is not widely used as a source of drinking water (because health codes usually require wells to be greater than 25 feet deep), the water table aquifer is an important source of water for streams, ponds, and lakes. Pollution plumes can contaminate seepage water discharging into these water features, particularly where they lie within 1000 feet of a pollution source (see Fig. 5.11).

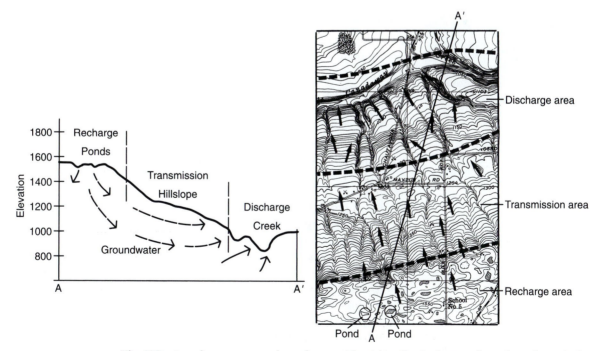

Fig. 7.9 A recharge area upslope from residential wells. Recharge takes place through the small water features at the top of the slope. Such recharge zones are common in glaciated landscapes.

Table 7.1   Suggested Criteria in Land Use Planning for Groundwater Protection

	Desirability	
Criteria	Worst	Best
Contaminant production	High	None
Handling and storage risk	High (e.g., nonsecured storage areas with open soil)	Low
Aquifer use	Drinking water (many wells within a radius of 1000–1500 ft)	None
Aquifer depth	Shallow (less than 200 ft)	Deep (greater than 1000 ft.)
Overlying material	Highly permeable (e.g., sands and gravels)	Impermeable (e.g., clayey soil and confining layer)
Aquifer system	Recharge zone	Beyond discharge (seepage) areas
Flow direction	Toward wells	Away from wells

Evaluation checklist

The planner's checklist for assessing the likelihood of groundwater contamination from a proposed land use should include the following (also see Table 7.1):

Part of a public awareness program on groundwater protection, Austin, Texas.

■ What are the contaminant production potential and the related risks in handling and storing hazardous waste?

■ Does the site lie over an aquifer that (a) presently serves as a source of drinking water; (b) could serve as a future source of drinking water; (c) feeds streams, lakes, and ponds; (d) is fed by a bedrock cavern system?

■ Is the aquifer deep or shallow and is the surface dominated by permeable material such as sand and gravel? Similarly, is the aquifer protected by a confining material, that is, a layer of impervious rock or soil material that limits penetration from sources directly above it?

■ What part of the aquifer system is the site associated with: recharge, transmission, withdrawal, or discharge (seepage)?

■ What is the direction of flow in the aquifer: toward or away from areas of concern? (This is often indicated on groundwater maps by a measure called the potentiometric surface, which is an indicator of the overall slope of the aquifer based on readings of well water levels.)

■ Will the proposed facilities fall into direct contact with groundwater because of deep footings, foundation work, underground utilities, and/or tunneling?

Management planning

For land uses already in place, the problem is to build a management plan to minimize the risk of groundwater contamination. The plan should address the three phases of the system leading to contamination: (1) *production* of contaminants, (2) *removal* from the production and/or disposal site, and (3) *diffusion* into aquifers (Fig. 7.10). Planning for contaminant management and groundwater protection carries the least risk when it is applied early in the system, in the production phase. Basically, three strategies can be employed to lower risk connected with production: reduce output rates, change technology to less dangerous contaminants, and reduce the risk of accidental spills.

At the removal phase, the objective is to limit the escape of contaminants from the site. The conventional strategy here is safely to contain the contaminants during storage, transportation, and disposal. The U.S. Environmental Protection Agency recommends various methods, including secure landfills of containerized waste (see Fig.

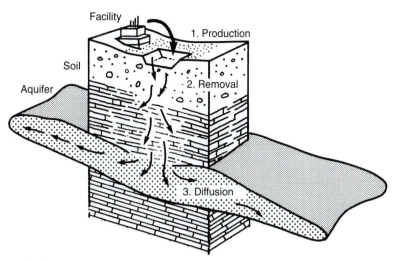

Fig. 7.10 The three-phase system associated with groundwater management: (1) on-site production, (2) removal from the production or disposal site, and (3) diffusion into the aquifer.

5.12). Soil conditions are extremely important inasmuch as compact clayey soils retard the advance of leachates because of their low permeabilities, whereas sandy soils allow the advance of leachates because of high permeabilities.

By the time fugitive contaminants have reached the diffusion phase, the likelihood of impact on aquifers and human water supplies has risen dramatically. However, the concentrations and chemical composition of the leachate may attenuate in passage as a result of filtration, adsorption, oxidation, and biological decay. *Final options* Once the contaminants have invaded an aquifer, preventive measures are no longer feasible and the only options left are (1) corrective pumping (heavy pumping to direct the plume away from wells) or (2) abandonment of vulnerable wells. Again, planning efforts for groundwater protection should be focused early in the system, at the production and site removal phases. Once contaminants escape the site, the environmental risk, technological difficulty of recovery, and recovery expense increase dramatically.

7.7 COMMUNITY WELLHEAD PROTECTION

To help protect community wells from contamination, the U.S. Environmental Protection Agency has mandated states to develop wellhead protection programs. The concept of wellhead protection involves managing the land uses and contamination sources in the contributing or recharge area for a community well (Fig. 7.11). Under this mandate, each state develops its own program plan that is submitted to the EPA for approval. Although programs can vary in design, each state's program must include a requirement for documentation of groundwater levels, flow directions, transmission rates, recharge/exchange times, recharge areas, and contaminant sources within the recharge area of a community well.

The plan for each community must include a program for testing the public water supply, a management plan for contaminant sources and areas, and a risk management plan for accidents such as chemical spills. Although the Wellhead Protection Program is generally viewed as an important step toward improved management of groundwater resources, the program is not supported by federal funds, and therefore is viewed with less than enthusiasm by many states and local communities.

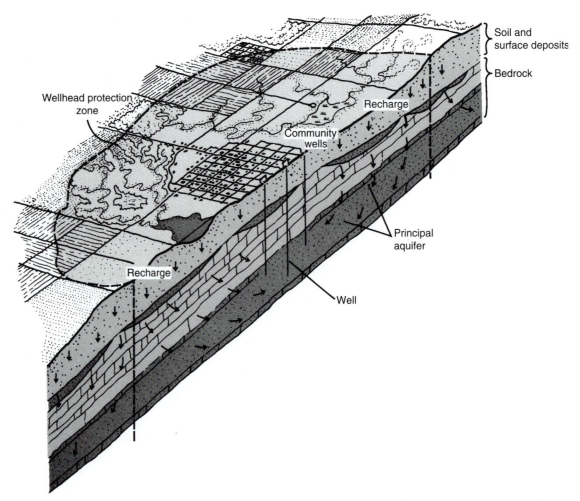

Fig. 7.11 The concept of wellhead protection, illustrating the protection zone around a set of community wells.

7.8 CASE STUDY

■ **Planning for Groundwater Protection Using a Geographic Information System**

Martin M. Kaufman

Residents of Tyrone Township in southeastern Michigan are dependent on groundwater for their drinking water supply. Almost all of this water comes from an aquifer in the glacial drift (i.e., the surface deposits of sands, gravels, and related materials that lie over the bedrock). Facing accelerated growth and overflow development from nearby urban centers, the Township became concerned about aquifer contamination and elected to initiate a groundwater vulnerability study to: (1) better understand its drinking water resource system; and (2) develop guidelines to control future development. The study employed a Geographic Information System (GIS) to map and synthesize the data.

Why was a GIS chosen to perform these tasks? Townships are large areas (36 square miles), and to manually map the factors contributing to this problem would be an enormous task. Moreover, the existing data, such as the soil reports and well logs, which would have to be used, come in different formats and geographic scales.

Finally, the limited time and resources available for this study ruled out extensive fieldwork and manual map redrafting. Therefore, a GIS was employed.

Geographic Information Systems are tools well suited to perform overlays of digital data that are spatially referenced by an established coordinate system, such as latitude and longitude. Within GISs, overlay operations combine data layers consisting of map geometries (points, lines, areas, grid cells), and their associated data attributes (e.g., soil types or textures) to produce new layers. Once this capability is operational within a GIS, the opportunities for timely (and cost-effective) data manipulation, analysis, and map display far exceed those possible with manual methods.

How was the GIS for this project implemented? In order to accurately depict groundwater vulnerability, it was necessary to obtain well log data, digital elevation data, measurements of groundwater levels, soils data, and current land use information. From these sources, additional data were generated including depth to water, percentage of sand and gravel, soil permeability, and land infiltration rates. These four variables were used in a map overlay model to identify areas of low, medium, and high groundwater vulnerability to contamination from sources such as stormwater, landfills, and chemical spills.

To accommodate the process model, a vector implementation of GIS was chosen. Vector GISs store data as points, lines, and polygons (geometric primitives), and have the capability to associate multiple attributes with each geometric primitive. Raster GISs, on the other hand, use grid cells to represent the earth's surface, with one attribute stored per grid cell. Thus, with four polygon layers of data used to model groundwater vulnerability, the capabilities of the vector GIS selected provided a good match for the processes requiring analysis. The final cartographic model is shown as follows:

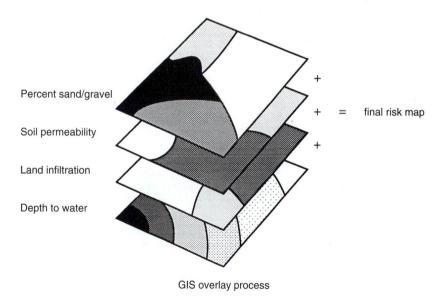

Percent sand/gravel

Soil permeability

Land infiltration

Depth to water

+ + = final risk map
+

GIS overlay process

What were the lessons learned from this project? The project team was aware that errors arising from GIS overlay operations could be minimized when data from similar sources and scales were used. Since digital data at a map scale of 1:24,000 were available from the same source for elevation, soils, and land use, the decision to use these data avoided a critical mistake often made in GIS applications. Too often, GIS developers concentrate on the images provided by the *Information and Systems* components of the software, but neglect the critical *Geographic* concepts, such as scale, throughout the project cycle.

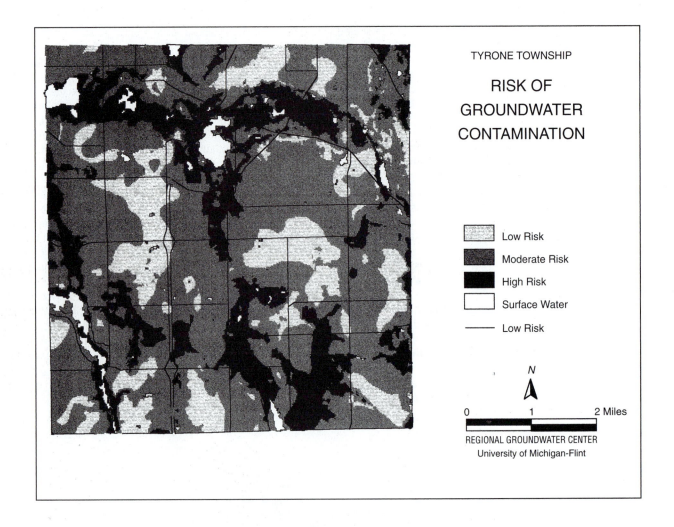

Good correspondence between the data layers also enhances the flexibility of the final GIS application and facilitates expansion. If the township requests additional geographic reference for the map locations, such as the addition of a digital road layer, also available at the 1:24,000 scale, they could be easily accommodated.

Another key aspect of GIS development is data preparation. In practice, over 50 percent of the costs associated with GIS development are related to data acquisition, conversion to a usable format, editing, and accuracy verification. Many GIS projects become "problems" because of their failure to accommodate the large labor expenditures related to data, so it is important to instill the GIS development process with a sound approach to data management. Prototyping the application with a small amount of data can help cut costs, and avoid larger expenditures later.

Before selecting an analytical tool, real-world knowledge must be used to identify the problem correctly, collect data, assess the physical processes at work and the different geographic scales at which they operate, and then accurately describe these processes and build a model depicting the system. Only then should the decision to use GIS or any other tool be made.

Martin M. Kaufman is chairman of the Earth and Resource Science Department at the University of Michigan—Flint. He specializes in water resources and GIS applications to land use planning.

7.9 SELECTED REFERENCES FOR FURTHER READING

Butler, Kent S. "Managing Growth and Groundwater Quality in the Edwards Aquifer Area, Austin, Texas." *Public Affairs Comment* 29:2, 1983.

DiNovo, Frank, and Jaffe, Martin. *Local Groundwater Protection: Midwest Region.* Washington, DC: American Planning Association, 1984.

Hill-Rowley, R., et al. *Groundwater Vulnerability Study for Tyrone Township.* Regional Groundwater Center, University of Michigan–Flint, 1995.

Josephson, Julian. "Groundwater Strategies." *Environmental Science and Technology* 14:9, 1980, pp. 1030–1035.

Moody, D. W. "Groundwater Contamination in the United States". *Journal of Soil and Water Conservation.* Vol. 45, 1990, pp. 170-179.

National Research Council. *Groundwater Contamination: Studies in Geophysics.* Washington, DC: National Academy of Sciences Press, 1984.

Page, G. W. *Planning for Groundwater Protection.* New York: Academic Press, 1987, 387 pp.

Pye, V. I., Patrick, Ruth, and Quarles, John. *Groundwater Contamination in the United States.* Philadelphia: University of Pennsylvania Press, 1983.

Rutledge, A. T. "Effects of Land Use on Ground-Water Quality in Central Florida—Preliminary Results: U.S. Geological Survey Toxic Waste—Ground-Water Contamination Program." *Water-Resources Investigations Report 86-4163*, Washington, DC: U.S. Government Printing Office, 1987.

Tripp, J. T. B., and Jaffe, A. B. "Preventing Groundwater Pollution: Towards a Coordinated Strategy to Protect Critical Recharge Zones." *Harvard Environmental Law Review* 3:1, 1979, pp. 1–47.

U.S. Council on Environmental Quality. *Contamination of Groundwater by Toxic Organic Chemicals.* Washington, DC: U.S. Government Printing Office, 1981.

STORMWATER DISCHARGE, WATER MANAGEMENT, AND LANDSCAPE CHANGE

8.1 INTRODUCTION

One of the most serious problems associated with land development is the change in the rate and amount of runoff reaching streams and rivers. Both urbanization and agricultural development effect an increase in overland flow, resulting in greater magnitudes and frequencies of peak flows on streams. The impacts of this change are serious, both financially and environmentally: property damage from flooding is increased, water quality is reduced, channel erosion is accelerated, and habitat is degraded.

Responsible planning and management of the landscape depend on accurate assessment of the changes in runoff brought on by land development. In the United States and parts of Canada this problem has reached epidemic proportions, and in most communities developers are now required to provide analytical forecasts of the changes in overland flow and stream discharge resulting from a proposed development. Such forecasts are used not only as the basis for recommending alternatives to traditional stormwater systems in order to reduce environmental impact, but also for evaluating the performance of an entire watershed that is subject to many development proposals.

This chapter is concerned with the stormwater runoff generated from rainstorms, the factors in the landscape that influence runoff rates, and how stormwater can be managed to minimize related environmental impacts. We examine a traditional model for forecasting stormwater from small watersheds, some alternative models, the selection of mitigation measures, and the issue of comprehensive water management.

8.2 OVERLAND FLOW

Infiltration and overland flow

Most precipitation reaching the landscape is disposed of in four ways (Fig. 8.1). Some is taken up on the surface of vegetation in a process known as **interception**. Some is absorbed directly by the soil in a process known as **infiltration**. Some, called **depression storage**, collects on the ground in small hollows and pockets. The remainder, called **overland flow**, runs off the surface, eventually joining streams and rivers. This is **stormwater**. In conventional stormwater management, infiltration and overland flow are the main concerns, and in most areas, the two are inversely related.

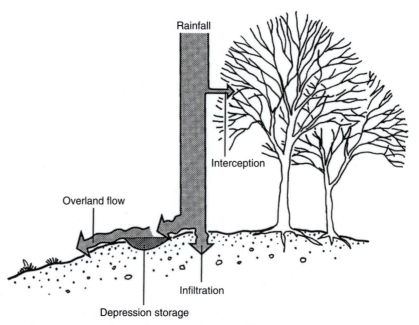

Fig. 8.1   The disposition of rainfall in natural or partially developed landscapes: interception, infiltration, depression storage, and overland flow.

Controls on overland flow

The percentage of precipitation going to infiltration and overland flow varies drastically in the modern landscape. In heavily vegetated terrain, infiltration can be so high that overland flow is practically negligible; instead, streams gain their discharge from subsurface sources, mainly groundwater and water within the soil, called *interflow*. This is especially pronounced in densely forested areas such as the Pacific Northwest and in areas of highly permeable soil such as the Nebraska Sand Hills. In dry areas, on the other hand, where vegetation is sparse or absent and in humid areas where forest and groundcover have been cleared and replaced by agriculture, settlements, and related land uses, overland flow is, by contrast, normally substantial. If we examine the ground to determine what factors control infiltration and overland flow, we will find that land cover (vegetation and land use), soil composition and texture, and surface inclination (slope) are usually the chief controls. Figure 8.2 illustrates the relationship among overland flow, soil, and vegetative cover on a hillslope.

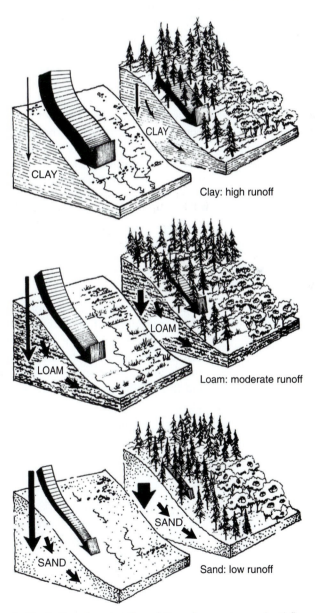

Fig. 8.2 Schematic illustration showing the relative changes in overland flow runoff with soil type and vegetation on sloping ground. The subsurface arrows represent water going to groundwater and interflow.

Table 8.1   Coefficients of Runoff for Rural Areas

Topography and Vegetation	Open Sandy Loam	Clay and Silt Loam	Tight Clay
Woodland			
Flat (0–5% slope)	0.10	0.30	0.40
Rolling (5–10% slope)	0.25	0.35	0.50
Hilly (10–30% slope)	0.30	0.50	0.60
Pasture			
Flat	0.10	0.30	0.40
Rolling	0.16	0.36	0.55
Hilly	0.22	0.42	0.60
Cultivated			
Flat	0.30	0.50	0.60
Rolling	0.40	0.60	0.70
Hilly	0.52	0.72	0.82

As a general rule, overland flow *increases* with slope, *decreases* with soil organic content and particle size, *increases* with ground coverage by hard surface material such as concrete and asphalt, and *decreases* with vegetative cover.

Coefficient of runoff

For a particular combination of these factors, a **coefficient of runoff** can be assigned to a surface. This is a dimensionless number between 0 and 1.0 that represents the proportion of a rainfall available for overland flow after infiltration has taken place. A coefficient of 0.60, for example, means that 60 percent of rainfall (or snowmelt) is available for overland flow, whereas 40 percent is lost to infiltration. Table 8.1 lists some standard coefficients of runoff for rural areas based on slope, vegetation, and soil texture. For urban areas, coefficients are mainly a function of the hard surface cover, and a number of values are given in Table 8.2.

Transient factors

Several transient factors may influence the coefficient of runoff and overland flow that are not standard considerations. As rainfall intensity increases, the coefficient of runoff temporarily rises for most surfaces. Extremely intensive rainstorms, such as those that yield 2 or 3 inches of rain in a 30-minute period, can override a soil's absorptive capacity, even those soils with high infiltration capacities. Prestorm soil moisture is another consideration. If an earlier rainstorm has already loaded a soil with water, or the level of groundwater has risen into the soil column, infiltration capacity falls and the coefficient of runoff rises substantially. In fact, naturally saturated soils such as those in wetlands behave as impermeable surfaces with coefficients of runoff approaching 0.90 or more. With no medium to take up the water, the result is overland flow leading to stormwater loading of wetlands and connecting streams.

Table 8.2   Coefficients of Runoff for Selected Urban Areas

Commercial	
Downtown	0.70–0.95
Shopping centers	0.70–0.95
Residential	
Single family (5–7 houses/ac)	0.35–0.50
Attached, multifamily	0.60–0.75
Suburban (1–4 houses/ac)	0.20–0.40
Industrial	
Light	0.50–0.80
Heavy	0.60–0.90
Railroad yard	0.20–0.80
Parks, Cemetery	0.10–0.25
Playgrounds	0.20–0.40

8.3 COMPUTING RUNOFF FROM A SMALL WATERSHED

Rational method

The runoff generated in the form of overland flow or stormwater from a small watershed (usually less than 1000 acres in area) can be computed by means of a simple manipulation called the **rational method**. This method is based on a formula that combines the coefficient of runoff with the intensity of rainfall and the area of the watershed. The outcome gives the **peak streamflow** (discharge) for *one rainstorm* at the *mouth of the watershed*:

$$Q = A \bullet C \bullet I$$

Where

Q = discharge in cubic feet per second
A = area in acres
C = coefficient of runoff
I = intensity of rainfall in inches or feet/hour

Design storm concept

To use the rational method, it is necessary to generate some essential data. Besides the obvious need to measure the area of the watershed and determine the correct coefficient of runoff, we must select an appropriate rainfall intensity value. Obviously, a wide variety of rainstorms deliver water to the watershed, and one storm that is likely to produce significant runoff, called the **design storm**, has to be selected in order to perform the computation. The design storm is defined according to local rainfall records for intensive storms of short duration (usually 1 hour or less), which occur, for example, on the average of once every 10 years, 25 years, or 100 years. These storms are called, respectively, the 10-year, 25-year, and 100-year storms of 1-hour duration. Which storm should be used usually depends on the recommendation of the local agency responsible for stormwater management. Figure 8.3 gives the values over the coterminous United States and the southern fringe of Canada for the 10-year, 1-hour storm. For more accurate values based on local records, local agencies such as the office of the city or county engineer should be consulted.

Once the desired storm has been selected, a second step must be taken to arrive at the actual value to be used in making the discharge computation. This step is based on two facts: (1) within the period of a rainstorm, say, 60 minutes, the actual

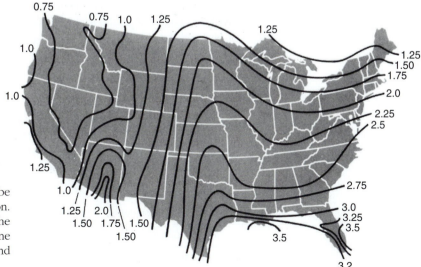

Fig. 8.3 The amount of rainfall that can be expected in the 10-year storm of 1-hour duration. Notice the dramatic differences between the Pacific Coast, the Rocky Mountain-Intermontane Region, and southern Canada on the one hand and the American South on the other.

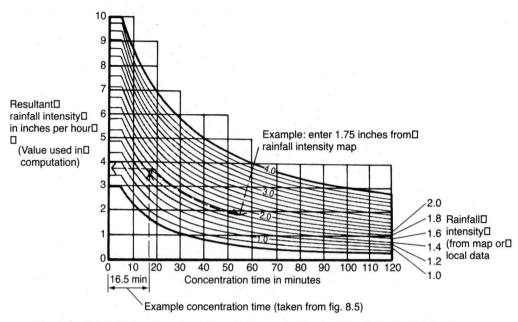

Fig. 8.4 Rainfall intensity curves. To use the graph, first find the desired rainfall value among the curves, then find the appropriate concentration time on the base of the graph. The rainfall intensity value is read from the intersection of the two on the left vertical scale. In the example, the resultant rainfall value is about 3.75 inches.

Concentration time

intensity of rainfall initially rises, hits a peak, and then tapers off; and (2) the time taken for runoff to move from the perimeter to the mouth of the watershed, called the **concentration time**, varies with the size and conditions of the watershed. Combining these two facts, we can see that in a small watershed the concentration time may be less than the duration of the storm so that the peak discharge is reached before the storm is over. Therefore, in order to accurately compute the peak discharge for the storm in question, we must use the value of rainfall intensity that corresponds to the time of concentration. The graph in Figure 8.4 gives representative curves for various storms and shows how to use the graph based on an example storm intensity of 1.75 inches per hour and a concentration time of 16.5 minutes.

Reliable estimates of the concentration time are clearly important in computing the discharges from small watersheds. The approach to this problem generally involves making separate estimates for (1) the time of overland flow; and (2) the time of channel flow, and then summing the two. The graph in Figure 8.5 can be used to estimate the time of overland flow if three things are known: (1) the length of the path (slope) from the outer edge of the watershed to the head of channel flow; (2) the predominant groundcover; and (3) the average slope of the ground to the head of channel flow. If these are not known, we must resort to an approximation based on representative overland flow velocities such as those offered in Table 8.3.

Channel flow velocity

Channel flow velocities can be estimated with the help of a formula (below) that is based on the channel's shape, roughness, and gradient. There are numerical values for each parameter and for most channels they can be approximated from maps

Table 8.3 Representative Overland Flow Velocities

Distance to Channel Head	Pavement	Turf	Barren	Residential
Around 100 ft	0.33 ft/sec	0.20 ft/sec	0.26 ft/sec	0.25–0.33 ft/sec
Around 500 ft	0.82/ft sec	0.25 ft/sec	0.54 ft/sec	0.50–0.65 ft/sec

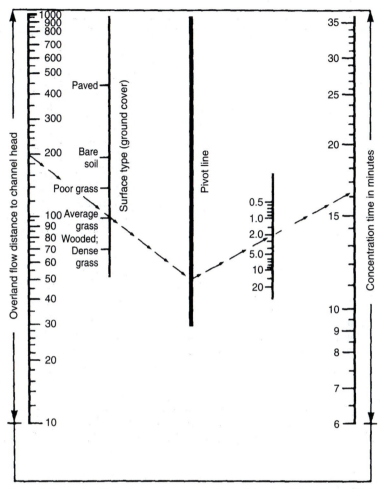

Fig. 8.5 A graph for estimating the concentration time of overland flow from a basin perimeter to a channel head or inlet. The string of arrows represents an example.

and field observations. For most natural channels when they are carrying a discharge at bankfull level, flow velocity falls in the range of 4 to 8 feet per second.

$$v = 1.49 \frac{R^{2/3}s^{1/2}}{n}$$

where

v = velocity, ft or m/sec
R = hydraulic radius (wetted perimeter of channel divided by cross-sectional area of channel)
s = slope or grade of channel
n = roughness coefficient (See p. 260 in Chapter 14 for roughness values.)

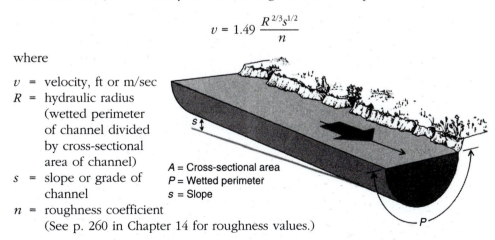

A = Cross-sectional area
P = Wetted perimeter
s = Slope

Because channel flow is so much faster than overland flow (typically ten times faster), it is important to delineate the full channel system, or drainage net, in a watershed. This means going beyond the stream channels shown on conventional topographic contour maps and mapping the small natural and manmade channels that carry water only intermittently. These include natural swales, road ditches, and farm drains, but only those that are clearly tied to the main channel system and contribute directly to its discharge (see Fig. 8.11b). As a rule, watersheds with an abundance of channels, such as those in urbanized areas, are much faster than those in undeveloped areas.

8.4 USING THE RATIONAL METHOD

As a forecasting device, the rational method is suited best not only to watersheds that are small, but also to those that are partially or fully developed. Beyond that, the reliability of the method depends on the accuracy of the values used for the coefficient of runoff, the concentration time, the drainage area, and the rainstorm intensity. With the exception of rainstorm intensity, the necessary data may be generated through field observations and surveys or from secondary sources, in particular topographic maps, soil maps, and aerial photographs. It is also necessary to add that many versions of the rational method have been devised that extend its utility and precision for various situations and geographic locations.

Computational procedure   Given that the pertinent data are in hand, the following *procedure* may be used to compute the **peak discharge (Qp)** resulting from a specified rainstorm. The result gives you the runoff leaving the watershed in a channel at the instant of peak flow (If the watershed or drainage area is not served by a channel, then the peak discharge cannot be computed using the rational method, but it may be possible to compute the volume of water available as overland flow. See number 8 below).

1. Define the perimeter of the watershed and measure the watershed area.
2. Subdivide the watershed according to cover types, soils, and slopes. Assign a coefficient of runoff to each, and measure each subarea.
3. Determine the percentage of the watershed represented by each subarea and multiply this figure by the coefficient of runoff of each. This will give you a coefficient adjusted according to the size of the subarea.
4. Sum the adjusted coefficients to determine a coefficient of runoff for the watershed as a whole.
5. Determine the concentration time using the graph in Figure 8.5 and the stream velocity formula.
6. Select a rainfall value for the location and rainstorm desired. Using this value and the concentration time, identify the appropriate rainfall intensity from the curves in Figure 8.4.
7. Multiply the watershed area times the coefficient of runoff times the rainfall intensity value to obtain peak discharge in cubic feet per second. (If you use inches for I, the answer is in acre inches per hour; however, these units are so close to cubic feet per second that the two are interchangeable.)
8. The total **volume of discharge (Qv)** produced as a result of a rainstorm can also be computed with the rational formula. The formula for volume of storm runoff is

$$Qv = A \bullet C \bullet R$$

where

A = drainage area, C = coefficient of runoff, and R = total rainfall.

In this case the full one-hour rainfall value is used, and we would solve for acre feet or cubic feet. This is a simpler computation because it does not involve deciphering concentration time and rainfall intensity. It is useful in estimating gross differences in pre- and post-development runoff.

8.5 OTHER RUNOFF MODELS AND CONCEPTS

Partial area concept

Scientists recognize a number of shortcomings in the rational method and the concepts on which it is based, and have devised some alternative models that deserve our attention. The principal alternative is based on the observation that most undeveloped and many partially developed watersheds do not produce basinwide overland flow in response to intensive precipitation. In other words, when it rains hard, only a fraction of the watershed develops overland flow and yields stormwater to streams. This observation has led to a model called the **partial area** or **variable source** concept.

According to this concept, the hydrologically active portion of a watershed in terms of stormwater discharge may be as little as 10 to 30 percent of the drainage area. Typically, the contributing area is found in collection zones at valley heads, in swales, along footslopes and low ground near streams (Fig. 8.6*a*). In addition, the size and shape of the contributing area may be different from event to event depending on soil moisture conditions, groundwater levels, and other factors.

Management implications

The partial area concept has serious implications for stormwater management. First, forecasts of stormwater discharge based on the rationale method may be larger than the actual discharges for many basins. Second, if large parts of a basin do not yield overland flow, then those areas apparently contain *natural buffers* that enable them to retain rainfall and limit its release to streams as stormwater. These buffers are usually high infiltration soils, forested areas, wetlands, and/or depression storage areas that take up all the water delivered to them by rainfall and surrounding runoff surfaces. If, in the course of planning and development, these noncontributing areas are equipped with stormwater facilities, as would be the logical assumption using the

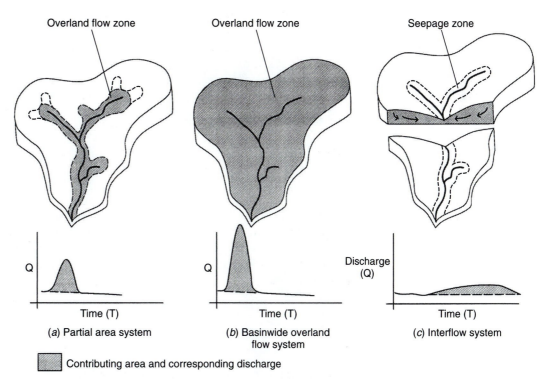

Fig. 8.6 Maps and corresponding hydrographs showing the differences in a basin's response according to (*a*) the partial area concept with contributions from overland flow at valley heads and channel fringes, (*b*) the rational method concept with contributions from overland flow over the entire basin and (*c*) the interflow concept with contributions limited principally to seepage along channels.

rational method, as is represented by Figure 8.6b, they then would be converted into contributing stormwater areas. Not only would basin stormwater discharge increase, but an existing stormwater mitigation opportunity would be lost.

Interflow concept

In many forested watersheds, field research has demonstrated that overland flow is effectively nonexistent. It appears that small streams in such watersheds rely entirely on subsurface water in the form of **interflow**. This water enters the ground at the point of rainfall reception and flows laterally within the soil mantle to the streambank. The corresponding hydrograph for stormflow in response to a rainstorm is slower, lower, and longer than ones for both partial area systems and full overland flow systems (Fig. 8.6c). Unfortunately, no methods comparable in ease of application to the rational method have been devised for computing discharge from an interflow system. The unit hydrograph method, which is described in Chapter 10, has potential in this regard, but it requires stream discharge measurements in response to a rainstorm, and for most basins, no discharge records are available.

8.6 TRENDS IN STORMWATER DISCHARGE

The clearing of land and the establishment of farms and settlements have been the dominant geographic changes in the North American landscape in the past 200 years. Almost invariably, these changes have led to an increase in both the amount and rate of overland flow, producing larger and more frequent peak flows in streams. With the massive urbanization of the twentieth century, this trend became even stronger and has led to increased flooding and flood hazard, to say nothing of the damage to aquatic environments.

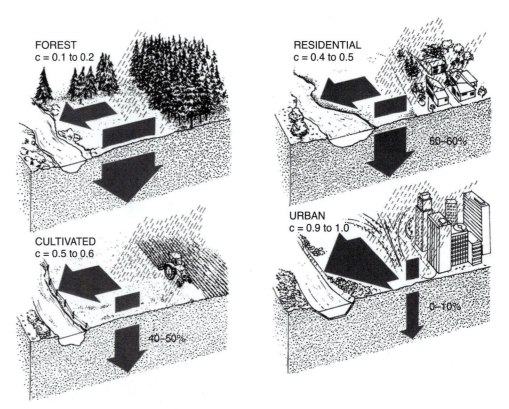

Fig. 8.7   Changes in the coefficient of runoff with land use and land cover: forest, cultivated, residential, and urban. Generally speaking, the coefficient of runoff and water lost to infiltration are inversely related.

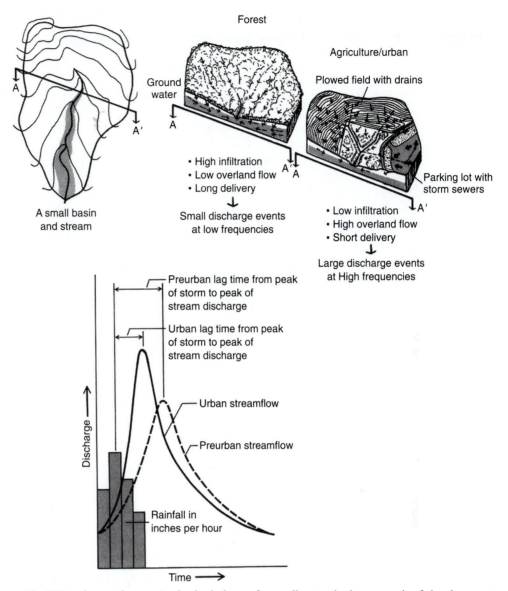

Fig. 8.8 Above, changes in the hydrology of a small watershed as a result of development. Below, the resultant change in stream discharge shown by a higher and quicker hydrograph after urbanization.

Primary causes   From a hydrologic standpoint, the source of the problem can be narrowed down to changes in two parameters: (1) the drastic increase in the *coefficient of runoff* in response to land clearing, deforestation, and the addition of impervious materials to the landscape (Fig. 8.7); and (2) the corresponding decrease in the *concentration time* (Fig. 8.8, upper diagram). In agricultural areas, this decrease is brought about by the construction of field drains and ditches as well as the straightening and deepening of stream channels. With urbanization, ditches are replaced with stormsewers, small streams are piped underground, and gutters are added to streets, all of which may reduce concentration times by more than tenfold. Together these changes produce a significant increase in the magnitude and frequency of peak discharges in receiving streams (Fig. 8.8, lower diagram). Changes in a third parameter, rainstorm intensity, may also contribute to discharge increases; storm magnitudes appear to be on the rise in metropolitan areas in the United States. At the present, however, it is

(a)   *(b)*

Fig. 8.9   Depression storage before and after development. (*a*) Abundant depression storage before development, but (*b*) very little is apparent (or possible) after development.

difficult to generalize about this trend because it is apparent only in certain metropolitan areas, but most large urban areas appear to be affected.

Other causes   Other changes brought on by development include reduction in depression storage and grafting of road drains onto drainage nets. *Depression storage* is the process by which overland flow is stored in low spots in the microtopography and thereby withheld from the stormflows (Fig. 8.9). In old-growth forests, for example, microtopography is typically diverse because of the presence of pits and mounds formed by windthrown trees, rotting stumps, burrowing animals, and so on. When forested land is cleared for agriculture, much of its depression storage capacity is eliminated by plowing away the microtopography. With residential and related development, the ground is not only graded smooth, but often mounded slightly or substantially to "improve" drainage.

Roads are a major source of increased runoff in many watersheds. Not only are they areas of much reduced infiltration, but they and their attendant stormwater drains increase drainage density and link or graft otherwise low runoff areas to stream systems. Many of these areas would normally be limited to low rates of runoff because of depression storage and/or high infiltration, but with roads superimposed on the watershed, they are spliced into the system of channels and lose their natural mitigative capacity (Fig. 8.10). This brings us to the concept of effective impervious cover.

In order for runoff from impervious cover to affect streams, the impervious surface has to be linked to a stream channel. If it is not linked to a stream channel, then the runoff it produces must reenter the landscape, where it is subject to infiltration, depression storage, and other means of reduction and disposal. This is the

Effective impervious cover   basic idea behind the term **effective impervious cover**, defined as impervious cover linked by ditch or pipe to a stream channel.

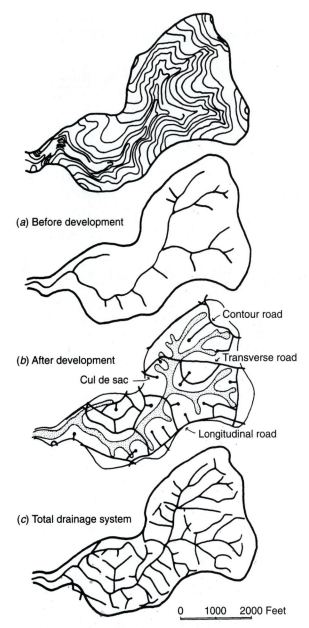

(a) Before development

(b) After development

Cul de sac —

Contour road

Transverse road

Longitudinal road

(c) Total drainage system

0   1000   2000 Feet

Fig. 8.10 The pattern of natural channels and roads (with stormdrains) before and after development in a small (320 acre) basin near Austin, Texas. Roads with curbs, gutters, and stormsewers are grafted onto the natural system of stream channels more than doubling the drainage density. The increase in drainage density drives up both the magnitude and frequency of stormflows.

With the exception of rural areas where impervious cover occurs in widely spaced islands, most impervious cover and related land use surfaces are linked to streams by engineered stormwater systems. The most common means of making this linkage, and thus creating effective impervious cover, is by building streets and roads because all roadways are designed with a paralleling set of stormdrains. Thus, as the density of roads increases with development and the drainage density (miles of channel per square mile of land) in a watershed increases, effective impervious cover rises

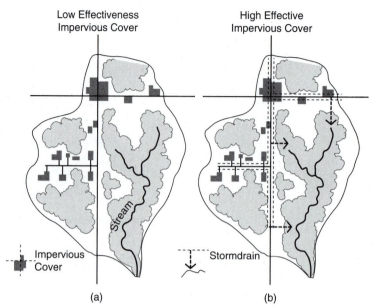

Fig. 8.11   An illustration of the concept of effective impervious cover: (*a*) no direct linkage between impervious cover and streams; (*b*) all impervious surfaces linked directly to streams.

until eventually all hard surfaces are serviced by stormdrains and the mitigating effects of the landscape on overland flow are eliminated (Fig. 8.11).

8.7 STORMWATER MITIGATION

Most communities have enacted stringent stormwater management standards on new development. Increasingly, these standards call for zero net increase in stormwater discharge from a site as a result of development. This means that the rate of release across the border of a site can be no greater after development than before. A basic dilemma thus arises because most forms of development effect an increase in runoff, but at the same time themselves require relief from stormwater.

Basic strategies   Three strategies may be used to achieve management of stormwater: (1) *store* the excess water on or near the site, releasing it slowly over a long time; (2) *return* the excess water to the ground, where it would have gone before development; (3) *plan* the development such that runoff is not significantly increased. The first, which is often referred to as "pipe and pond", is the most common strategy. It involves directing stormwater to a holding basin and then releasing it slowly over an extended period of time, thereby reducing or "shaving" the peak discharge. The main objective is to reduce the rate of stormwater delivery to streams, and it usually involves the con-

Storage basin strategy   struction of storage facilities such as detention basins (Fig. 8.12*a*).

Detention basins are ponds sized to store the design storm and then allow it to discharge at a specified rate. For instance, if a design storm generates a total runoff of 4.2 acre feet (182,952 cubic feet) in one hour, and the allowable release rate is 12 cubic feet per second (43,200 cubic feet per hour), then about 3.2 acre feet of gross storage are needed. Basins that store water on a long-term basis with the objective of releasing it to the ground and to evaporation are called *retention basins*, and they are generally used in areawide stormwater management. Increasingly, the pipe and pond strategy is falling into disfavor. Among other things, receiving streams often suffer loss of summer low flows. In addition, the absence of channel-shaping peak flows

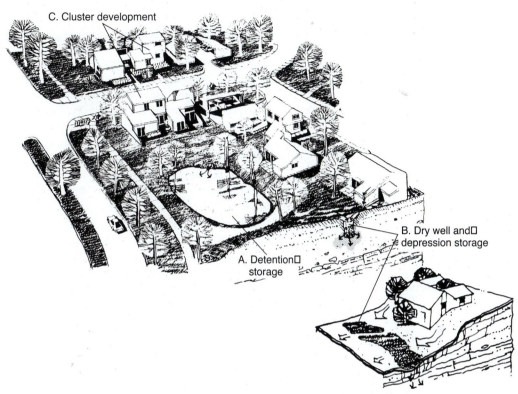

Fig. 8.12 Examples of stormwater mitigation measures: (*a*) detention storage, (*b*) on-site infiltration, and (*c*) cluster development. The lower drawing illustrates a version of (*b*) based on depression storage.

has been shown to be disruptive to aquatic and riparian habitat by reducing stress, and allowing invasion by alien plant species.

Source control strategy   The *second strategy* relies on **source control** of stormwater, that is, disposing of it at its point of origin. This strategy utilizes some form of soil infiltration and is usually accomplished onsite. Stormwater is directed into vegetated areas, shallow depressions, troughs, or pits, from which it percolates into the ground (Fig. 8.12*b*). Since infiltration rates are often slower than rainfall rates, this strategy is usually most effective for small storms, low intensity, long duration storms, or the water produced by the first part of large rainstorms. It is particularly useful on the North American West Coast where the majority of rain comes in events producing 1 cm or less. The lower diagram in Fig. 8.12 illustrates a source-control concept based on depression storage in residential development. For a 0.25-acre lot (11,000 square feet), the entire volume of a 0.5-inch rainfall can be held in an area of 1400 square feet at a depth of 4 inches. In dry regions this method may prove to be an effective means of recharging local soil moisture supplies and reducing yard and parkway irrigation. But to use depression storage effectively, many communities, their engineers, and landscape architects would have to change their standards and practices that now dictate smooth and/or mounded grading of yards and parkways, ruling out any possibility for depression storage.

Site planning strategy   The *third strategy* reduces the need for corrective measures and relies on **site adaptive planning** to resolve the stormwater problem. Attention is first paid to locating development on sites with good hydrologic performance. Next, attention is paid to surface materials, avoiding impervious surfaces wherever possible, and to density ratios, that is, the balance between developed land and open space.

Clustering is one means of reducing impervious areas by grouping buildings and related facilities to improve the per person impervious cover ratio and allow for more comprehensive approaches to stormwater mitigation (Fig. 8.12c). Although it is possible to achieve zero net increase in runoff solely through careful site planning, especially when previously cleared land is involved, most successful stormwater management requires a combination of site planning, source control, and storage strategies. (These ideas are developed further in Chapter 13.)

Selecting mitigation measures

Which strategies and mitigation measures should be used depends on many factors, including local topographic and soil conditions and the character of the development program and its layout. In some areas there is no choice in the matter because local ordinances dictate what is to be used and many of these ordinances are based strictly on traditional engineering methods. Where there is a choice, however, several criteria can be used to guide the selection of stormwater mitigation measures. The *first criterion* is discharge magnitude. The question is where within the flow system to focus mitigation efforts: (a) on small flows in and around planning sites, or (b) on large flows downstream? This is significant because the force of running water, that is, the stress it exerts on the channel and riparian environment, increases substantially

The flow magnitude criterion

with discharge. As a rule, therefore, large flows are inherently more difficult to contain and regulate than small ones. It follows that large flows usually require substantial engineering treatments, whereas small flows can often be handled as a part of landscape design treatments. Unfortunately, stormwater management in most North American communities is limited to traditional engineering approaches involving gutters, pipes, and basins scaled to street and road networks and tied to local streams. Any mitigation of this system usually involves regional stormwater basins that inevitably lead to more engineering and higher mitigation costs.

Other criteria

The *second criterion* is environmental compatibility. This addresses the questions of (1) environmental sacrifices and tradeoffs such as wetland impacts and habitat loss and (2) compatibility with the landscape design scheme as an aesthetic and land use issue (Fig. 8.13). The second question includes the potential for multiple uses, such as recreation and habitat enhancement as well as neighborhood character and social efficiency.

Fig. 8.13 Retention ponds in Iowa farm country built for flood control. While compatible with agricultural needs, such measures may conflict with other objectives, such as habitat conservation and downstream water supply.

The *third criterion* is cost, both construction and maintenance, and the *fourth criterion* is liability. The risks of children around stormwater ponds and the hazards associated with downstream flooding are the principal liability concerns. Risk management and liability are becoming colossal issues on many fronts in environmental planning, often demanding inordinate attention from engineers, planners, and facility managers.

8.8 THE CONCEPT OF PERFORMANCE

The performance concept

Any effort to plan and manage the environment rests on a concept about how the environment should perform. When plans are formulated, the objective is either to guide and structure future change in order to avoid undesirable performance and/or to improve performance in an environment, setting, or system whose existing performance is judged to be inadequate. The phrase *judged to be inadequate* is important because any judgment on performance is based on human values. "That stream floods too often and it poses a danger to local inhabitants," is a value judgment by someone about environmental performance. Similarly, an ordinance restricting development from wetlands reflects a societal value about either the performance of wetlands, the performance of land use, or both.

Performance goals

In watershed planning and management, **performance goals** must be set early in the program to determine the *best management practices* (BMPs) to be used in the planning process. Generally, the larger the watershed, the more difficult is the task because large watersheds usually involve many communities and interest groups with different views and policies on the matter. The task is usually less complex for a small watershed, not only because fewer players are involved, but also because the watershed itself is usually less complex.

The process of formulating performance goals and BMPs usually begins with the definition of local (watershed land users') values, attitudes, and policies. Next, regional factors are considered, in particular, policies pertaining to development density, stormwater retention, wetlands, open space, and the like. In addition, the values and needs, both articulated and apparent, of downstream riparians must be given serious consideration. Where, for instance, upstream users depend on a watershed and its streams to carry stormwater away, and downstream riparians have set a goal around the maintenance of habitat and quality residential land near water courses, the two goals are in conflict, and unless modified, the downstream group stands little chance of success.

Watershed hydrologic versatility

A third consideration in setting performance goals is the *carrying capacity* of the watershed based on its biophysical character. It is important to recognize the diversity, or **hydrologic versatility** of watersheds, even small ones. This means that there is often a wide range of land types within a basin which calls for different types, densities, and configurations of development. In other words, watersheds cannot be treated as uniform entities, ignoring fundamental differences in terrain, soils, and water processes from place to place.

Each watershed must be evaluated as a biophysical system, including existing land use, and weighed against the proposed goals based on local and regional values and policies. A watershed and streams that could not support a trout population under predevelopment conditions certainly cannot support one after development; therefore, a performance goal of habitat quality suitable to sustain trout would be physically unrealistic. It is usually the role of the environmental specialist in hydrology, soils, ecology, or forestry, for example, to examine the watershed, determine its condition and potentialities, and recommend appropriate modifications in proposed performance goals.

Performance standards

Once performance goals have been formulated, **performance standards** and controls have to be defined. Performance standards are the specific levels of performance that must be met if goals are to be achieved. For example, a stormwater discharge standard might call for zero net increase in peak discharges on first-order streams after a particular date or development density has been reached. This means that in planning new development it is necessary to take into account all changes in land use activities in the watershed at this time and determine from the balance of both positive and negative changes whether special measures or strategies are needed. If, for instance, cropland is being converted to woodland at the same time woodland is being cleared for new residential development, analysis may show that one change offsets the other, thereby maintaining the performance standard of zero net change in stormwater discharge.

Performance controls

Performance controls are the rules and regulations used to enforce the standards and goals. These may be specific ordinances limiting the percentage of impervious surface, requirements for site plan review and approval, or incentives such as tax breaks for restoring open space. Controls are necessary because without them the plan has no real "teeth" and thus no regulatory strength.

8.9 COMPREHENSIVE WATER MANAGEMENT PLANNING

The problem

The tendency in modern land planning and engineering is to isolate the various components of the environment and address them individually and separately. This is fostered by community ordinances that typically address stormwater, wetlands, grading, landscaping, roads, parking, and utilities as though they are more or less mutually independent. As a result, not only do the rules and guidelines often lack balance and coordination, but some are actually incompatible. For example, in one Midwestern community site grading specifications for development projects call for mounding lawn areas in parkways, which tends to increase stormwater runoff whereas stormwater management guidelines call for reduction of runoff through onsite detention. In a California community, residential property owners are urged and sometimes required to practice water conservation but are restricted by ordinance from storing and recycling stormwater for garden and lawn irrigation.

Unifying theme

Water is the one theme that is central to virtually all landscape planning and through which site planning can be unified. Water management planning provides an opportunity to weave a common thread through a land use plan which can help reduce both construction and landscape management costs, improve risk management, and provide for more sustainable landscapes and facilities in the long run. Water management planning must begin with the system that is universal to all terrestrial environments: the *watershed* and its *drainage net*. Every planning site is nested within a drainage system. A first consideration is simply its location and the implications of location in terms of upstream and downstream opportunities and constraints such as flood risk, water supply, and stormwater management. A second consideration is the search for complementary systems and how to integrate them. Wetlands and stormwater are examples.

Complementary systems

Although wetland mapping is standard practice in planning projects today, little or no attention is usually paid to wetland watersheds as a part of the exercise. As a result, land use development in wetland watersheds rarely makes allowances for maintenance of wetland water supplies as a part of stormwater management. Consequently, some wetlands are overloaded with redirected stormwater whereas others are deprived of natural runoff supplies. If stormwater systems could be coordinated with wetland water supply needs, there is an opportunity, with proper allowances for changes in discharge rates and water quality, to achieve better performance in both systems at lower costs. The same sorts of complementary relations can be found among wetlands and groundwater systems, stormwater and irrigation

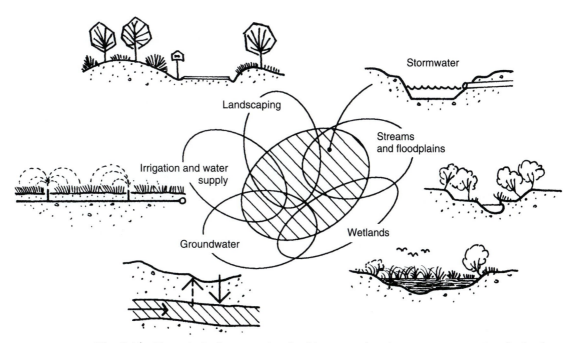

Fig. 8.14 The principal systems involved in comprehensive water management for land use development projects.

systems, and landscaping (grading, soil treatment, ground covers, etc.) and stormwater runoff (Fig. 8.14).

8.10 CASE STUDY

A Case for Green Infrastructure in Stormwater Management, Surrey, British Columbia

Patrick Condon

In 1998, farmers sued the City of Surrey, British Columbia, for flood damages on their agricultural lands and received a settlement of an undisclosed amount in the tens of millions. Farmers blamed the flooding on suburban development upstream, which had cleared forests and grasslands, added impervious cover, funneled runoff into stormdrains, which in turn ran unmitigated and unabsorbed to streams, and flooded farmland downstream.

Low-density, highly paved residential communities are a major threat to natural drainage systems. Conventional stormwater management requires that rainwater be drained along street curbs, into grates, and then conveyed by pipes directly into waterways. This stormwater surges to streams at velocities and volumes many times greater than predevelopment rates, scouring streambeds, and sometimes causing flooding.

In addition to being environmentally damaging, building and maintaining standard infrastructures with stormdrains and arterial roads is costly to communities, developers, and, ultimately, homeowners. The high price of new homes in low-density development reflects the cost of the inefficient and overbuilt street networks that are part and parcel of the stormwater infrastructure. Furthermore, the replacement and upkeep costs of infrastructure in areas of conventional low-density residential development are consuming a large and growing share of municipal budgets.

The City of Surrey recognized the need to change its basic approach to stormwater management. In 1998, the City Council voted to collaborate with the James Taylor Chair in Landscape and Liveable Environments at the University of

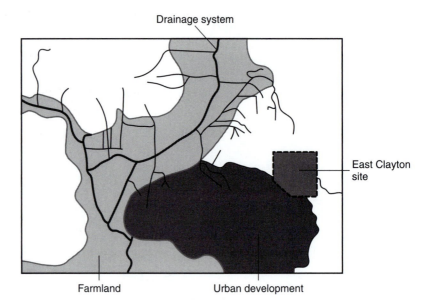

British Columbia to build a model community that would apply sustainable planning principles and alternative development standards "on the ground." The preservation and promotion of stream corridors and their drainage regimes was a key sustainability goal. The City chose a 250-hectare site called East Clayton for a demonstration project.

The approach we chose employed a green infrastructure strategy. We saw this not only as a way of managing stormwater and reducing the community's liability to landowners downstream, but as a way to combat unsustainable conventional development practices. Green infrastructure relies on source control of stormwater; that is, it encourages infiltration of rainwater at or close to its point of origin where it can be filtered into the soil before either being taken up by trees, recharging groundwater, or flowing slowly to streams as interflow. This approach helps support stream baseflows, reduces steam peak flows, and eliminates flooding downstream, ultimately bringing watershed performance close to its predevelopment levels.

In addition, organizing development around green infrastructure creates more walkable communities. The excessive paving and hierarchical pattern of suburban streets (typically five times the amount of paving per capita than in traditional, higher density, urban neighborhoods) creates an unfriendly and dysfunctional pedestrian environment. A green infrastructure system, by contrast, uses narrow streets arranged in an interconnected network, which makes walking and cycling easier while reducing stormwater runoff. Conversely, the need for automobile travel is reduced, especially high-frequency short trips, thereby reducing per capita energy use and lowering green house gas emissions. Street networks that encourage walking and cycling also provide an opportunity to improve the health of residents, for researchers have linked obesity with the sedentary, auto-dependent lifestyles of suburban development.

In 2003, the first phase of the East Clayton neighborhood was completed on a 20-hectare parcel. Integral to the East Clayton design is a green infrastructure system that builds on existing waterways to create an integrated and multifaceted network of streams, green streets, greenways, self-mitigating parcels, parks, and riparian areas. The stormwater system at East Clayton has the capacity to retain 2.4 cm (1 inch) of rainfall per day, which accounts for 90 percent of the 60 plus inches of annual rain falling on the site.

The integration of green infrastructure into our built environments is a viable solution for maintaining the ecological health of watersheds and building affordable and sustainable communities. The cost of green infrastructure can be less than half the

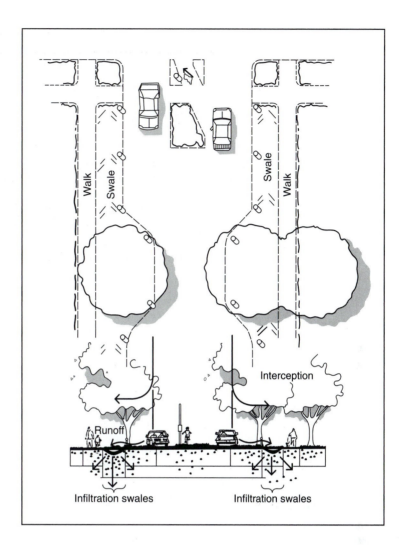

cost of conventional infrastructure, a savings in Surrey of over $12,000 per household. Lower infrastructure costs in turn lead to lower housing and long-term maintenance costs. Finally, green infrastructure creates more walkable neighborhoods, giving us safe and more socially sustainable communities. East Clayton provides a demonstration of the links between affordability, sustainability, and walkability.

Patrick Condon is professor of landscape architecture at the University of British Columbia. He holds the James Taylor Chair in Landscape and Liveable Environments. ∎

8.11 SELECTED REFERENCES FOR FURTHER READING

Burchell, Robert, and Lostokin, David. *Land, Infrastructure, Housing Costs and Fiscal Impacts Associated with Growth: The Literature on the Impacts of Sprawl Versus Managed Growth*, Lincoln Institute of Land Policy Research Paper, Cambridge, MA, 1995.

Condon, Patrick and Gonyea, Angela. "Status Quo Versus an Alternative Standard, East Clayton Two Alternative Development Standards Compared." *Technical Bulletin No. 2*. James Taylor Chair, University of British Columbia, 2000.

Dunne, Thomas, and Black, R. D. "Partial-Area Contributions to Storm Runoff in a New England Watershed." *Water Resource Research*, 1970, pp. 1296–1311.

Ferguson, Bruce K. "The Failure of Detention and the Future of Stormwater Design." *Landscape Architecture*, 81, 1991, pp. 76–79.

Hewlett, J. D., and Hibbert, A. R. "Factors Affecting the Response of Small Watersheds to Precipitation in Humid Regions." In *Forest Hydrology*. Oxford: Pergamon Press, 1967, pp. 275–290.

Horton, Robert E. "The Role of Infiltration in the Hydrologic Cycle." *American Geophysical Union Transactions* 14, 1933, pp. 446–460.

Marsh, W. M., and Marsh, N. L. "Hydrogeomorphic Considerations in Development Planning and Stormwater Management, Central Texas Hill Country, USA." *Environmental Management* 19:5, 1995, pp. 693–702.

Poertner, Herbert G. *Practices in Detention of Urban Stormwater Runoff.* Chicago: American Public Works Association, 1974.

Seaburn, G. E. "Effects of Urban Development on Direct Runoff to East Meadow Brook, Nassau County, Long Island, New York." *U.S. Geological Survey Professional Paper 627-B*, 1969.

U.S. Natural Resources Conservation Service. *Urban Hydrology for Small Watersheds.* Technical Release No. 55. Washington, DC, U.S. Department of Agriculture, 1975.

Whipple, W., et al. *Stormwater Management in Urbanizing Areas.* Englewood Cliffs, NJ: Prentice–Hall, 1983.

9

WATERSHEDS, DRAINAGE NETS, AND LAND USE

9.1 INTRODUCTION

Most overland flow moves only a short distance over the ground before it gathers into minute threads of water. These threads merge with one another, forming rivulets capable of eroding soil and shaping a small channel. The rivulets in turn join to form streams, the streams join to form larger streams, and so on. This system of channels, characterized by streams linked together like the branches of a tree, is called a **drainage network**, and it represents nature's most effective means of getting liquid water off the land. The area feeding water to the drainage network is the **drainage basin**, or **watershed**, and for a given set of geographic conditions, the size of the main channel and its flows increase with the size of the drainage basin.

Early in the history of civilization, humans learned about the advantages of channel networks, for both distributing water and removing it from the land. The earliest sewers were actually designed to carry stormwater, and they were constructed in networks similar to those of natural streams. Whether or not they knew it, the ancients followed the *principle of stream orders* in the construction of both stormsewer and field irrigation systems. This principle describes the relative position, called the *order*, of a stream in a drainage network and helps us to understand the relationships among streams in a complex flow system.

Modern land development often alters drainage networks by obliterating natural channels, adding artificial channels, or changing the size of drainage basins. Such alterations can have serious environmental consequences, including increased flooding, loss of aquatic habitats, reduced water supplies during low flow periods, and lower water quality. In land use planning generally little attention is paid to drainage networks as geographic entities. Site planners and designers typically ignore questions related to site location in drainage networks and their implications for risk, liability, and water management. Likewise, community planners often ignore the implications of retention basins and other stormwater facilities in terms of streamflow, water quality, and flood management at the regional scale.

9.2 THE ORGANIZATION OF NETWORKS AND BASINS

Stream orders

The **principle of stream orders** is built on a classification system based on the rank of streams within the drainage network. First-order streams are channelized flows with no tributaries. Second-order streams are those with at least two first-order tributaries. Third-order streams are formed when at least two second-order streams join together, and so on (Fig. 9.1).

Bifurcation ratio

The number of streams of a given order that combine to form the next higher order generally averages around 3.0 and is called the **bifurcation ratio**. From a functional standpoint, this ratio tells us that the size of the receiving channel must be at least three times the average size of the tributaries. For drainage nets in general, a comparison of the total number of streams in each order to the order itself reveals a remarkably consistent relationship, which defines the principle of stream orders, in which stream numbers decline progressively with increasing order. Therefore, first-order streams are the most abundant streams in every drainage network.

Given a classification by order of the streams in a drainage net, we can examine the relationship between orders and other hydrologic characteristics of the river system such as drainage area, stream discharge, and stream lengths. This provides a basis for comparing drainage nets under different climatic, geologic, and land use conditions; for analyzing selected aspects of streamflow, such as changes due to urbanization; and for defining zones with different land use potentials. Table 9.1 lists and defines several factors involved in drainage network and basin analysis.

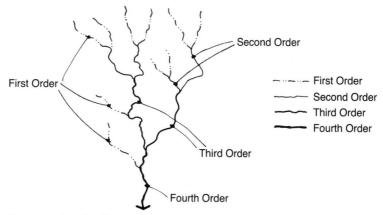

Fig. 9.1 Stream order classification according to rank in the drainage network. This follows the scheme originally defined by American hydrologist Robert Horton.

Basin order

Drainage basins can also be ranked according to the stream order principle. First-order drainage basins are those emptied via first-order streams; second-order basins are those in which the main channel is of the second order; a fifth-order basin would be one in which the trunk stream is of the fifth order. Just as all high-order streams are products of a complete series of lower streams, high-order (large) basins are comprised of a complete series of lower-order basins, each set inside the other. This is sometimes referred to as a *nested hierarchy*, as is illustrated in Figure 9.2.

Nonbasin area

In accounting for the combined areas of the basins that make up a larger basin, however, not all the land area is taken up by the lower-order basins. Invariably a small percentage of the land drains directly into higher-order streams without first passing through the numerical progression of lower-order basins (Fig. 9.3). This land is called the *nonbasin drainage area*, and it generally constitutes 15 to 20 percent of the total drainage area in basins of second order or larger.

Table 9.1 Factors Important to the Analysis of Drainage Networks

Number of Streams—The sum total of streams in each order.
Bifurcation Ratio—(branching ratio) Ratio of the number of streams in one order to the number in the next higher order.

$$BR = \frac{N}{N_u}$$

where

BR = bifurcation ratio
N = the number of streams of a given order
N_u = the number of streams in the next highest order

Drainage Basin Order—Designated by highest order (trunk) stream draining a basin.
Drainage Area—Total number of square miles or square kilometers within the perimeter (divide) of a basin.
Drainage Density—Total length of streams per square mile or square kilometer of drainage area.

$$\text{Density} = \frac{\text{sum total length of streams (mi or km)}}{\text{drainage area (mi}^2 \text{ or km}^2\text{)}}$$

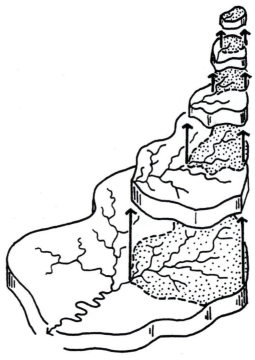

Fig. 9.2   Illustration of the nested hierarchy of lower-order basins within a large drainage basin.

Drainage density

Drainage density is the total length of channels per unit area (mi^2 or km^2) of land. It is an important parameter in watershed management because it indicates how efficiently land is served by channel systems. The more channels per square mile, the faster surface water is removed from the land, because channels are the most efficient means of dispatching runoff. Virtually everywhere drainage density increases

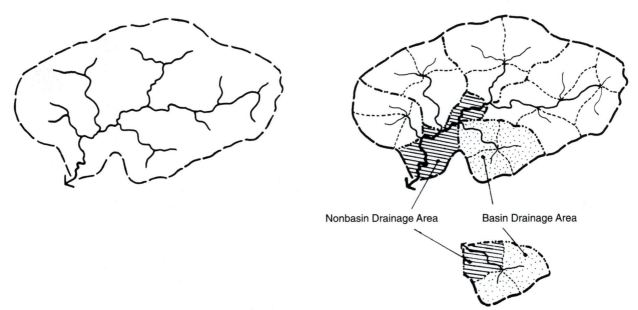

Nonbasin Drainage Area   Basin Drainage Area

Fig. 9.3   Two types of drainage areas can be defined in basins larger than the first order: basin and nonbasin areas. Nonbasin area borders the trunk stream(s) and releases its water directly to the main channel.

with land clearing and development as drainage ditches are cut along roads and farmfields and stormsewers are added to communities. Unless mitigated by water management measures, stormwater loading and delivery rates in such landscapes increase with drainage density while water quality decreases.

9.3 MAPPING THE DRAINAGE BASIN

Mapping basins and networks

The process of mapping the individual drainage basins within a large drainage net requires finding the **drainage divides** between channels of a particular order. This is best accomplished with the use of a topographic contour map; however, field inspection may be needed in developed areas in order to find culverts, diversions, and other drainage alterations. As a first step, channels should be traced and ranked, taking care to note the scale and the level of hydrographic detail provided by the base map. This is important because maps of large scales will show more first-order streams, owing to the fact that they are usually drawn at a finer level of resolution than smaller scale maps.

In areas of complex terrain, the task of locating basin perimeters can be tedious, but it can be streamlined somewhat if the pattern of surface runoff (overland flow) is first demarcated. This can be done by mapping the direction of runoff using short arrows drawn perpendicular to the contours over the entire drainage area. Two basic runoff patterns will appear: divergent and convergent. Where the pattern is divergent, a drainage divide (basin perimeter) is located; where it is convergent, a watershed and drainageway form (Fig. 9.4). These patterns are also illustrated schematically in Figure 9.10.

Runoff patterns

By delineating all the drainage divides between first-order streams, a watershed can be partitioned into first-order basins. From this pattern, second-order basins can be traced, and they are comprised of two or more first-order basins plus a fraction of nonbasin area. This will account for roughly 80 to 85 percent of the total drainage area, the remainder being nonbasin (i.e., non first-order) drainage.

Since perennial streams are fed by both surface water and groundwater, it is also important to be aware that the drainage basins for these two sources may be differ-

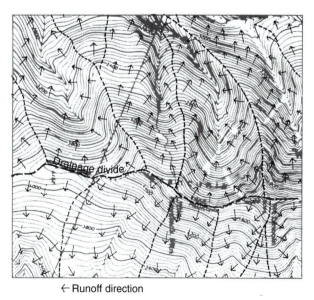

← Runoff direction

Fig. 9.4 Mapping and partitioning the watershed using vectors of overland flow. Flow is divergent on divides and convergent in the basins.

ent. In other words, an aquifer may extend beyond the topographic perimeter. In most instances, however, it is impossible to identify such an arrangement without the aid of an areawide groundwater study. A clue to the existence of a significant difference in the two can sometimes be found in first- or second-order streams that have exceptionally large or small baseflows for their surface drainage area relative to similar streams in the same region.

9.4 TRENDS IN THE DEVELOPMENT OF SMALL DRAINAGE BASINS

Clearing and development of land often have a pronounced influence on drainage networks and basins. Deforestation and agriculture may initiate soil erosion and gully formation. As gullies advance, they expand the drainage network, thereby increasing the number of first-order streams and the drainage density. And as we pointed out in Chapter 8, road development in rural areas can have the same effect. The main hydrologic consequence of these changes is shortened concentration times, that is, faster response times, because the distance that water must travel as overland flow or interflow is reduced (Fig. 9.5). Discharges are in turn larger because, for small basins, concentration times fall closer to the peaks of rainstorm intensity. It follows that large flows occur with higher frequencies, erosion is greater, channel drainage more prevalent, and water quality can be expected to decline.

Urbanization   Urbanization also leads to considerable change in the shape and density of a drainage network. One of the first changes that takes place is a "pruning" of natural channels, that is, removal of parts of the channel network. These channels are often replaced by ditches, usually where no channel or swale existed previously, and underground channels in the form of stormsewers. Although the natural network may be pruned, the total drainage network is usually enlarged and intensified. The *net effect* of urbanization is usually an increase in total channels and in turn an increase in the overall drainage density (Fig. 9.6; also see Fig. 8.10). Coupled with the lower infiltration rates and extensive effective impervious cover in urbanized

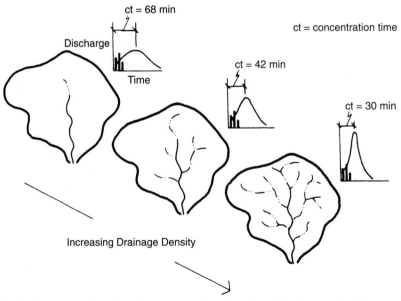

Fig. 9.5 Hydrographs showing the changes brought about by increased drainage density: concentration time is reduced, causing greater peak discharges for a given rainstorm.

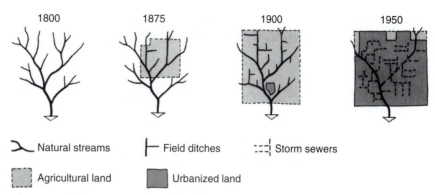

Natural streams   Field ditches   Storm sewers

Agricultural land   Urbanized land

Fig. 9.6 Pruning, grafting, and intensification of a drainage network with community development.

areas, this leads to increased amounts of runoff and shorter concentration times for the drainage basin, both of which produce larger peak discharges. As a result, both the magnitude and frequency of peak discharges are increased for receiving streams and rivers (see Fig. 8.8).

Stormsewers   **Stormsewers** are underground pipes that conduct surface water by gravity flow from streets, buildings, parking lots, and related facilities to streams and rivers. The pipes are usually made of concrete, sized to the area they serve, and capable of conducting stormwater at a very rapid rate. Studies show that stormsewers and associated impervious surfaces increase the frequency of floodflows on streams in fully urbanized areas by as much as sixfold (Fig. 9.7).

The graph in Figure 9.7 assumes that the size of the drainage basin has remained unchanged as sewering advanced. In many instances, however, the size of the basin is also increased with sewering, as is illustrated in the case of the Reeds Lake watershed (Fig. 9.8). The additional drainage area is completely urbanized, laced with stormsewers, and produces peak discharges perhaps four to five times larger than drainage areas of comparable size without stormsewers and such heavy development.

Modern agriculture   Modern agriculture is also responsible for altering drainage basins and stream networks. In humid regions, farmers often find it necessary to improve field drainage to facilitate early spring plowing and planting. In addition to cutting ditches through and around fields and deepening and straightening small streams, networks of drain tiles are often installed in fields. Drain tiles are small perforated pipes (originally ceramic but now plastic) buried just below the plow layer. They collect water that

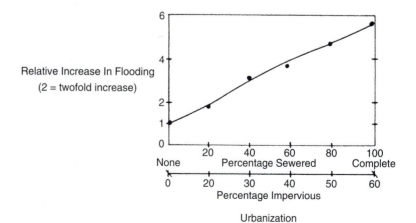

Fig. 9.7 Increased frequency of floodflows related to stormsewering and impervious surface.

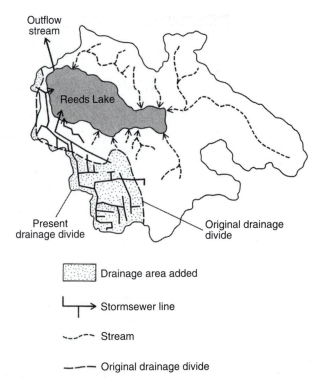

Fig. 9.8   The Reeds Lake watershed showing the drainage added (in tint) through storm-sewering with urban development.

has infiltrated the soil and conduct it rapidly to an open channel. Although the effects of tile systems on streamflow have not been documented, these systems undoubtedly serve to increase the magnitude and frequency of peak flows locally in much the same way as yard drains and stormsewers do in cities. Another serious hydrologic change associated with agriculture is the draining of wetlands. Wetlands are important in the storage of runoff and floodwater, but where they are eliminated, stream peak flows increase in turn.

Lakes and reservoirs   The watersheds that serve lakes and reservoirs (impoundments) follow the same organizational principles as river watersheds, with the impoundment itself representing a segment of some order in the drainage network or a node linking streams of lower orders. On the other hand, the land use patterns are different in one important respect: The heaviest development is usually found in the nonbasin drainage area (see Fig. 11.11). This is a discontinuous belt of shoreland that encircles the waterbody which is highly attractive to residential, recreational, and commercial development. As development takes place, the need for storm drainage usually arises and yard drains, ditches, and stormlines are constructed where, under natural conditions, channels never existed. In addition, the shoreland zone is often expanded inland with stormsewer construction, thereby capturing additional runoff and rerouting it directly to the waterbody.

9.5 PLANNING AND MANAGEMENT CONSIDERATIONS

Basin components   Since small drainage basins are the building blocks of large drainage systems, it is essential that watershed planning and management programs address the small basin. Most small basins (primarily first, second, and third orders) are comprised of

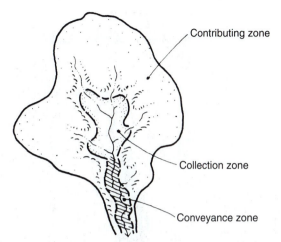

Fig. 9.9 The three main hydrologic zones of a small drainage basin. The lower two zones are the least suited to development but essential to the basin's operation.

three interrelated parts: (1) an outer, contributing zone that receives most of the basin's water and generates runoff in the form of groundwater, interflow, and overland flow; (2) a low area or collection zone in the upper basin where runoff from the contributing zone accumulates; and (3) a central conveyance zone represented by a valley and stream channel through which water is transferred from the collection zone to higher order channels (Fig. 9.9). The hydrologic behavior of each zone is different, and each in turn calls for different planning and management strategies.

Hydrologic zones   The **contributing zone** is generally least susceptible to drainage problems. It provides the greatest opportunity for site-scale stormwater management because most sites in this zone have little upslope drainage to contend with and because surface flows are generally small and diffused. By contrast, the **collection zone** in the upper basin is subject to serious drainage problems. Seepage is common along the perimeter, and groundwater saturation can be expected in the lower central areas during much of the year. During periods of runoff, this zone is prone to inflooding, caused by stormwater loading superimposed on an elevated groundwater surface. Inflooding is very common in rural areas of modest local relief; in fact, it is probably the most common source of local flood damage today in much of the Midwest and southern Ontario.

Edge forms   The edge separating the upland zone and the collection zone is one of the most critical borders in landscape planning. This border is made of two edge forms, convex and concave (Fig. 9.10). The *convex forms* create divergent patterns of runoff and therefore tend to be dry along their axes—indeed very dry in some instances. By contrast, *concave forms* create convergent patterns of runoff, concentrating flows downslope, and therefore tend to be wet along their axes. The concave edges are the functional links between the upland surface, where runoff is generated, and the lowland collection areas where water accumulates and streams head up. As such they are vital management points in the drainage basin, serving as the gateways to the lowland and riparian environments of the collection zone. It follows that the concave edges also represent important habitat corridors between upland and lowland surfaces.

Conveyance zone   The central **conveyance zone** contains the main stream channel and valley, including a small floodplain. The flows in this zone are derived from the upper two zones as well as from groundwater inflow directly to the channel. Groundwater contributions provide the stream baseflow and constitute the vast majority of the stream

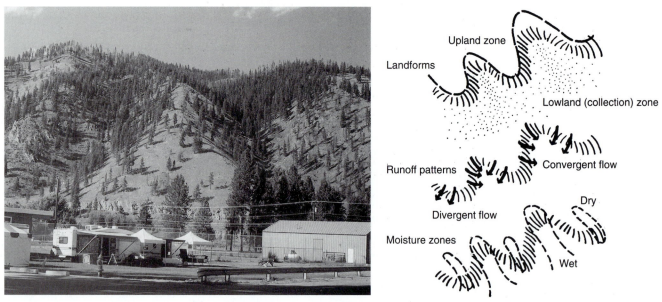

Fig. 9.10 Concave and convex edges and associated runoff patterns. Vegetation (photograph) reveals the contrast in moisture conditions.

discharge over the year. Stormwater, on the other hand, is derived mainly from the upper zones, and though small in total, it constitutes the bulk of the largest peak flows in developed or partially developed basins. The central conveyance zone is also subject to flooding, but in this case it is outflooding caused by the stream over-topping its banks. Because this zone is prone to comparatively large floods, especially if the upper basin is developed, it is generally least suited to development and most difficult to manage hydrologically. In hilly and mountainous terrain, such as coastal California, where slopes tend to be unstable during wet periods, it is the upper fringe of the conveyance zone and the upper collection zone that receive the landslide and mudflow debris from slope failures (see Fig. 4.12).

Land use implications   Recognition of the constraints and opportunities associated with each of these drainage zones is an important step toward forming development guidelines for small basins, including buildable land, open space, and special use areas. What it does not provide, however, is a rationale for establishing density guidelines. *Density* is a measure of the intensity of development, defined, for example, as percentage impervious surface, total building floor space per acre, or population density. For water and land management, percentage impervious surface is commonly used to define density. As a rule of thumb, both the magnitude and the pollution load of stormwater flow increase with impervious cover.

Carrying capacity   Establishing the appropriate density for a drainage basin should be based on performance goals and standards (see Chapter 8 for a discussion of performance concepts) and **basin-carrying capacity**. The carrying capacity of a drainage basin is a measure of the amount and type of development it is able to sustain without suffering degradation of water features, water quality, biota, soils, and land use. While this concept is easy to envision, determination of a basin's carrying capacity may be difficult to derive. As a general rule, 30 percent development may be the advisable maximum for most basins, but this can vary with the style of development and the character of the basin, especially its hydrologic versatility. For residential development, it is significant that the per capita impervious cover, that is, the ratio of hard surface to persons inhabiting the basin, is greatest for large-lot single-family dwellings. Therefore, where the basin land use density is measured by impervious cover, single-family large lots support the smallest population, and it follows, produce the largest per capita stormwater runoff rate. Cluster development, on the other hand, supports a larger population with less impact on runoff rates and water quality for a comparable impervious cover.

Basin physiography

But this is only part of the picture because the basin's *physiography* and *hydrologic versatility* exert a strong control on carrying capacity. Basins in hilly or mountainous terrain often have the lowest carrying capacity because of the abundance of steep, unbuildable slopes, rapid rates of stormwater transfer and the paucity of opportunities for source control of stormwater. The soil mantle may be thin, which not only minimizes the basin's capacity to filter and retain infiltration water, but often increases the tendency for slope failure during wet periods (see Section 4.6 in Chapter 4). For basins with less relief, lower gradients, deep soil mantles, and good soil drainage, the carrying capacity may be considerably higher. Curiously, few communities take basin physiography and hydrologic versatility into account when formulating land use planning and engineering policy. Basins with steep terrain, complex drainage systems and few opportunities for stormwater management are usually treated no differently in terms of land use design, road alignments, and stormwater engineering than low relief basins with simple drainage systems and lots of management opportunities.

Land use and engineering

Besides fundamental differences in basin physiography, carrying capacity should also take into account the nature of the development and the proposed strategy for stormwater management within the basin. With respect to the latter, higher land use densities may be allowable where stormwater controls such as onsite detention or dry well infiltration can be employed effectively (see Fig. 8.12 in Chapter 8). Where curbs, gutters, and stormsewers are required by community ordinance, a formal basinwide stormwater mitigation program is a virtual necessity because these drains facilitate pollutant transfer to and increased flooding in receiving streams and related waterbodies. Ideally, a multiple pronged approach should be taken; first aimed at reducing onsite stormwater production and effective impervious cover followed by detention and/or disposal of excess runoff.

9.6 LAND USE PLANNING IN THE SMALL DRAINAGE BASIN

Step one

The procedure for building land use plans for small drainage basins begins with a definition of the stream system, patterns of runoff, and the three hydrologic zones. Each zone should be analyzed for soils, slopes, vegetation, and existing land use to determine its limitations for development. Steep slopes, runoff collection areas, unstable and poorly drained soils, forested areas in critical runoff zones, and areas prone to flooding and seepage should be designated as nonbuildable. The remaining area is more or less the developable land, and the bulk of it usually lies in the upland zone of the basin separated from the lowlands by the concave and convex edges described earlier (Fig. 9.11*a*).

Step two

The next step is to define land use units. These are physical entities of land, with the fewest constraints to development in general, that set the spatial scale and general configuration of development (Fig. 9.11*b*). Within this framework, development schemes may be evaluated and either discarded or assigned to the appropriate land unit. To achieve the desired performance, different options for the siting of different activities should be exercised; for instance, high-impact activities may be assigned to land units that are buffered by forest and permeable soils from steep slopes and runoff collection areas.

Guidelines

In general, the recommended guidelines for site selection and site planning in small drainage basins are: (1) maximize the distance of stormwater travel from the

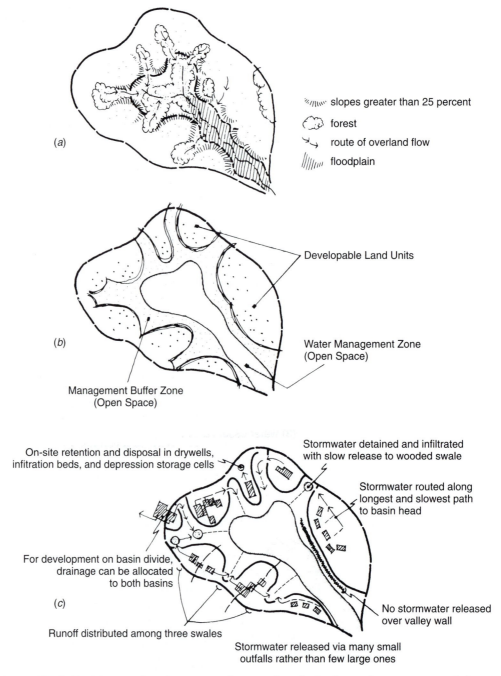

Fig. 9.11   Maps produced as a part of a procedure for land use planning in a small drainage basin: (*a*) The three hydrologic zones and the significant features of each. (*b*) The land use units. (*c*) Some strategies and guidelines in site selection and site planning.

site to a collection area or stream; (2) maximize the concentration time by slowing the rate of stormwater runoff; (3) minimize the volume of overland flow per unit area of land; (4) utilize or provide buffers such as forests and wetlands to protect collection areas and streams from development zones; and (5) divert stormwater away from or around critical features such as steep slopes, unstable soils, or valued habitats (Fig. 9.11*c*).

9.7 CASE STUDY

■ **Watershed Concepts Applied to Industrial Site Management, Saltillo, Mexico**

Jack Goodnoe

With the shift in industrial operations from the United States to Mexico, many corporations are facing the challenge of siting and designing new manufacturing facilities. But unlike the industrial site planning and design practiced earlier in the twentieth century, the exercise today calls for more than muscling engineered facilities onto a parcel of ground. This was the case with an industrial site near Saltillo in northeastern Mexico where General Motors built a set of automobile plants. The program not only called for state-of-the-art manufacturing systems, but for water and landscape management systems quite unlike those of the humid, temperate settings where most GM facilities were situated in the United States.

Rimmed by the Sierra Madre Mountains, the Saltillo Valley is extremely arid. Average annual rainfall amounts to about 10 inches (25 cm) per pear, and most of it is given up to evaporation. On the face of it, stormwater management would seem to be a fairly insignificant part of planning an industrial site in this location. But that is far from the case, first, because most rain comes each year in a dozen or so intensive downpours, and when it hits the desert landscape, there is practically nothing to hold it back. As a result, rainstorms can produce not only massive stormwater runoff, but dramatic erosion of soil unprotected by vegetation. Consequently, stormwater management is a major concern in such arid regions because facilities must, among other things, be protected from flooding and sedimentation. Using conventional stormwater methods of collecting runoff in pipes and ditches, such as shown in alternative A, has serious limitations because (1) such systems only concentrate the runoff into even bigger flows that produce even greater erosion and flooding downstream; (2) the pipes and ditches become clogged with sediment; and (3) water essential to the maintenance of vegetation is discarded from the site.

Given a development program that called for an engine plant, an assembly plant, and an administrative building (covering about 150 acres including parking and storage areas), a strategy for stormwater management and landscape development was needed that would serve multiple purposes. These included reduction of:

- stormwater runoff and downstream flooding
- catastrophic onsite flooding that would impair manufacturing operation and wash away spill debris;
- soil erosion and sedimentation of pipes, ditches, and stream channels;
- airborne dust, a threat not only to human health, but to air supplies used for industrial cooling and automotive painting operations;
- infrastructure construction and maintenance costs; and
- the harsh industrial character of the site, making it more attractive and functional as a working landscape.

Our approach to water management and landscape stabilization, which is shown in alternative B, began by viewing the site as several small watersheds rather than one large one. Because upslope drainage flowed toward the operations area, it was necessary to separate runoff into subbasins and invoke source controls as the main line of defense. With this approach the stormwater system is designed to hold runoff as close to its point of origin as possible and thereby prohibit rapid transfer and concentration downslope. It has the advantage of reducing runoff and soil erosion by keeping flows small and promoting local infiltration, which in turn reduces the need for stormwater piping and irrigation infrastructure. Because of the size (550 acres) of the industrial site and the availability of soil material from building excavations, we seized the opportunity to create several small subbasins to capture upslope drainage before it could reach the central operations area. Within each sub-

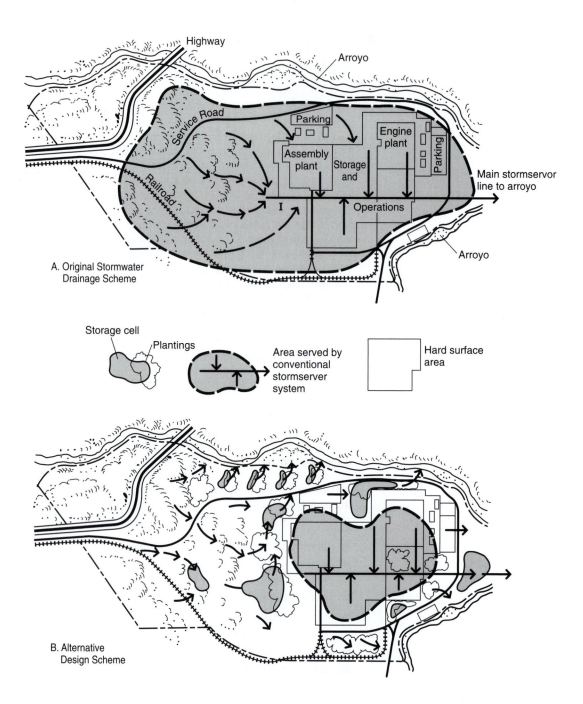

basin, we created a broad depression storage cell. Each subbasin and storage cell formed a "water management unit," a parcel of land designed to integrate water, soils, plants, and landscape aesthetics into a manageable and, ultimately, sustainable subsite or landscape cell. In this manner, the site was organized into a number of individual water management units that were, in turn, framed by a larger landscape design scheme organized around a micro watershed theme.

Each unit functioned more or less independently, and the need for piping upslope stormwater through the center of the site was virtually eliminated. Exceptionally large storms were allowed to overflow, with the discharge redirected laterally to the arroyos on the site perimeter, but only after substantial quantities of water had already been detained in the depression storage cells and introduced to the soil. Risks of disruption to manufacturing operations and facilities were significantly

The effects of runoff from one rainstorm on the piping system proposed in Alternative A. This inlet was located near the point marked I.

reduced (1) by breaking down the stormwater system into smaller drainage units (rather than concentrating into a single large system as is customary in conventional stormwater engineering), (2) by reducing the need to move stormwater over long distances, (3) by diverting stormwater away from manufacturing facilities at the heart of the site, and (4) by reducing the magnitude and frequency of damaging flows in the lower site and beyond.

The depression storage cells took the form of shallow basins (averaging 0.5 to 1.5 meters in depth). Their small scale and simple design allowed construction with uncomplicated engineering, using readily available excavators and untrained labor. The cells were planted with grasses and groves of trees using species, such as *Pinus resinosa*, adapted to sporadic water supplies. The cells were designed to place the groves of trees where they would help to screen and soften views of the industrial buildings and storage yards. In addition, worker recreation fields were designed to retain stormwater and serve as a "no-cost" water management feature with self-irrigating capacity.

This strategy of integrated stormwater management and landscape design helped create a multifunctional industrial landscape in which water, soil, vegetation, and land use are not only complementary but mutually supportive. It improved the performance and the success of each of the individual landscape systems while sharing costs, enhancing aesthetic quality, and reducing environmental and economic risks. Among other things the plan demonstrated the appropriate application of an alternative approach to industrial site management, one that does not rely solely on conventional stormwater and site engineering. Finally, the concept of water management units was easily translated into a site management plan. Because of their simple design and operation, the water management units can be serviced and maintained by conventional grounds crews that find it easy to grasp the purpose behind the system and how it functions for the site as a whole.

Jack Goodnoe, a landscape architect, is president of Land Planning and Design, Ann Arbor, Michigan. He specializes in strategic approaches to institutional planning and landscape design.

9.8 SELECTED REFERENCES FOR FURTHER READING

Copeland, O. L. "Land Use and Ecological Factors in Relation to Sediment Yield." U.S. Department of Agriculture, *Misc. Publication 970.* Washington, DC, 1965, pp. 72–84.

Dietrich, W. E., et al. "Analysis of Erosion Thresholds, Channel Networks, and Landscape Morphology Using a Digital Terrain Model." *The Journal of Geology* 101, 1993, pp. 259–278.

Dunne, Thomas, and Leopold, L. B. "Drainage Basins." In *Water in Environmental Planning.* San Francisco: Freeman, 1978, pp. 493–505.

Dunne, Thomas, Moore, T. R., and Taylor, C. H. "Recognition and Prediction of Runoff-Producing Zones in Humid Regions." *Hydrological Sciences Bulletin* 20: 3, 1975, pp. 305–327.

Gregory, K. J., and Welling, D. E. *Drainage Basin Form and Process.* New York: Halsted Press, 1973.

Horton, R. E. "Erosional Development of Streams and Their Drainage Basins: Hydrophysical Approach to Quantitative Morphology." *Geological Society of America Bulletin* 56, 1945, pp. 275–370.

Leopold, L. B. "Hydrology for Urban Land Planning—A Guidebook on the Effects of Urban Land Use." *U.S. Geological Survey Circular* 554, 1968.

Leopold, L. B., and Miller, J. P. "Ephemeral Streams: Hydraulic Factors and Their Relation to the Drainage Net." *U.S. Geological Survey Professional Paper 282-A,* 1956.

Meganck, Richard. *Multiple Use Management Plan, San Lorenzo Canyon* (summary). Organization of American States and Universidad Autonoma Agraria, 1981.

Riggins, R. E., et al. (eds.). *Watershed Planning and Analysis in Action.* New York: American Society of Civil Engineers, 1990.

Strahler, A. N. "Quantitative Geomorphology of Drainage Basins and Channel Networks." In *Handbook of Applied Hydrology* (Van te Chow, ed.). New York: McGraw–Hill, 1964.

10

STREAMFLOW, RIVER VALLEYS, AND FLOOD HAZARD

10.1 INTRODUCTION

Prior to the nineteenth century, it was widely believed that streamflows could not possibly be produced by precipitation. After all, if this were so, then how was it that streams could continue flowing when no rain was falling? Moreover, it was argued, there was not nearly enough rainfall to account for the flows seen in streams, especially large ones like the Seine and the Rhine. In the 1700s, a Frenchman named Pierre Perrault decided to test this supposition by comparing rainfall and stream discharge in the Seine Basin of France. He found that the annual volume of precipitation exceeded streamflow by sixfold.

But how rainwater got to the river and what controlled its rate of delivery remained largely unanswered until the twentieth century. We now understand that streams are supplied water from various sources—some large, some small; some slow, some fast; some variable, some steady. But all begin as precipitation and many factors—including numerous human factors as we saw in previous chapters—influence the disposition of this water and the path it takes on its way to stream channels. In some instances, more water is delivered to a channel than it can carry, and the result is a flood.

Floods have been part of the human experience for thousands of years. Our understanding of the nature and causes of flooding has increased dramatically over the past century or so, yet property damage and loss of life continue to rise. Much of this trend is related to global population growth and to the shift of people toward river valleys, coastal areas, and cities throughout the developing world. But flood damage is also rising in the developed world. In the United States, flooding ranks near the top of the national environmental agenda. About 7 percent of the country lies within a 100-year floodplain and these areas have been favorite places for rural and urban development throughout the twentieth century. By 1980, between 3.5 and 5.5 million acres of floodplain had been developed for urban land uses, including more than 6000 communities with populations of 2500 or more. Since then floodplain development has been growing by about 2 percent a year. Today the total list of vulnerable land uses is truly massive.

The nation's traditional response to the flood problem has been to "engineer the threat away." The federal government's first involvement in flood projects began with the Flood Control Acts of 1928 and 1936. Congress assigned the U.S. Army Corps of Engineers the responsibility of constructing reservoirs, levees, channels, and diversions to prevent flood damage along the Mississippi and its major tributaries. Since the late 1930s, the Corps has built 76 reservoirs and built or improved 2200 miles of levees in the upper Mississippi Basin alone. Added to that are an estimated 5800 miles of nonfederal levees in the upper basin, built mainly by state and local governments. In addition, there are more than 3000 reservoirs on the smaller tributaries of the upper Mississippi Basin constructed under the auspices of another federal agency, the U.S. Natural Resources Conservation Service. But despite all these efforts flood damage in the Mississippi and other watersheds has climbed since 1900. We now realize that the problem requires strategies beyond those offered by engineering alone, a point driven home by the shocking flood disasters of the 1990s, especially the Mississippi flood of 1993 and the Red River (North Dakota/Minnesota) flood of 1997.

10.2 SOURCES OF STREAMFLOW

Discharge

The quantity of water carried by a stream is termed **discharge**. It is measured in volumetric units of water passing a point on a stream, such as a bridge, over time. The conventional units in English-speaking countries are *cubic feet per second*, abbreviated cfs. In recent years, *cubic meters* have begun to replace cubic feet, but the latter are still widely used in science, engineering, and land planning.

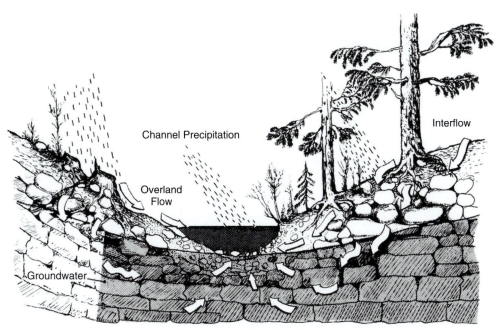

Fig. 10.1 Sources of streamflow: channel precipitation, overland flow, interflow, and groundwater in order of their arrival at the channel.

Sources of discharge

Discharge is derived from four sources: channel precipitation, overland flow, interflow, and groundwater (Fig. 10.1). *Channel precipitation* is the moisture falling directly on the water surface, and in most streams, it adds very little to flow. *Groundwater*, on the other hand, is a major source of discharge. Groundwater enters the streambed where the channel intersects the water table and provides a steady supply of water, termed **baseflow**, during both dry and rainy periods. Owing to the large supply of groundwater available to streams and the slowness of the response of groundwater to precipitation events, baseflow changes only gradually over time.

Interflow is water that infiltrates the soil and then moves laterally to the stream channel in the zone above the water table. Much of this water is transmitted within the soil itself, some of it moving along the soil horizons. Until recently, the role of interflow in streamflow was not understood, but field research has begun to show that next to baseflow, it is the most important source of discharge for streams in forested lands. This research also reveals that overland flow in heavily forested areas makes negligible contributions to streamflow (see Fig. 8.6).

Stormflow

In dry regions, cultivated lands, and urbanized areas, however, overland flow is usually a major source of streamflow, particularly in small streams. **Overland flow** is stormwater (or snowmelt) runoff that begins as a thin layer of water that moves very slowly (typically less than 0.25 ft/sec) over the ground. Under intensive rainfall in the absence of barriers such as rough ground, vegetation, and absorbing soil, however, it can mount up rapidly reaching stream channels in minutes and causing sudden rises in discharge. The quickest response times between rainfall and streamflow occur in urbanized areas where yard drains, street gutters, and stormsewers collect overland flow and route it to streams straightaway.

Factors influencing runoff

It is important to understand that the relative balance of contributions from overland flow, interflow, and groundwater changes dramatically with (1) the nature of rainfall and (2) the conditions of the drainage basin. As we illustrated in Chapter 8, land clearing and development tend to reverse the natural balance between overland flow and infiltration. In addition, the rate of stormwater delivery changes with land use development as the area served by slow-moving overland flow is reduced and

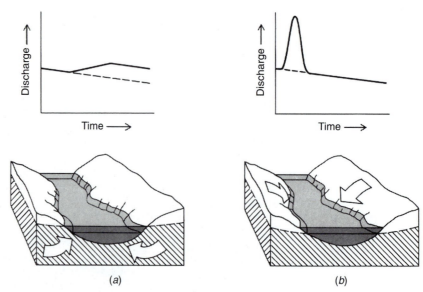

Fig. 10.2 (*a*) Stream response to a long steady rainfall that produces groundwater recharge and a rise in baseflow. (*b*) Stream response to an intensive rainstorm that produces only overland flow.

replaced by a much faster delivery system of gullies, gutters, ditches, and stormsewers. Overland flow increases in volume, and travel time quickens with development, resulting in much greater magnitudes and frequencies of peak discharges in receiving streams.

The amount and rate of water delivered to streams also vary with rainfall intensity and duration. Light rains rarely produce overland flow but, if extended over many days, may produce enough percolation through the soil to cause an increase in baseflow (Fig. 10.2*a*). Thus discharge may rise without appreciable contributions from surface runoff. In striking contrast to this situation is the streamflow from a short, heavy rainstorm in which rainfall intensity exceeds soil infiltration capacity. The result is immediate overland flow and a sudden rise in stream discharge. If coupled with a "fast" watershed, that is, one that has been developed and laced with artificial drains, the stormwater load amassed in the stream may be extreme indeed (Fig. 10.2*b*).

The 1993 Mississippi Flood

The 1993 Mississippi River flood, the largest ever recorded on the river, was a response to heavy, long-duration spring and summer rainfalls. Early rains saturated the soil over more than 300,000 square miles of the upper watershed, greatly reducing infiltration and leaving soils with little or no storage capacity. As rains continued, surface depressions, wetlands, ponds, ditches, and farmfields filled with overland flow and rainwater. With no remaining capacity to hold water, additional rainfall was forced from the land into tributary channels and thence to the Mississippi. For more than a month, the total load of water from hundreds of tributaries exceeded the Mississippi's channel capacity, causing it to spill over its banks onto adjacent floodplains (Fig. 10.3). Where the floodwaters were artificially constricted by levees and unable to spill onto the large sections of floodplain, the floodwaters were forced even higher.

10.3 METHODS OF FORECASTING STREAMFLOW

If the drainage area of a stream is relatively small and no discharge records are available, we must make discharge estimates using the rational method (see Chapter 8).

Fig. 10.3 The 1993 Mississippi flood, the largest in the recorded history of the river, was a response to prolonged spring and summer rainfall.

Unit hydrograph method

However, if chronological records of discharge are available for a stream, a short-term forecast of discharge can be made for a given rainstorm using a hydrograph. This method involves plotting the discharge from a watershed over time as the watershed and its streams respond to the rainstorm. The method is called the **unit hydrograph method** because it addresses only the runoff produced by a particular rainstorm in a specified period of time; the time taken for a river to rise, peak, and fall in response to the storm. Once the rainfall–runoff relationship is established, then rainfall data can be used to forecast streamflow for selected storms, called standard storms. A *standard rainstorm* is a high-intensity storm of some known magnitude and frequency. One method of unit hydrograph analysis involves expressing the hour-by-hour or day-by-day increase in streamflow as a percentage of total runoff. Plotted on a graph, these data form the unit hydrograph for that storm, which represents the runoff added to the prestorm baseflow (Fig. 10.4).

Magnitude and frequency approach

To forecast the flows in a large drainage basin using the unit hydrograph method would prove to be difficult because in a large basin, such as the Ohio, Hudson, or the Columbia, conditions may vary significantly from one part of the basin to another. This is especially so with the distribution of rainfall, because the basin is rarely covered evenly by an individual rainstorm. As a result, the basin does not respond as a unit to a given storm, making it difficult to construct a reliable hydrograph. In such cases, we turn to the **magnitude and frequency method** and calculate the

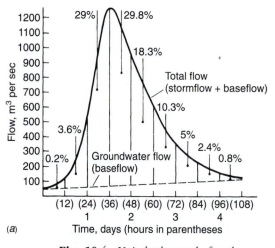

Flow period	Total flow	Storm-flow
0-12 hrs	63m³/s	63m³/s
12-24	192	127
24-36	1065	991
36-48	1101	1019
48-60	714	623
60-72	453	354
72-84	275	170
84-96	194	85
96-108	144	28

Peak = 1253; 1175 at hour 36

Flow period	Percentage of total flow
0-12 hrs	0.2%
12-24	3.6
24-36	29.0
36-48	29.8
48-60	18.3
60-72	10.3
72-84	5.0
84-96	2.4
96-100	0.8

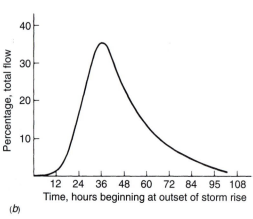

Fig. 10.4 Unit hydrograph for the uppermost part of the Youghiogheny River near the Pennsylvania-West Virginia border. Streamflow (*a*) is expressed as a percentage of total flow (*b*).

probability of the recurrence of large flows based on records of past years' flows. In the United States, these records are maintained by the Hydrological Division of the U.S. Geological Survey for most rivers and large streams. For a basin with an area of 5000 square miles or more, it is not uncommon for the river system to be gauged at five to ten places. The data from each gauging (gaging) station apply to the part of the basin upstream of that location.

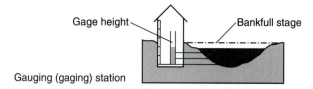

Since we are concerned mainly with the largest flows, only *peak annual flows* are recorded for most rivers. With the aid of some statistical techniques, we can use these flow data to determine the probability of recurrence of a given flow on a river. Key data available for each gauging station include:

- *Peak Annual Discharge:* the single largest yearly flow in cubic feet per second (cfs).
- *Date of Peak Annual Discharge:* the month and year in which this flow occurred.
- *Gage Height:* the height of water above channel bottom.
- *Bankfull Stage:* gage height when channel is filled to bank level.

The following steps outline a procedure for calculating the recurrence intervals and probabilities of various peak annual discharges:

Computing discharge probability

■ First, *rank* the flows in order from highest to lowest; that is, list them in order from biggest to smallest values. For two of the same size, list the oldest first.

■ Next, determine the *recurrence interval* of each flow:

$$t_r = \frac{n + 1}{m}$$

where

t_r = the recurrence interval in years
n = the total number of flows
m = the rank of flow in question

■ Third, determine the probability (p) of any flow:

$$p = \frac{1}{t_r}$$

This will yield a decimal, for instance 0.5, which can be converted to a percent (50 percent) by multiplying by 100.

■ In addition, is it also useful to know *when* peak flows can be expected; accordingly, the monthly frequency of peak flows can be determined based on the dates of the peak annual discharges. Furthermore, the percentage of peak annual flows that actually produce floods can be determined by comparing the stage (elevation) represented by each discharge with the bankfull elevation of the gauging station, that is, any flow that exceeds bankfull elevation is a flood.

Given several decades of peak annual discharges for a river, it is possible to make limited *projections* to estimate the size of some large flow that has not been experienced during the period of record. The technique most commonly used involves projecting the curve (graph line) formed when peak annual discharges are plotted against their respective recurrence intervals. In most cases, however, the curve bends rather strongly, making it difficult to plot a projection accurately using all the points (Fig. 10.5). This problem can be overcome by plotting the discharge and/or the recurrence interval data on logarithmic graph paper or logarithmic/probability paper. Once the plot is straightened, a line can be ruled through the points.

Making floodflow projections

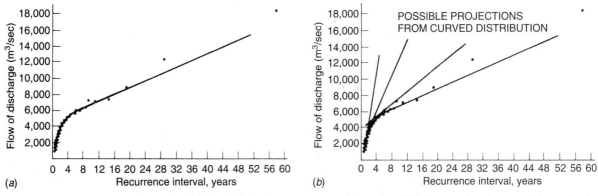

Fig. 10.5 (*a*) Graph based on recurrence interval and discharge of peak annual flows for the Eel River, California, during the period of 1911–1969. (*b*) This graph shows why it is impossible to plot one straight graph line from a curved distribution.

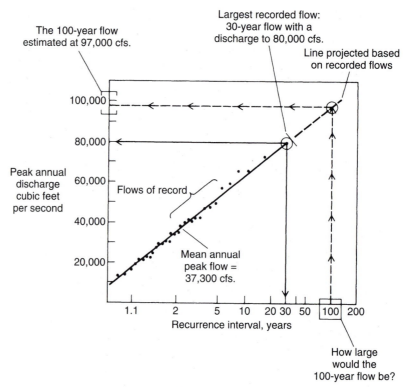

Fig. 10.6 Straightened curve of the distribution of discharges for the Skykomish River at Gold Bar, Washington. An estimate of the magnitude of flows at intervals beyond 30 years is now possible as shown by the broken line.

The process of making a projection can be accomplished by merely extending the line beyond the points and then reading the appropriate discharge for the recurrence interval in question. In Figure 10.6 the curve was straightened by using a logarithmic scale on the recurrence interval (horizontal) axis. The period of record covers 30 years, but we would like to know the magnitude of the 100-year flow. Assuming that the 100-year (or larger) flow is not represented within the 30 years of record, we can project the curve beyond the 30-year event and read the magnitude of the 100-year event, which is at 97,000 cfs, as shown in Figure 10.6. This figure represents a best approximation of the 100-year flow.

Forecasting constraints   Making forecasts of riverflows based on discharge records is extremely helpful in planning land use and in engineering bridges, buildings, and highways in and around river valleys. On the other hand, this sort of forecasting technique has several distinct limitations. First, the period of record is very short compared to total time the river has been flowing; therefore, forecasts are based only on a glimpse of the river's behavior. For a stream with records covering, say, the period 1900–2000, there is no guarantee that the true 100-year event actually occurred in the period of record. Second, throughout much of North America, watersheds have undergone such extensive changes because of urban and agricultural development, forestry, and mining that discharge records for many streams today have a different meaning than those several decades ago. Third, climatic change in some areas, for example, near major metropolitan regions, may be great enough over 50 or 100 years to produce measurable changes in runoff. Together, these factors suggest that forecasts based on projections of past flows should be taken as approximations of future flows, and for greatest reliability, they should be limited to events not too far beyond the years of record, say, 50-year, 100-year, and 200-year flows.

10.4 THE SIZE AND SHAPE OF VALLEYS AND FLOODPLAINS

Misfit stream concept

Flood analysis and forecasting also require an understanding of the geomorphology of stream valleys because valley size, shape, and topography influence the distribution, movement, and force of floodwaters. As a rule, stream valleys are carved by the streams that occupy them, and it follows that large streams flow in large valleys and small ones in small valleys. There are exceptions, called **misfit streams**, that are either too large or too small for their valleys: for example, midlatitude streams that carried huge discharges and carved broad valleys during the last glaciation but today carry much smaller flows. But most of the streams we encounter, especially those in the rural and predeveloped landscape, more or less fit the scale of their valleys.

In the developed landscape, particularly in and around urban areas, however, many streams have become misfit. Stormwater loading in urban streams (and some agricultural ones as well) has so changed the magnitude and frequency of peak flows that these streams often behave as much larger streams. Not only is flooding more frequent, but because it occurs in relatively small valleys, its stage is higher, that is, floodwater is much deeper over the valley floor. When streamflow exceeds the channel capacity, it flows onto the valley floor and the valley more or less takes on the function of a huge channel. The height that the floodwaters reach over the valley floor depends mainly on the valley's size and shape. If it is wide, the water spreads over a large area in a thin layer; however, if it is narrow, it builds up into a deep layer. As we noted earlier, the levees along the Mississippi reduced the size of the river's effective valley, restricting and elevating floodwaters in the 1993 flood. At St. Louis the width of the Mississippi's channel in 1849 was 4300 feet; today it is only 1900 feet.

Valley size and shape

Valley formation involves two essential geomorphic processes: downcutting or deepening and lateral cutting or widening. Downcutting generally leads to *V-shaped valleys* with little or no flat ground or floodplain between the stream channel and the valley walls. V-shaped valleys are characteristic of many headwater streams and streams in mountainous terrain. In extreme cases, the valleys are so sharp that they merge with the channel without a topographic break (Fig. 10.7a). However, most stream valleys, even those in deep gorges, have an area of flat ground, or **floodplain**,

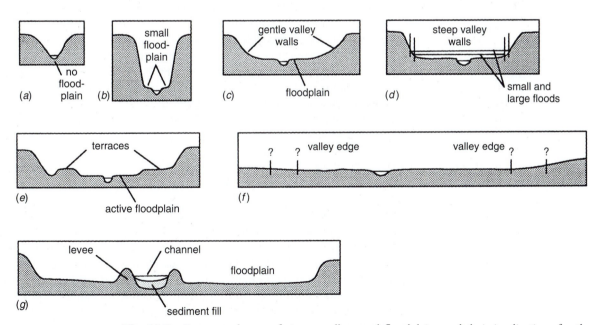

Fig. 10.7 Common shapes of stream valleys and floodplains and their implications for the depth and extent of flooding.

on the valley floor (Fig 10.7*b*). This landform contains the stream channel, and within it the stream shifts its course, erodes and deposits sediments, shapes soils and habitats, and overflows with floodwaters. These processes and related features, which are important to understanding the local dynamics of flooding, are part of the topographically diverse landscape of floodplains and they are discussed in detail later in Chapter 14.

V-shaped and U-shaped valleys

Besides V-shaped valleys and gorges with small floodplains, there are five other basic valley forms. If we widen a V-shaped valley, the next form is a *U-shape* (Fig. 10.7*c*). This valley has a central corridor of flat ground bordered by gentle slopes leading up to steeper valley walls. The gentle slopes are comprised of various deposits, such as alluvial fans, built by small tributary streams and debris from slides and mudflows. Where the valley walls are lower and/or more resistant, or where the stream has been in contact with them, these gentle footslopes may be absent and the valley has more of a *rectangular shape* with straighter and often steeper valley walls (Fig. 10.7*d*).

Terraced valleys contain benchlike areas situated above the valley floor (Fig. 10.7*e*). They are usually remnants of former floodplains when the channel and valley floor were both at a higher elevation in the valley. Terraces are typically found along the sides of valleys, next to the valley walls but occasionally they occur as isolated parcels ("islands") within the modern floodplain. They are, of course, less prone to flooding, and if large enough, are attractive ground for settlement.

Broad, flat valleys

Broad, flat floodplains without distinct valley walls are characteristic of many streams crossing lowlands such as coastal plains (Fig. 10.7*f*). These valleys are best described as hydrologic rather than geomorphic because their geographic limits are defined more or less by the areal extent of various flood events rather than by landforms. In fact their internal topography represented by levees, terraces, backswamps, and other features is more important to understanding flood management than topography on the valley margins. The lower third or so of the Mississippi (the area called the Embayment) is the strongest example of this kind of valley. Many other streams, including the Trinity, Red, Tombigbee, and Savannah, also exhibit hydrologic valleys where they cross the Gulf and Atlantic Coastal Plain.

Decidedly the most dangerous valley/floodplain type is the one we shall call *inverted*. In these valleys, the streambed lies at or above the valley floor, and is held in place by natural or constructed levees (Fig. 10.7*g*). The inverted condition is caused by sediment buildup in the channel, which can be caused by various conditions including excessive soil erosion higher in the watershed. Not many streams fall into this class; the lower Mississippi in the Embayment section does where tributary basins parallel the main channel. Perhaps the most extreme case is the Yellow River of China whose elevated channel is dangerously high above the floodplain and poses a threat to millions of people living on the valley floor should the confining levees fail under the force of a floodflow.

10.5 APPLICATIONS TO LAND PLANNING

Relating discharge to topography

Property damage from floods in the United States and Canada has increased appreciably in this century (Fig. 10.8). In order to curb this trend, improved understanding of the hydrologic behavior of streams and rivers is necessary. This is not simply a matter of relating a stream's general discharge regime to large sections of valley; rather it requires detailed segment by segment analysis taking into consideration changes in floodplain area and elevation, tributary discharges, and valley constrictions such as bridges, levees, and natural narrows. Analysis of a river's flood potential involves translating a particular discharge into a level (elevation) of flow and then

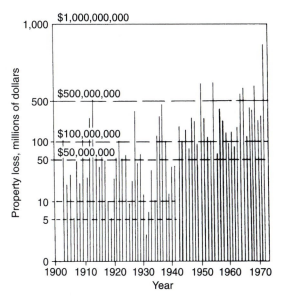

Fig. 10.8 Property loss (in millions of dollars) as a result of floods in the United States over 70 years. (Data from U.S. National Weather Service.)

relating that flow level to the topography in the river valley (Fig. 10.9*a*). In this manner we can determine what land is inundated and what is not. This procedure requires accurate discharge and topographic data as well as a knowledge of channel conditions. The latter is necessary in order to compute flow velocity, which is the basis for determining the water elevation in different reaches of the channel, that is, just how high a given discharge will reach above the channel banks.

For planning purposes, however, neither the time nor money is usually available for such detailed analysis. We must, instead, turn to existing topographic maps and existing discharge data and make a best estimate of the extent of various frequencies of flow. The accuracy of such estimates depends not only on the reliability of the data, but also on the shape of the river valley. Where the walls of a river valley rise sharply from the floodplain, the limits of flooding are often easy to define, as the profile in Figure 10.7*d* suggests. And as the lower diagram in Figure 10.9*b* shows, there would not be much difference in the extent of the 25- and 100-year floods because of the stream valley's box-shape. However, where a river flows through broad lowlands the problem can be very difficult and defining the difference in the extent of the 25-, 50-, or 100-year event might be a very hazy enterprise indeed (Fig. 10.9*b*).

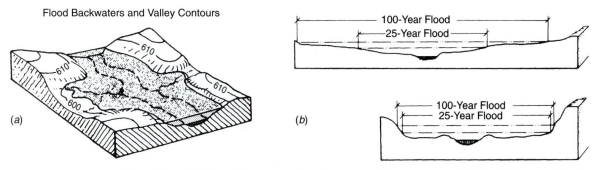

Fig. 10.9 (*a*) Flood backwaters in relation to valley contours. (*b*) The influence of valley shape on the extent of the 25-year and 100-year floods.

National Flood Insurance

The **U.S. National Flood Insurance Program** is based on a definition of the 100-year "floodplain." In most areas the boundaries of this zone are delineated by relating discharge data and flow elevations to the topography of the stream valley. According to this program, two zones are actually defined: (1) the regulatory *floodway*, the lowest part of the floodplain where the deepest and most frequent floodflows are conducted; and (2) the *floodway fringe*, on the margin of the regulatory floodway, an area that would be lightly inundated by the 100-year flood. Buildings located in the regulatory floodway are not eligible for flood insurance, whereas those in the floodway fringe are eligible provided that a certain amount of floodproofing is established (Fig. 10.10). The flood insurance program is intended to serve as one of the controls to limit development in floodplains; however, it can be argued that it has actually produced the opposite effect and encouraged development in the floodway fringe by offering government insurance in exchange for adding certain "floodproofing" measures to buildings such as elevated foundations and waterproofing. Experience has shown that these floodproofing measures are ineffective in curbing damage from truly large floods, which are, of course, inevitable in the areas labeled as floodway fringe.

Other management strategies

Other controls designed to prevent floodplain development include zoning restrictions against vulnerable land uses and educational programs to inform prospective settlers of the hazards posed by river valleys. In the case of river valleys where development is already substantial, the only means of reducing damage is to relocate flood-prone land uses or reduce the size of hazardous riverflows. Because of the high costs and the sociological problems associated with relocation, decision makers have traditionally turned to the flow-reduction and flood-control options. This calls for structural (engineering) changes such as building reservoirs, dredging channels, diverting flows, and/or constructing embankments (levees) to confine flow. All of these measures are expensive, deleterious to the environment, and for large

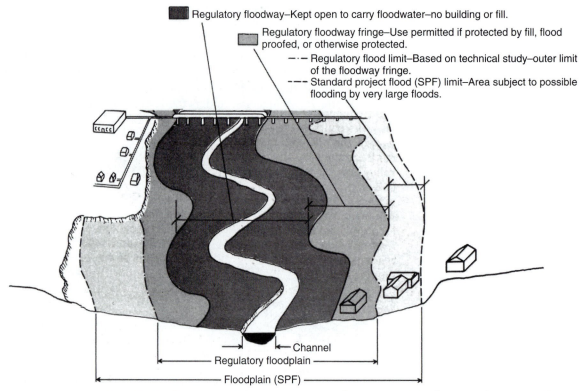

Regulatory floodway–Kept open to carry floodwater–no building or fill.

Regulatory floodway fringe–Use permitted if protected by fill, flood proofed, or otherwise protected.

—·— Regulatory flood limit–Based on technical study–outer limit of the floodway fringe.

—--— Standard project flood (SPF) limit–Area subject to possible flooding by very large floods.

Channel
Regulatory floodplain
Floodplain (SPF)

Fig. 10.10   Illustrations defining the regulator floodway and the floodway fringe according to the U.S. National Flood Insurance Program.

flood events, some measures have been shown to actually increase local flood magnitudes and damages. However, owing to the high costs of construction, tighter environmental regulation, and the growing specter of public liability, public agencies are now giving more consideration to various sorts of nonstructural programs.

Risk management   Increasingly, **risk management** programs offer a promising alternative to the problem or a supplement to traditional approaches. This approach recognizes the inevitability of floods and the risks they pose in floodplains and coastal areas. Risk management planning endeavors to build response plans to minimize disruption, property damage, and loss of life. Such plans include public information programs on disaster preparation, early warning systems, evacuation plans, and plans and programs for disaster recovery.

10.6 FLOODPLAIN MAPPING

Small stream valleys   Many streams are too small to qualify for an official gauging station, yet they merit serious consideration in local land use planning. Communities bordering on such streams are often pressed for information not only on flood hazards, but also on many other features of the stream valley that are important to planning decisions. Typical concerns include soil suitability for foundations and basements, siting of sanitary landfills and wastewater disposal facilities, the role of wetlands in the floodplain environment, and the distribution of forests and wildlife habitats. Much of this information can be generated from published sources, in particular, topographic contour maps, aerial photographs, and soil maps.

Floodplain significance   Of all the features of the stream valley, the floodplain is the most important from a planning standpoint. *First*, excluding the stream channel itself, it is generally the lowest part of the stream valley and thus is most prone to flooding. *Second*, floodplain soils are often poorly drained because of the nearness of the water table to the surface and saturation by floodwaters. *Third*, floodplains are formed by incremental erosion and deposition associated with the lateral migration of streams in their valleys. Therefore, the borders of the floodplain can be taken as a good indicator of the extent of alluvial soil. These features are the key indicators for floodplain mapping (Fig. 10.11).

Floodplain mapping criteria   Floodplains can be delineated in four ways: according to (1) topography, (2) vegetation, (3) soils, and (4) the extent of past floodflows. Topographic *contour maps* can be used to delineate floodplains if the valley walls are high enough to be marked

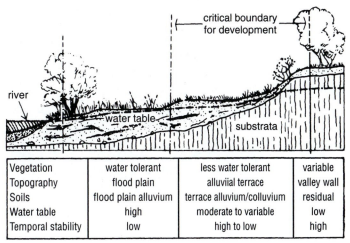

Vegetation	water tolerant	less water tolerant	variable
Topography	flood plain	alluviial terrace	valley wall
Soils	flood plain alluvium	terrace alluvium/colluvium	residual
Water table	high	moderate to variable	low
Temporal stability	low	high to low	high

Fig. 10.11 A good illustration of the abrupt change in topography, soils, drainage, and vegetation from the floodplain to the valley walls.

Data sources

by two or more contour lines. However, those that carry only a single contour, or fall within the contour interval, cannot be delineated, even though they may be topographically distinct to the field observer. In the latter case, aerial photographs can be helpful in mapping this feature. Viewed with a standard stereoscope, topographic features appear exaggerated, thereby making identification of valley features is a relatively easy task in many instances.

Vegetation and land use

Aerial photographs also enable us to examine vegetation and land use patterns for evidence of floodplains and wet areas. In the prairies and the major farming regions, such as the Great Plains and the Corn Belt, the floodplains of streams and rivers are often demarcated by corridors of forest. Similarly, in forested areas the floodplain tree covers are sometimes different in floristic composition, and once these differences are known, they can be used to help identify floodplains. (Floodplain habitat and vegetation are discussed in Section 14.7 of Chapter 14.)

Soils

Because river floodplains are built from river deposits, the soils of floodplains often possess characteristics distinctively different from those of the neighboring uplands. They are generally described in the U.S. Natural Resources and Conservation Service (NRCS) reports as soils having high water tables, seasonal flooding, and diverse composition. Therefore, *soil maps* prepared by the NRCS can also be used in the definition of floodplains. A note of caution is called for in using these maps, however, because the NRCS bases much of its own decisions for drawing the boundaries of different soil types on topography, vegetation, and land use patterns. Therefore, we must be aware that a good correlation between floodplains based on NRCS soil boundaries and floodplains based on observable vegetation, topography, and land use may constitute circular reasoning.

Past floodflows

Finally, floodplains may be defined according to the extent of *past floodflows*. Evidence of past flows may be gained from firsthand observers who are able to pinpoint the position of the water surface in the landscape and from features such as organic debris on fences and deposits on roads that can be tied to the flood. In addition, geobotanical indicators such as scars on tree trunks from ice damage or flood-toppled and partially buried trees are helpful evidence (see Fig. 14.12).

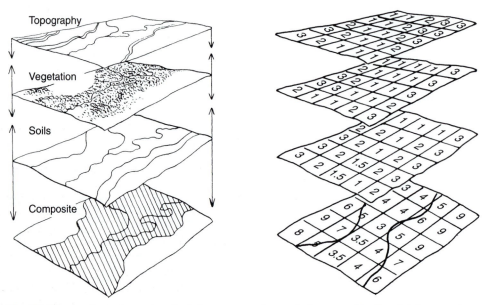

Fig. 10.12 A schematic portrayal of the map overlay technique used in floodplain mapping. *Left*, maps delineating the pertinent features; *right*, these features codified into a numerical scheme based on land use suitability.

Understandably, this method usually demands extensive field work, including interviews with local residents and tramping through brushy bottomlands.

Map synthesis   The map overlay technique is most commonly employed in synthesizing all these data. Although it has been long used by geographers, the application of this technique has been advanced by landscape architects and planners in recent decades for problems involving complex spatial arrangements among multiple sets of data. One overlay scheme involves assigning numerical values to the various classes of each component of the landscape (vegetation, drainage, soils, and the like) according to their land use suitability. The values are then summed and the results used to delineate areas or zones of different development potentials (Fig. 10.12). We should be cautioned against attaching too much meaning to the numerical results, however, because the values are not intrinsic quantities but rather arbitrary numbers assigned to qualitative classes. GIS techniques are ideally suited to floodplain mapping; however, you face the same problems as in manual techniques, namely, data acquisition, data reliability, and the meaning of weighted overlays.

10.7 CASE STUDY

Flood Risk and the Impacts of Fire in a Small Forested Watershed, Northern Arizona

Charlie Schlinger, Cory Helton, and Jim Janecek

Communities downstream of large earthen dams are potentially at risk of disastrous flooding. Indeed, catastrophic floods associated with earthen dam failures, such as the 1889 "rainy day" Jonestown, Pennsylvania, flood, or the 1976 "sunny day" Teton Dam failure, collectively have moved legislators and regulators toward a focus on worst-case scenarios. Although such scenarios have an exceedingly low probability of occurring, no one involved is willing to suffer the consequences.

In addition to flood hazards posed by dams, many areas experience infrequent high-magnitude storm events with the potential to cause severe flooding. "Flashy" ephemeral watercourses in arid and semiarid regions generate stormflows that rapidly peak following intense storms of brief duration. This is counterintuitive to many unfamiliar with hydrologic processes. To make matters more interesting, the effects of fire on a watershed can magnify the "flashiness" of these systems and greatly increase peak flows, particularly for low-to moderate-intensity precipitation events. High intensity fires not only destroy water-absorbing vegetation, but change the soil structure, creating a phenomenon called hydrophobicity, a highly impermeable surface condition that promotes rapid overland flow. Successful management of a watershed with an earthen dam, flashflood potential, and forest fire impacts requires a strategy that considers watershed and stream channel conditions as well as the performance and reliability of engineered structures.

The Black Canyon Lake reservoir and watershed cover 5.6 square miles in remote semi-arid ponderosa pine country above the Mogollon Rim in north-central Arizona. The dam is nearly 80 feet tall, was built in the early 1960s and has a capacity of 1580 acre-feet. Downstream lies the community of Heber, Arizona, population of nearly 2500, which partially occupies the floodplain. A failure of Black Canyon Dam could be catastrophic to Heber. Therefore, in the summer of 2002, when the Rodeo-Chedski fire burned most of the Black Canyon watershed, regulatory agencies became concerned. Approximately 84 percent of the watershed was affected by the fire; nearly 25 percent was severely burned (see map).

The Black Canyon dam is regulated by the Arizona State Department of Water Resources. Given the downstream risk to Heber from a dam failure at Black Canyon, the Department stipulated that the reservoir and spillway must accommodate the probable maximum flood (PMF). The PMF arises from a hypothetical worst-case

Burned over area of upland timber.

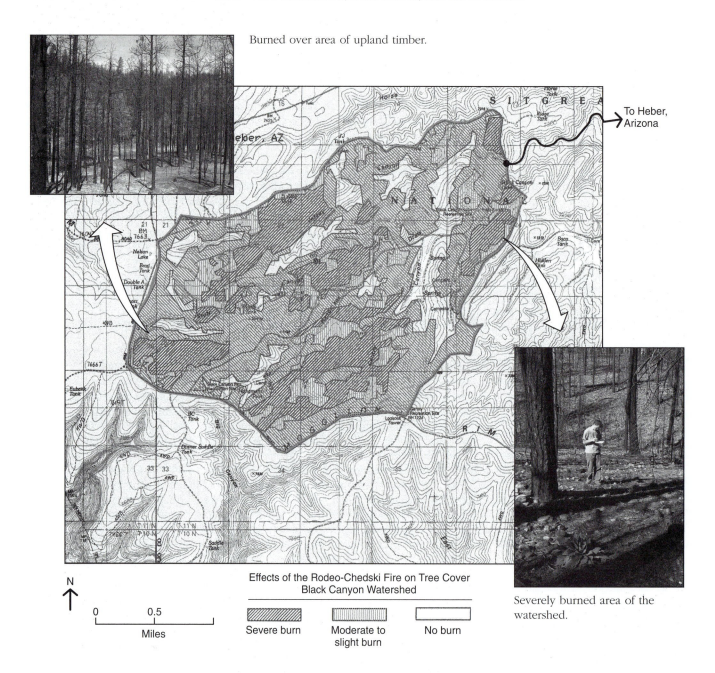

Effects of the Rodeo-Chedski Fire on Tree Cover
Black Canyon Watershed

Severe burn   Moderate to   No burn
               slight burn

N

0   0.5

Miles

Severely burned area of the
watershed.

rainfall event, known as the probable maximum precipitation (PMP). The concern is that overtopping of the earthen dam during a PMF event could lead to catastrophic failure of the dam and downstream propagation of a flood wave not only driven by the PMF but augmented by the nearly 1600 acre-feet of reservoir water.

The 6-hour PMP is estimated at around 10.4 inches, with nearly 5.5 inches falling in a 15-minute period midway through the storm. (The spillway at Black Canyon was originally designed based on a 6-hr 100-yr storm with total precipitation of only 4.5 inches, considerably less than the estimated PMP!) Using a unit hydrograph method developed by the US Natural Resources and Conservation Service, and a multibasin model with numerous subcatchments to account for varying degrees of the burn and spatial variability of other watershed parameters, discharge projections were calculated for the PMP event. For a watershed condition representing recovery one year after the fire, peak discharge was estimated at

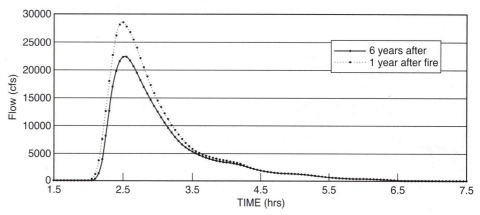

Hydrographs illustrating the change in discharge from the watershed largely in response to vegetation conditions 1 year after the fire and 6 years after the fire.

28,000 cfs. For a watershed condition representing recovery six years after the fire, estimated discharge had dropped 5500 cfs to 22,500 cfs (see lower curve above).

For events such as the 2- or 25-year storm, it is generally important to construct a hydrological model with attention to detail, meaning the subdivision of a watershed into multiple catchments. However, this is less of an issue for a PMP event. The impact of fire on runoff from a watershed can be extreme especially for small watersheds and low-recurrence-interval storms. Yet, recovery of understory vegetation (grasses and shrubby vegetation) and topsoil conditions from fire usually occurs over several years to as many as 5 to 10 years. A final consideration is that the PMF is not overly sensitive to soil type or vegetation condition; nonetheless the runoff is extreme.

To handle a flow generated by the PMP will require making changes in the dam and/or watershed. Alternatives for the Black Canyon Dam include: increasing storage by raising the dam crest elevation; increasing discharge capacity by widening the spillway channel or lowering the spillway crest elevation; hardening the earthen embankment so that an overtopping flow can pass over it without significant damage, or devising some combination of the three.

Finally, although PMF study results often lead to a focus on structural remedies, such as those listed above, watershed and forest conditions deserve equal consideration. Management of forested watersheds to mitigate risks associated with catastrophic burns can alleviate the need for conservatism in PMF evaluations, and thus reduce the required magnitude, and cost, of structural remedies.

Charlie Schlinger is associate professor of civil and environmental engineering and Jim Janecek is a research hydrologist at Northern Arizona University, where they work on water management problems. Cory Helton is an engineer with Jon Fuller Hydrology & Geomorphology, Inc.

∎

10.8 SELECTED REFERENCES FOR FURTHER READING

Bhowmik, N. G., et al. *The 1993 Flood on the Mississippi River in Illinois* (Miscellaneous Publication 151). Champaign: Illinois State Water Supply Survey, 1994.

Burby, Raymond J., and French, S. P. "Coping with Floods: The Land Use Management Paradox." *Journal of the American Planning Association* 47:3, 1981, pp. 289–300.

Dunne, Thomas, and Leopold, Luna B. "Calculation of Flood Hazard." In *Water in Environmental Planning.* San Francisco: Freeman, 1978, pp. 279–391.

Glassheim, Eliot, "Fear and Loathing in North Dakota." *Natural Hazards Observer,* 21:6, 1997, pp. 1–4.

Interagency Floodplain Management Review Committee. *Sharing the Challenge: Floodplain Management into the 21st Century.* Washington, DC: U.S. Government Printing Office, 1994.

Kochel, R. Craig, and Baker, Victor R. "Paleoflood Hydrology." *Science* 215:4531, 1982, pp. 353–361.

Linsley, R. K. "Flood Estimates: How Good Are They?" *Water Resources Research.* 22:9, 1986, pp. 1595–1645.

Natural Hazards Research and Applications Information Center. *Floodplain Management in the United States: An Assessment Report.* Boulder: University of Colorado, 1992.

Platt, R. H. "Metropolitan Flood Loss Reduction Through Regional Special Districts." *Journal of the American Planning Association* 52:4, 1986, pp. 467–479.

Rahn, Perry H. "Lessons Learned from the June 9, 1972, Flood in Rapid City, South Dakota." *Bulletin of the Association of Engineering Geologists* 12:2, 1975, pp. 83–97.

Schneider, William J., and Goddard, J. E. "Extent of Development of Urban Flood Plains." *U.S. Geological Survey Circular 601-J,* 1974.

Thompson, A., and Clayton, J. "The Role of Geomorphology in Flood Risk Assessment". *Proceedings of the Institution of Civil Engineers.* Paper 12771, 2002, pp. 25-29.

White, Gilbert F. *Flood Hazard in the United States: A Research Reassessment.* Boulder: University of Colorado, Institute of Behavior Science, 1975.

Wolman, M. Gordon, "Evaluating Alternative Techniques for Floodplain Mapping." *Water Resources Research* 7, 1971, pp. 1383–1392.

11

WATER QUALITY, RUNOFF, AND LAND USE

11.1 INTRODUCTION

The quality of water in lakes and streams has been a national issue in the United States and Canada for more than a quarter century. Both countries have enacted complex bodies of law calling for nationwide pollution control programs. In the United States the past 30 years have produced several significant trends. The Water Pollution Control Act of 1972 (now amended and known as the Clean Water Act) provided federal funding for new and improved sewage treatment facilities, which resulted in reduced fecal bacteria and nutrient loading and decreased BOD (biological oxygen demand) in the nation's inland and coastal waters. Lead levels also declined with the change to unleaded gasoline for automobiles.

Despite these strides, the overall quality of the nation's surface waters has not met forecasts based on the programs implemented under the Clean Water Act. Sewage from many municipalities is still inadequately treated. More than 1000 communities, including a number of large cities, rely on combined sewer overflow systems that discharge raw sewage directly into streams, lakes, and the ocean via storm-sewers when treatment plants become overloaded during wet weather. These cities are reluctant to build separate systems for sewage because of the high costs involved—estimated around $20 million per linear mile.

Little progress has been made on managing stormwater pollution, although we understand the problem better today than we did 10 or 20 years ago. Stormwater contributions have increased significantly with economic development and population growth. BOD loadings from both agriculture and urban stormwater sources have risen in most parts of North America, offsetting the gains made through improved municipal and industrial sewage treatment facilities. Stormwater is also responsible for increases in nitrogen, phosphorus, sediment, and chloride. The nitrogen increase is attributed primarily to a substantial rise in fertilizer applications from agriculture. Agriculture is also responsible for most of the increase in sediment loading, whereas road deicing is mainly responsible for the increase in chloride.

11.2 POLLUTION TYPES, SOURCES, AND MEASUREMENT

Water pollutants can be classified in different ways according to several different criteria, including environmental effects, influence on human health, types of sources, and pollutant composition. Based on their influences on the environment and human health, pollutants can, for our purposes, be grouped into eight classes.

Types of water pollutants   **Oxygen-demanding Wastes**   As organic compounds in sewage and other organic wastes are decomposed by chemical and biological processes, they use up available oxygen in water that is essential to fish and other aquatic animals. Biological oxygen demand (BOD) is the most common measure of this form of water pollution.

Plant Nutrients   Water-soluble nutrients such as phosphorus and nitrogen accelerate the growth of aquatic plants, causing the buildup of organic debris from these plants that can lead to increased BOD and the elimination of certain organisms in streams, lakes, and other water features.

Sediments   Particles of soil and dirt eroded from agricultural, urban, and other land uses cloud water, cover bottom organisms, eliminate certain aquatic life forms, clog stream channels, and foul water-supply systems.

Disease-causing Organisms   Parasites such as bacteria, viruses, protozoa, and worms associated mainly with animal and human wastes enter drinking water and cause diseases such as dysentery, hepatitis, and cholera.

Toxic Minerals and Inorganic Compounds Substances such as heavy metals (e.g., lead and mercury), fibers (e.g., asbestos), and acids from industry or various technological processes are harmful to aquatic animals and cause many diseases in humans, including some forms of cancer.

Synthetic Organic Compounds Substances, some water-soluble (e.g., cleaning compounds and insecticides), some insoluble (e.g., plastics and petroleum residues manufactured from organic chemicals), cause various conditions in animals and humans, including kidney disorders, birth defects, and probably cancer.

Radioactive Wastes Byproducts of nuclear energy manufacturing from both commercial and military sources emit toxic radiation, a cause of cancer.

Thermal Discharges Heated water discharged mainly from power plants and some industrial facilities causes changes in species and increased growth rates in many aquatic organisms.

There are two conventional ways of describing water pollution sources. One is based on the activity that produces the pollutant, and the other on the way the pollution is discharged into the environment. The latter uses only two categories, generally called point sources and nonpoint sources. **Point source pollution** is that from a specific source, usually a facility, and is released at a known discharge point or outfall, usually a pipe or ditch. Chief among point sources are municipal sewer systems, industry, and power plants. While point source pollutants are very important, the majority of stream pollution in the United States and Canada today comes from nonpoint sources.

Point and nonpoint sources

Nonpoint source pollution represents spatially dispersed, usually nonspecific, sources that are released in various ways at many points in the environment. Stormwater in both urban and rural areas is the principal source of nonpoint pollution. It is generated from various land uses over large areas and is discharged into streams, lakes, and other water features at countless places along channels and shores. In most regions, agriculture is the chief contributor of nonpoint source pollution and sediment is the main contaminant. However, the most geographically extensive source of water pollution is fallout (in both wet and dry forms) from the atmosphere.

In general, stormwater pollution systems consist of three essential parts: (1) onsite production, followed by (2) removal from the production site, and ending with (3) delivery to a stream, lake, or wetland. *Production* represents the amounts and types of contaminants generated by a land use. For example, on streets and highways cars and trucks are the principal contaminant producers (e.g., oil, gasoline, and tire residues) and the rate of output is directly tied to traffic levels. *Removal* is accomplished mainly by runoff. Roads are built with crowns in the middle so that runoff moves efficiently to the shoulders and collects with its load of contaminants in the drains along the road edge. *Delivery* entails routing the stormwater in a pipe or ditch from the road edge to a stream or water body (Fig. 11.1). Understanding this system is important because it is the framework within which planning and management actions take place.

Production–delivery system

In order to describe and analyze water pollution problems, we must be able to measure pollution levels. For most pollutants, such as sediments and plant nutrients, measurement involves separating the pollution from the water and measuring the total units or parts of pollutant per parts of water, usually given in *parts per million* (ppm).

Units of measure

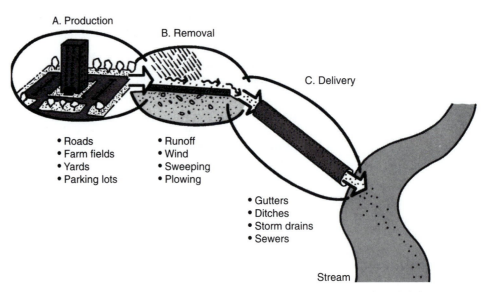

A. Production

B. Removal

C. Delivery

- Roads
- Farm fields
- Yards
- Parking lots

- Runoff
- Wind
- Sweeping
- Plowing

- Gutters
- Ditches
- Storm drains
- Sewers

Stream

Fig. 11.1 Three-part stormwater pollution system: production, removal, and delivery.

A more widely used measure is *milligrams per liter*, which is the weight of the pollutants in a liter of water and is denoted by mg/l. Although they are different units, ppm and mg/l are equivalent and both are standard measures of pollution concentration.

Concentration and load Important to the understanding of pollution data is the distinction between pollution concentration and total pollution load or discharge. **Pollution concentration** is the amount of contaminants in water at a single moment, such as during a pollution spill. It is measured in ppm or mg/l and varies with both changes in streamflow and changes in the amount of pollutants discharged. When there is a lot of water, pollution is diluted and the concentration is low. When there is a lot of pollution and little water, concentration is high. The **pollutant load**, on the other hand, is measured by adding up the total pollution discharge into a stream over an extended period of time, such as a year. Unlike pollution concentration, load is measured irrespective of the amount of water involved.

Background levels In most cases, the problem of measuring water pollution usually involves first determining background levels of the contaminant in question, because many substances labeled as pollutants (e.g., sediment, nitrogen, and certain heavy metals) are found naturally in water. In a stream where sediment is the issue, this can be difficult because sediment levels naturally fluctuate with changes in discharge. Therefore, defining background levels may require considerable sampling. But what if sediment from pollution sources is already present? Then there is really no way to define background levels precisely. As a result, background levels of naturally occurring pollutants often represent estimates, but for manufactured pollutants such as synthetic organic compounds, which are *not* found in the environment, detection at any level constitutes pollution, strictly speaking.

NPDE System Stormwater quality is getting serious attention as part of environmental regulation. Most municipalities recognize that the stormwater detention basins required for new development also help reduce sediment and some other contaminants in runoff. The U.S. EPA has implemented stormwater management rules and guidelines including a permitting requirement under the **National Pollutant Discharge Elimination System (NPDES)**. This system addresses mainly point sources of water pollution but can also include contaminated stormwater, such as that from industrial areas, which is released from pipes. In 1987 this requirement was extended to communities with populations over 100,000 mandating them to obtain an NPDES permit for their stormwater management systems. Unfortunately, agriculture remains untouched by stormwater regulations despite its enormous contributions to water pollution.

11.3 STORMWATER, LAND USE, AND WATER QUALITY

Urban and rural nonpoint

The first question in the relationship between water quality and land use is the balance of contributions between urban and agricultural land uses. This is not an easy question to answer, and the general response is that stormwater pollution is very large in both urban and agricultural lands. Based on one measure, BOD, it is estimated that: (1) in 35 percent of the watersheds in the United States, urban nonpoint sources (principally stormwater) are greater contributors than point sources; and (2) in 54 percent of the watersheds, agricultural nonpoint sources (also principally stormwater) are greater contributors than point sources. The concentration of BOD pollutants in stormwater is generally equal to or greater than that in effluents discharged from modern sewage treatment facilities. However, the total loading from stormwater may be significantly greater than that from point sources because of the huge volumes of stormwater discharged from the land.

The automobile connection

Our concern is mainly with urban stormwater, including its relationship to the automobile. The rise of **urban stormwater pollution** more or less parallels the growth of cities and the use of automobiles since 1920 or so. In the United States the number of operating motor vehicles has grown to nearly 200 million in only 75 years. Between 1970 and 1995 the total miles traveled annually by Americans has more than doubled, from 1.1 trillion miles to 2.25 trillion miles. There are now more than a million miles of paved roads and countless acres of parking lots, driveways, garages, and other hard surfaces for automobiles (Fig. 11.2). In most metropolitan areas, the total length of paved streets is equivalent to a two-lane highway stretching from Michigan to Florida.

As impervious surfaces, these facilities generate huge amounts of stormwater containing a host of automobile contaminants—oil, paint, lead, organic compounds, and many other residues. Today the quantity of petroleum residue washed off streets, highways, parking lots, and industrial sites each year exceeds the total spillage from oil tankers and barges worldwide. Between 1950 and 1980, highway salt applications in the United States increased 12 times. The salt content in expressway drains during spring runoff may be a hundred times greater than natural levels in freshwater. Unfortunately, we have not demanded of transportation planners the same standards demanded of development (land use) planners, and it remains conventional engineering practice to release contaminated stormwater directly to streams and other water features in the quickest and cheapest way.

Land use density relations

In urban areas, the pollutant loading of stormwater increases with *land use density* (Fig. 11.3). Land use density is measured by the percentage of land covered by impervious materials, and it reflects the level of pollution-producing activities such as automobile traffic, spills, leakage, atmospheric fallout, and garbage production per acre, or hectare. It also reflects the efficiency of surface flushing by stormwater runoff, which is nearly 100 percent in heavily built-up areas. The heaviest pollution loads are typically carried by the so-called *first-flush flow*, representing the first half-inch or so of a rainstorm.

Although impervious cover is an important indicator of pollution loading, it, of course, is not a direct cause of stormwater pollution. Rather, the amount and types of land use activity associated with impervious cover are the sources of pollution, for example, commercial activity, automobile traffic, air pollution, and garbage production. Other measures of land use may also be used as indicators of pollution loading, for example, dwelling units per acre and people per acre. As the data in Table

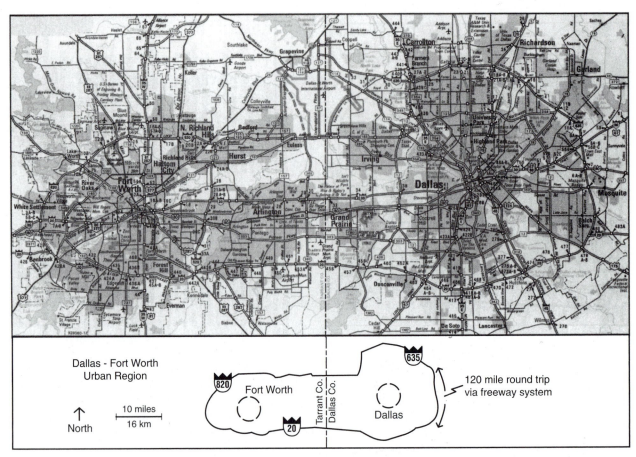

Fig. 11.2 The system of highways in the Dallas-Fort Worth urban region is similar to that in most large urban areas. Add city streets to this system and the road density more than doubles.

11.1 indicate, pollution loading in residential areas increases with both dwelling units and persons per acre. This is very significant, but equally significant for planning purposes is the *rate* of loading per person or dwelling units at different density levels.

Table 11.1 Annual Stormwater Pollution Loading for Residential Development

Density	Phosphorus[a] (per person rate)	Nitrogen[a] (per person rate)	Lead[a] (per person rate)	Zinc[a] (per person rate)	Sediment[b] (per person rate)
0.5 unit/ac	0.8	6.2	0.14	0.17	0.09
(1.25 person/acre)	(0.64)	(4.96)	(0.11)	(0.14)	(0.07)
1.0 unit/ac	0.8	6.7	0.17	0.20	0.11
(2.5 persons/acre)	(0.32)	(2.68)	(0.07)	(0.08)	(0.04)
2.0 units/ac	0.9	7.7	0.25	0.25	0.14
(5 persons/acre)	(0.18)	(1.54)	(0.05)	(0.05)	(0.03)
10.0 units/ac	1.5	12.1	0.88	0.50	0.27
(25 persons/acre)	(0.06)	(0.48)	(0.04)	(0.06)	(0.01)

[a] Pounds per acre per year.
[b] Tons per acre per year.
Source: Guidebook for Screening Urban Nonpoint Pollution Management Strategies, Northern Virginia Planning District Commission, 1979.

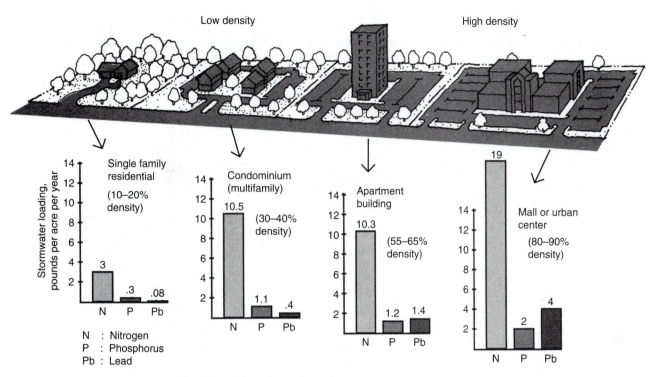

Fig. 11.3 The relationship of stormwater pollution and land use density in an urban region. Pollution loading increases with density from the suburban fringe toward the city center.

Per capita loading

The *per capita loading* rate, that is, the amount of stormwater pollution per person, actually *decreases* with higher residential densities. By way of example, let us consider residential densities of 0.5, 2.0, and 10 dwelling units per acre. At a density of 0.5 house per acre (2-acre lots), which is equivalent to about 1.25 persons per acre, the per person loading rate for phosphorus is 0.64 pound per year (i.e., 0.8/1.25) and lead is 0.11 pound per year (i.e., 0.14/1.25). At a density of 2 houses per acre (0.5-acre lots), the per person loading rate is 0.18 pound phosphorus and 0.05 pound lead. For town house apartments at 10 units per acre, the person loadings are 0.06 pound phosphorus and 0.035 pound lead.

The trend is clear. Measured on a per capita basis, large residential lots in the range of 1 to 2 acres tend to be the most damaging to water quality. The reasons are related to the large size of lawns, the large numbers of automobiles per household, and the relatively great lengths of roads and drives, among other things. The fact that the per capita loading rate decreases with higher densities suggests that communities should discourage large lot development on the urban fringe and promote cluster development as a means of reducing water pollution.For communities faced with the dilemma of how to accomodate growth with minimal water quality degradation, the answer is clearly not large lot single family for residential development.

11.4 WATER QUALITY MITIGATION ON LAND

Broadly speaking, the first objective in mitigating stormwater quality is to *slow the overall rate of response* of the stormwater flow system (Fig. 11.4). Decelerating the system will induce infiltration and settling of contaminants and also reduce rates of surface flushing and erosion. One measure of system response is concentration time. If

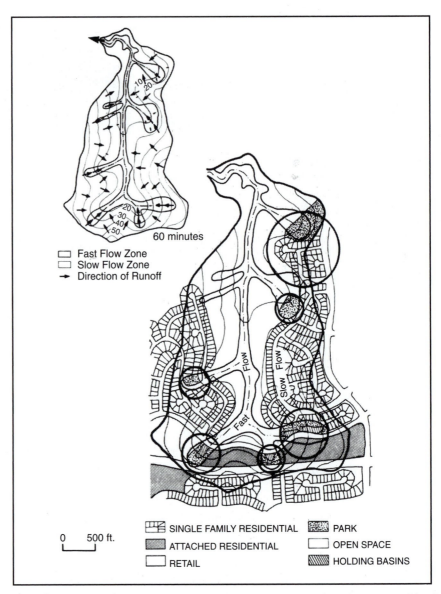

Fig. 11.4 The pattern of runoff travel times in a small basin (inset) and a proposed land use plan designed to minimize acceleration of runoff in fast flow zones by limiting development near channel heads and using stormwater basins and parkland to slow delivery rates to channels.

the concentration times of stormflow, especially those that flush surfaces frequently—for instance flows from storms in the range of an 0.5- to 1.0-inch rainfall—can be extended rather than decreased with development, an important step can be taken toward water quality control. Slower flow systems encourage infiltration, settling, and sequestering of nutrients and other contaminants. In addition, slow flows lower the risk of channel erosion and the sediment it releases into the delivery system itself.

Mitigation strategies   With regard to other mitigation strategies, we can approach the problem at three levels: (1) control of pollutant production; (2) control of pollutant removal from the site; and (3) control of pollutant transfer through the delivery system. The first level includes the regulation of land use types, development density, lawn fertilizing, garbage burning, and so on. Considering the per capita pollutant loading rates associated with large lot residential development, we see that cluster development is clearly

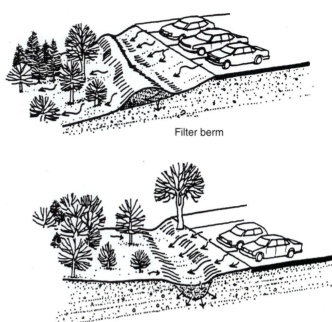

Fig. 11.5 Illustrations (right) of a filter berm and an infiltration trench, two simple, small-scale mitigation-devices to reduce contaminant levels in stormwater. Left, an elaborate infiltration swale, or stormwater garden.

an important planning strategy of minimizing pollution loading of stormwater in residential areas. (These approaches are developed further in Chapter 13, which discusses best management practices.)

Soil absorption

Measures for limiting the transfer of pollutants from the site into the stormwater system are aimed largely at regulating the volume of runoff. The most common strategy relies on increasing **soil absorption** by, for example, increasing the ratio of vegetated-to-impervious groundcover, using porous pavers, and diverting runoff into infiltration beds or dry wells. According to a study conducted in the Washington, D.C., region, soil-absorption measures are the most effective means of removing pollutants from stormwater. This study found that, for soil with average permeability, expected removal capacities are in the range of 35 to 65 percent for total annual phosphorus; 40 to 85 percent for annual biochemical oxygen demand (BOD); and 80 to 90 percent for annual lead, depending on land use type.

There is wide variety of soil-absorption/filter systems in use today and several are described below. We have already mentioned dry wells and depression storage cells. Dry wells are vertical shafts filled with stones or coarse aggregate, designed to store and infiltrate stormwater. They have the advantage of not taking up much surface area and releasing water at depth. Infiltration beds come in various forms and names, with and without plants. Those that come with plants lend themselves to ready integration into landscape design schemes. Among the most interesting are stormwater gardens or rain gardens (Fig. 11.5).

Filter berms

Some communities have experimented with the use of soil material as a filter medium for stormwater. Austin, Texas, for example, has favored two such measures in new developments: filter berms and filtration basins. **Filter berms** are elongated earth mounds constructed along the contour of a slope. They are usually constructed of soil containing different grades of sand and a filter fabric and are designed to function in the same fashion as soil infiltration trenches, which have been shown to

Fig. 11.6 Filter basin used to reduce contaminant loads in stormwater. Water is filtered through layers of sand that lie between the baffles.

be highly effective in contaminant removal (Fig. 11.5). With berms and related measures, treatment is limited to small flows, either those from individual lots or small groups of lots. Soil-filtering efficiency is very high according to tests based on the application of treatment plant effluent to soil in various parts of the United States.

Filtration basins
                Filtration basins (also called water quality basins or filtering ponds) are concrete structures floored with several grades of sand and a filter fabric through which the stormwater is conducted (Fig. 11.6). Filtration basins are generally used for higher density land uses, such as shopping centers, than would be appropriate for filter berms. They are designed to filter the first 0.5 inch of runoff, the so-called first flush, which is heaviest in contaminants. The performance of filtration basins based on the city of Austin's experience is good for small stormflows (less than 0.5 inch runoff). For first-flush flows Austin reported the following filtering rates: Fecal coliform 76 percent, total suspended solids 70 percent, total nitrogen 21 percent, total Kjeldahl nitrogen 46 percent, nitrate nitrogen 0, total phosphorus 33 percent, BOD 70 percent, total organic carbon 48 percent, iron 45 percent, lead 45 percent, and zinc 45 percent.

Vegetated buffer
                Another filtering measure strongly favored by some communities is the **vegetated buffer**. Several experimental studies show that vegetative buffers can be extremely efficient in sediment removal (up to 90 percent or more) if they meet the following design criteria: (1) continuous grass/turf cover, (2) buffer widths generally greater than 50 to 100 feet, (3) gentle gradients, generally less than 10 percent, and (4) shallow runoff depths, generally not exceeding the height of the grass. In hilly terrain, vegetative buffers should, to the greatest extent possible, be located on upland surfaces and integrated with depression storage and soil-filtration measures such as berms and dry wells (see Fig. 8.12b). It is doubtful that vegetated buffers alone are effective in reducing contaminants such as nitrogen and phosporus, especially where stormwater is discharged over wooded hillslopes.

Holding basins
                The final class of mitigation measures are those placed in the delivery system. These measures are **holding basins**, usually detention ponds and retention basins. These features are designed to withhold stormwater from the flow system in order to reduce peak discharge. By holding stormwater, water quality can also be improved, especially the early runoff from a storm. Investigators generally point to the importance of sediment settling in stormwater basins and its role in the overall

Table 11.2 Representative Removal Efficiencies for Stormwater Holding Basins

Pollutant	Percentage Removal
Suspended sediment	40–75%
Total phosphorus	20–50%
Total nitrogen	15–30%
BOD	30–65%
Lead	40–90%
Zinc	20–30%

Sources: Based on a compilation from various sources by Michael Sullivan Associates, Austin, Texas.

removal of pollutants from the water. Table 11.2 offers representative removal values reported by various studies from retention basins and detention basins. Values tend to be higher for retention basins because they hold water on a permanent basis and have a correspondingly higher potential for sediment settling and biochemical synthesis than detention basins. Retention basins are often larger than detention basins and employ infiltration and evaporation rather than outflow to dispose of stormwater, except in the case of massive runoff events which overflow the basin.

11.5 EUTROPHICATION OF WATERBODIES

Nutrient loading

Among the many problems caused by water pollution, **nutrient loading** is one of the most serious and widespread in North America. Nutrients are dissolved minerals that nurture growth in aquatic plants such as algae and bacteria. Among the many nutrients found in natural waters, nitrogen and phosphorus are usually recognized as the most critical ones, because when both are present in large quantities they can induce accelerated rates of biological activity. Massive growths of aquatic plants in a lake or reservoir will lead to (1) a change in the balance of dissolved oxygen, carbon dioxide, and microorganisms; and (2) an increase in the production of total organic matter. These changes lead to further alterations in the aquatic environment, most of which are decidedly undesirable from a human use standpoint:

- Increased rate of basin in-filling by dead organic matter.
- Decreased water clarity.
- Shift in fish species to rougher types such as carp.
- Decline in aesthetic quality; for example, increase in unpleasant odor.
- Increased cost of water treatment by municipalities and industry.
- Decline in recreational value.

Eutrophication

Together, the processes of nutrient loading, accelerated biological activity, and the buildup of organic deposits are known as **eutrophication**. Often described as the process of aging a waterbody, eutrophication is a natural biochemical process that works hand in hand with geomorphic processes to close out waterbodies. Driven by natural forces alone, an inland lake in the midlatitudes may be consumed by eutrophication within several thousand years, but the rate varies widely with the size, depth, and bioclimatic conditions of the lake. In practically every instance, however, land development accelerates the rate by adding a surcharge of nutrients and sediment to the lake. So pronounced is this increase that scientists refer to two eutrophication rates for waterbodies in developed areas: natural and cultural (Fig. 11.7).

For plants to achieve a high rate of productivity, the environment must supply them with large and dependable quantities of five basic resources: heat, light, carbon

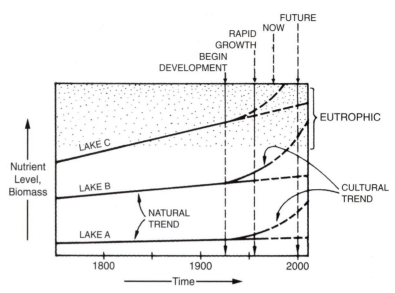

Fig. 11.7 Graph illustrating the concepts of natural and cultural eutrophication. The three sets of curves represent waterbodies at three stages of natural eutrophication as they enter the cultural stage of their development in the 20th century.

dioxide, water, and nutrients. According to the biological *principle of limiting factors*, plant productivity can be controlled by limiting the supply of any one of these resources. In inland and coastal waters, heat, light, carbon dioxide, and water are abundant on either a year-round or seasonal basis. Nutrients on the other hand, often tend to be the most limited resource, not just in terms of quantities but also in terms of types. If nitrogen and phosphorus are available in ample quantities, productivity is usually high; however, if either one is scarce, productivity may be retarded.

Phosphorus and nitrogen   When introduced to the landscape in the dissolved form, **phosphorus and nitrogen** show different responses to runoff processes. Nitrogen, which is generally more abundant, tends to be highly mobile in the soil and subsoil, moving with the flow of soil water and groundwater to receiving waterbodies. If introduced to a field as fertilizer, for example, most of it may pass through the soil in the time it takes infiltration water to percolate through the soil column—as little as weeks in humid climates. In contrast, phosphorus tends to be retained in the soil, being released to the groundwater very slowly. As a result, under natural conditions most waters tend to be phosphorus-limited, and when a surcharge of phosphorus is directly introduced to a waterbody, accelerated rates of productivity can be triggered. Accordingly, in water management programs aimed at limiting eutrophication, phosphorus control is often the primary goal. In the study of inland lakes, a classification scheme for levels or stages of eutrophication has been devised based on the total phosphorus content (both organic and inorganic forms) of lake water (Table 11.3).

11.6 LAKE NUTRIENT LOADING AND LAND USE

Nutrient budget concept   For any body of water it is possible to compute the **nutrient budget** by tabulating inputs, outputs, and storage of phosphorus and nitrogen over some time period. *Inputs* come from four main sources: point sources, surface runoff (streamflow, stormwater, and the like), subsurface runoff (chiefly groundwater), and the atmosphere. *Outputs*

Table 11.3 Levels of Eutrophication Based on Dissolved Phosphorus

Level or Stage	Total Phosphorus, mg/l[a]	Water Characteristics
Oligotrophic (pre-eutrophic)	Less than 0.0.25	No algal blooms or nuisance weeds; clear water; abundant dissolved oxygen
Early eutrophic	0.025–0.045	
Middle eutrophic	0.045–0.065	
Eutrophic	0.065–0.085	
Advanced eutrophic	Greater than 0.085	Algal blooms and nuisance aquatic weeds throughout growing season; poor light penetration; limited dissolved oxygen

[a] Representative mean annual values of phosphorus in phosphorus-limited waterbodies.

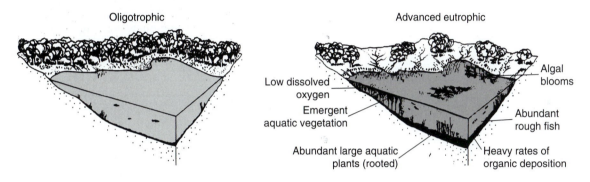

take mainly three forms: streamflow, seepage into the groundwater system, and burial of organic sediments containing nutrients (Fig. 11.8). Storage of nutrients is represented by plants and animals, both living and dead, which, if not buried by sediment,

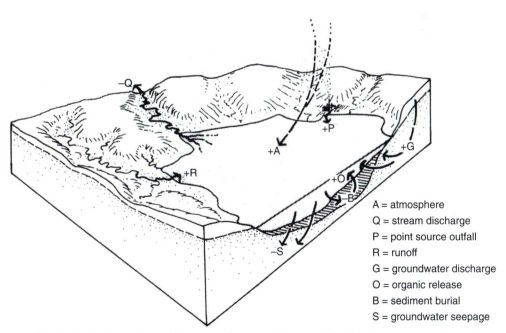

Fig. 11.8 Main inputs and outputs of nutrients to a waterbody. Computation of a nutrient budget over some time period tells us the trend of the nutrient concentrations in the water.

release synthesized nutrients upon decomposition. A formula for the nutrient budget may be written as follows:

$$P + R + O + G + A - Q - S - B = 0$$

where

P = point source contributions
R = surface runoff contributions
O = organic sediment contributions
G = groundwater contributions
A = atmospheric contributions

Q = losses to streamflow
S = losses to groundwater
B = losses to organisms and
    sediment burial

Although the nutrient budget is easily described, it has proven very difficult to compute accurately for most waterbodies. The primary reasons for this are: (1) some of the pertinent data, such as nutrient losses to groundwater seepage, are often difficult to generate; and (2) exchanges of nutrients among water, organisms, and organic sediments are difficult to gauge. As a result, most nutrient budgets are based on a limited set of data, usually those representing streamflow, septic drainfield seepage, point sources, and atmospheric fallout. However, data have been produced from field studies relating land use and surface cover to the nutrient content of runoff.

Nutrient loading of runoff   In a study of the nitrogen and phosphorus contents of streams throughout the United States, the U.S. Environmental Protection Agency and various state water quality programs made several interesting findings. *First*, the export of these nutrients from the land by streams tends to vary widely for different runoff events and for different watersheds with similar land uses. *Second*, although regional variations in nutrient export do appear in the United States for small drainage areas, they do not correlate very well with rock and soil type. Instead, nutrient export by small streams tends to correlate best with land use and cover, in particular to the proportion of agricultural and urban land in a watershed. Nitrogen and phosphorus concentrations in streams draining agricultural land, for example, are typically five to ten times higher than those draining forested land.

Based on these findings, it is possible to estimate the nitrogen and phosphorus loading of streams and lakes in most areas. The loading values are applicable to surface runoff, principally channel flow. The values in Table 11.4 are given in kilograms per square kilometer for seven basic land use/cover types. (The definition of each land use/cover type is given on page 218.) Figure 11.9 gives the loading values in milligrams per liter of water (runoff) for agriculture and forest use/cover types for the eastern, middle, and western regions of the coterminous United States. For compar-

Table 11.4   Nutrient Loading Rates for Six Land Use/Cover Types

Use/Cover	Nitrogen (kg/km²/yr)	Phosphorus (kg/km²/yr)
Forest	440	8.5
Mostly forest	450	17.5
Mostly urban	788	30.0
Mostly agriculture	631	28.0
Agriculture	982	31.0
Mixed	552	18.5
Golf Course	1500	41.0

Source: J. Omernik, *The Influence of Land Use on Stream Nutrient Levels.* U.S. Environmental Protection Agency, 1977.

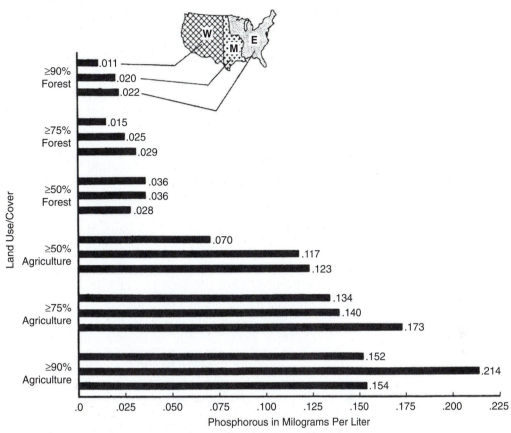

Fig. 11.9 Phosphorus loading values per liter of runoff based on agricultural and forest land use/cover for the coterminous United States.

ative purposes, examine these values with those representative of the water systems listed in Table 11.5.

Estimating nutrient loading

To determine the nutrient contributions to a waterbody, we must first delineate the drainage system and define its drainage areas. This may be a difficult task in developed areas because of the complexity of the land uses and the artificial drainage patterns that are superimposed on the natural drainage system. As a first step, it is instructive to identify the various land uses and major cover types in the area, noting their relationship to the drainage system and water features, what kinds of pol-

Table 11.5 Representative Levels of Phosphorus and Nitrogen in Various Waters

Water	Total P, mg/l	Total N, mg/l
Rainfall	0.01–0.03	0.1–2.0
Lakes without algal problems	Less than 0.025	Less than 0.35
Lakes with serious algal problems	More than 0.10	More than 0.80
Urban stormwater	1.0–2.0	2.0–10
Agricultural runoff	0.05–1.1	5.0–70
Sewage plant effluent (secondary treatment)	5–10	More than 20

Source: John W. Clark et al., *Water Supply and Pollution Control*, 3rd ed. (New York: IEP/Dun-Donnelley, 1977); and American Water Works Association, "Sources of Nitrogen and Phosphorus in Water Supplies," *Journal of the American Water Works Association* 59, 1967, pp. 344–366.

lutants they are apt to contribute (both nutrients and other types), and the locations of critical entry points. The remaining steps are as follows:

■ Determine the percentage of each drainage area occupied by forest, agriculture, and urban development.

■ Classify each area according to the relative percentages of forest, agriculture, and urban land uses based on the following percentages:

Forest	> 75% forested
Mostly forest	50–75% forested
Agriculture	> 75% active farmland
Mostly agriculture	50–75% active farmland
Mostly urban	> 40% urban development (residential, commercial, industrial, institutional)
Mixed	Does not fall into one of the above classes; for example, 25% urban, 30% agriculture, and 45% forest

■ Using the nutrient-loading values given in Table 11.4, multiply the appropriate value for each area by its total area in square kilometers.

■ To calculate the loading potential from septic drainfield seepage, count the number of homes within 100 yards of the shore or streambank for each drainage area and multiply this number by the following nutrient loading rates:

Phosphorus/Year	*Nitrogen/Year*
0.28 kg/home	10.66 kg/home

■ Combine the two totals for each drainage area for the total input from the watershed. If the *grand* total input from all sources is called for, then atmospheric contributions must also be considered. Atmospheric input need only be considered for water surfaces, since it is already included in the values given for land surfaces. Given a fallout rate for the area in question, this value (in mg/m^2) should be multiplied by the area of the waterbody; this quantity would be added to those from the watershed to obtain the grand total input (less groundwater contributions, if any) to the waterbody.

11.7 PLANNING FOR WATER QUALITY MANAGEMENT IN SMALL WATERSHEDS

Planning for residential and related land uses near water features presents a fundamental dilemma: People are attracted to water, yet the closer to it they build and live, the greater the impact they are apt to have on it. As impacts in the form of water pollution and scenic blight rise, the value of the environment declines, which in turn usually lowers land values (Fig. 11.10). Moreover, the greater the proximity of development to water features, generally the greater the threat to property from floods, erosion, and storms. In spite of these well-known problems, the pressure to develop near water features has not waned in recent years. In fact, it has increased, and in areas where there are few natural water features, developers are often inclined to build artificial ones to attract buyers.

Corrective approach   In planning and management programs aimed at water quality, basically two approaches may be employed: preventive and corrective. The *corrective approach* is used to address an unsatisfactory condition that has already developed—for example, cleaning beaches and surface waters after an oil spill. On small bodies of water with weed growth problems, corrective measures may involve treating the water with chemicals that inhibit aquatic plants, harvesting the weeds, and/or dredging the lake basin

(*a*)   (*b*)

Fig. 11.10 Examples of water quality degradation that can lead to deterioration of land values: (*a*) Long Beach, California, at the mouth of the Los Angeles River, and (*b*) Montreal, Canada, along the St. Lawrence Seaway.

to remove organic sediments. Such measures are generally used as a last resort for waterbodies with serious nutrient problems and/or advanced states of eutrophication.

Preventive approach   A *preventative approach* is generally preferred for most bodies of water, though it may actually be more difficult to employ successfully. This approach involves limiting or reducing the contributions of nutrients and other pollutants from the watershed by controlling onsite sources of pollution, limiting the transport of pollutants from the watershed to the lake, or both. Measures to control nutrient sources include improving the performance of septic drainfields for sewage disposal, replacing septic drainfields with community sewage treatment systems, reducing fertilizer applications to cropland and lawns, controlling soil erosion, eliminating the burning of leaves and garbage and controlling animal manure. Measures to limit nutrient transport to the lake include filtering water through the soil or a soil medium, source control of stormwater, eliminating stormsewers in site planning and design, disconnecting farmfields from the drainage network, maintaining wetlands, floodplains, and natural stream channels, and constructing wetlands such as the one described in the case study at the end of this chapter.

Management planning   The formulation of a water quality *management program* for a waterbody usually begins with an estimate of the nutrient budget. Nutrient contributions are placed in two categories: those than can be managed using available technology and funding, and those that cannot. The latter usually includes groundwater and atmospheric contributions, whereas the former includes primarily surface and near surface runoff, that is, stormwater, septic drainfield seepage, and the like.

Setting goals   The second step entails defining management goals, such as "to slow the rate of eutrophication," or "to improve on the visual character" of a certain part of the waterbody. This process is important because the goals must be realistic and attainable. Generally speaking, the more ambitious the goals (for example, "to reverse the trend in eutrophication"), the more difficult and expensive they will be to achieve. Once goals are established, a plan (or set of plans) is formulated that identifies the actions

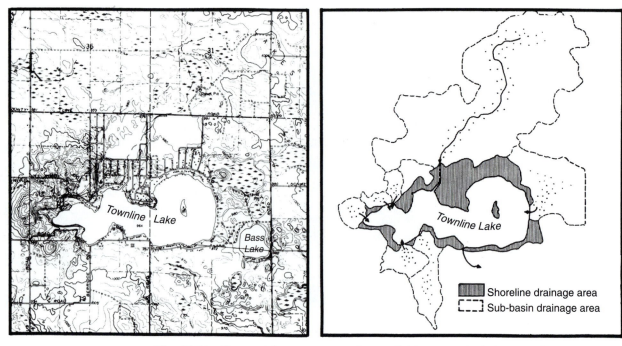

Fig. 11.11 The spatial organization of a small lake watershed showing the two types of drainage subareas: shoreland and subbasin.

and measures that need to be implemented. Some actions, such as prohibiting the use of phosphorus-rich fertilizers, may apply to the entire watershed, whereas others may apply only to a specific subarea that the nutrient budget data show to be a large contributor. In addition, measures are proposed to limit nutrient production and transport as related to future development by defining appropriate development zones, and enacting guidelines on sediment control, stormwater drainage, and sewage disposal.

Implementation and evaluation

Finally, after the plan is implemented, the waterbody must be monitored and the results weighed against the original goals, financial costs, and public support. Evaluation is a difficult task because it often requires comparing the results with forecasts about what conditions would have been without the plan. Uncertainty often arises relating to the validity of the original forecasts and what the effects of natural perturbations in climatic, hydrologic, or biotic systems may have had in masking or enhancing the efforts of the plan.

The watershed framework

In water quality planning for lakes and ponds, we need to understand the drainage system with which we are dealing. As we noted in Chapter 9, every waterbody is supported by a watershed, which is comprised of many water systems including stream networks, groundwater, and precipitation and evaporation. The watershed runoff system, particularly that on the surface, provides the spatial framework within which land use planning related to water quality management takes place.

Drainage areas and Land use

The watershed illustrated in Figure 11.11 is typical of that around most impoundments. The bulk of the area is taken up by *subbasins* that drain all but the narrow belt of land along the shore, called *shoreland*. Subbasins are small watersheds with integrated drainage systems that usually empty via a single stream, whereas shorelands, being too narrow to develop streams, lack integrated drainage systems, and drain directly to the waterbody by overland flow, interflow, and groundwater. Land uses in the watersheds of impoundments tend to correlate with these two types of drainage areas. Water-oriented development (recreational activities and residential uses) is concentrated in the shoreland, whereas nonwater-oriented development (agriculture, suburban residential, commercial, and so on) is located mainly in the subbasins. The proportion of the watershed occupied by these two groups is usually very lopsided in favor of the subbasin users, which in turn can create a stakeholder's dilemma. Occupants of both shorelands and subbasins are watershed stakeholders, but only shoreland occupants with lake frontage property are usually recognized as such.

Management of subbasins and shorelands

In the shoreland zone management efforts usually involve dealing with individual property owners because each site drains directly into the waterbody (see the diagrams in Case Study 21.8). In the subbasins, on the other hand, many land uses are linked together by a runoff system that connects to the waterbody at a single point. Management of a subbasin, therefore, is in three respects a more difficult task. First because a great many players may be involved, second, the system is often larger and more complex, and third, land users lack a sense of connection, and hence stakeholder responsibility, for the lake and its management problems. Subbasins do, however, provide management opportunities that shorelands do not. For example, runoff can be managed by various source control techniques in the upper basin to reduce nutrients, sediment, and other impurities as the following case study illustrates.

11.8 CASE STUDY

Sediment and Nutrient Trapping Efficiency of a Constructed Wetland Near Delavan Lake, Wisconsin

John F. Elder and Gerald L. Goddard

Delavan Lake is a highly valued and heavily used sport fishery and recreational lake in southeastern Wisconsin which for many years has been plagued by water quality degradation and algal growth. By the late 1970s, the 2072-acre lake had reached a eutrophic state. Despite efforts in 1981 to reduce nutrient loading by diverting sewage and septic tank effluent from the lake basin, a severe algal bloom occurred in 1983 that signaled the need for additional action which led to hydrologic and water quality studies and a detailed rehabilitation plan. Among other recommendations, the plan called for construction of a wetland at the confluence of Jackson Creek and its two tributaries, which combine to form the principal inflow to Delavan Lake. The purpose of this wetland was to reduce sediment and nutrient inflow to the lake and thus contribute to its long-term protection and maintenance.

Jackson Creek Wetland was constructed in the fall of 1992 when a 15-acre wetland was enlarged to 95 acres. Nearly all surface-water inflow to the wetland is through three streams, which have a combined drainage area of 16.6 square miles. This area is used principally for agriculture with increasing areas of residential and commercial development. In addition to areas of sedge meadow, wet prairie, and shallow-water marsh, the wetland contains three retention ponds along its upstream edge at the mouths of the inflowing streams. Surface areas of these ponds are about 3.4, 1.2, and 1.2 acres, respectively. Each pond has 1 to 4 outlet swales that distribute runoff into the wetland. The primary objective of this study was to assess the effectiveness of the wetland as a retention system for suspended sediment and nutrients. The study was conducted between 1993 and 1995.

Sediment accumulated in the ponds each year of the study. Total accumulation over two years was 752 tons in all three ponds. This represents a retention efficiency of 58 percent of the suspended sediment input to the ponds from stream discharge. Retention was greater than 20 percent at all times of the year except during the winter/early spring period. During the growing season, retention was frequently greater than 80 percent. For nutrients, however, the trapping effectiveness of the wetland was much less consistent. With the exception of ammonia, total nutrient loads generally decreased substantially between the inflows and the outflow on an annual basis. The inflow/outflow ratio, however, was highly variable seasonally and was frequently less than 1, an indication that more nutrients were given up by the ponds than retained.

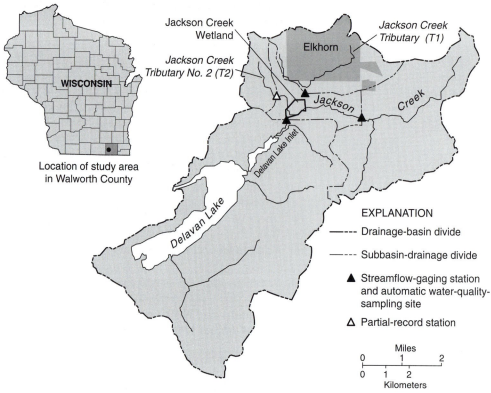

Delavan Lake, its watershed, and the Jackson Creek Wetland.

In 1994, the season of greatest transport, by a large margin, was the winter/early spring period. That period was also the time of greatest fractional retention of total and dissolved phosphorus. This combination of high transport and uptake accounted for nearly all of the annual net retention; it overshadowed later small-volume releases during low-flow periods. In fact, virtually all retention in 1994 occurred during a single month: February. Nearly all other months were actually periods of net phosphorus release, reflected by the large positive changes during the third and fourth quarters. Phosphorus mobilization during spring and summer might be due to increased solubility in anaerobic conditions that could, in turn, be caused by higher rates of microbial respiration in warmer temperatures.

What do these seasonal variations imply for aquatic ecosystems downstream? Although no direct evidence of their significance is available, the fact that phosphorus releases tended to occur in late spring and summer makes them coincident with the likely timing of algal blooms downstream in Delavan Lake Inlet and Delavan Lake. Further indication of a potential problem is the fact that the proportional

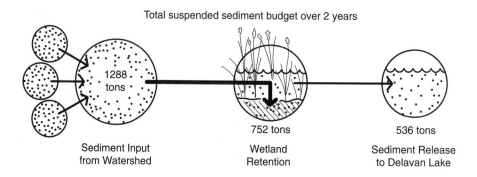

release of dissolved orthophosphate—the fraction that is available for algal uptake—was even greater than that of total phosphorus.

The lack of coupling between suspended sediment and nutrient transport dynamics is also notable. The suspended material presumably is retained because of the reduced velocity of water flow, which allows the material to settle out. One might expect that the nutrient loads would be largely associated with oxidized suspended sediments and, therefore, would be entrained together with the sediments. However, this was clearly not the case; net release of nutrients often coincided with substantial retention of sediments. This result indicates that biogeochemical processes mobilized the sediment-associated nutrients and thus prevented their effective retention in the wetland, at least periodically.

Climate is also a consideration on nutrient load characteristics. Precipitation was 3.49 inches above normal during the 1993 study period, 12.19 inches below normal in water year 1994, and 4.74 inches below normal in water year 1995. The high water flows in 1993 produced heavy nutrient fluxes into and out of the wetland. The large flows may also have reduced the nutrient retention effectiveness of the wetland.

In summary, although the wetland functioned as a sink for both sediments and, at times, for some forms of nutrients, its efficiency was greater and more consistent for sediment than nutrients. The wetland retained 46 percent of the input sediment load over the full period of study. This finding, though illustrating some limitation of the wetland as nutrient sink, does not necessarily conflict with the general concept of nutrient retention in the wetland. As observed in other wetlands, the variability reflects the capacity of the system to function not only as a sink but also, periodically, as a facilitator of nutrient transformation and transport. The results of this study thus illustrate the complexities of biogeochemical cycles in any ecosystem, wetlands included. Understanding these complexities may help us to avoid placing unrealistic expectations on natural or constructed wetlands as systems with unlimited filtering capacity.

John F. Elder (retired) is a research limnologist and Gerald L. Goddard is a hydrologic technician. Both are with the U.S. Geological Survey-Wisconsin District Lake-Studies Team. ∎

11.9 SELECTED REFERENCES FOR FURTHER READING

Clark, John W., Viessman, Warren, Jr., and Hammer, Mark J. *Water Supply and Pollution Control.* 3rd ed. New York: IEP/Dun-Donnelley, 1977.

Dillon, P. J., and Vollenweider, R. A. *The Application of the Phosphorus Loading Concept to Eutrophication Research.* Burlington, Ontario: Environment Canada; Center for Inland Waters, 1974.

Elder, J. F. "Factors Affecting Wetland Retention of Nutrients, Metals, and Organic Materials." In Kusler, J. H., and Brooks, G. (eds). *Wetland Hydrology.* Chicago: Association of State Wetland Managers, 1988.

Marsh, William M., and Hill-Rowley, Richard. "Water Quality, Stormwater Management, and Development Planning on the Urban Fringe." *Journal of Urban and Contemporary Law* 35, 1989, pp. 3–36.

Naiman, R. J., and Turner, M. G. "A Future Persprctive on North America's Freshwater Ecosystems." *Ecological Applications* 10: 4, 2000, pp. 958–970.

National Academy of Sciences. *Eutrophication: Causes, Consequences, and Correctives.* Washington, DC: National Academy of Science Press and the National Research Council, 1969.

OECD. *Environmental Impact Assessment of Roads.* Paris: Organization for Economic Cooperation and Development, 1994.

Omernik, James M. *Nonpoint Source–Stream Nutrient Level Relationships: A Nationwide Study.* Corvallis, OR: U.S. Environmental Protection Agency, 1977.

Smith, R. A., Alexander, R. B., and Wolman, M. G. "Water Quality Trends in the Nation's Rivers." *Science* 235, 1987, pp. 1607–1615.

Tilton, Donald L., and Kadlec, R. H. "The Utilization of a Fresh-Water Wetland for Nutrient Removal from Secondarily Treated Waste Water Effluent." *Journal of Environmental Quality* 8:3, 1979, pp. 328–334.

U.S. Soil Conservation Service. *Ponds for Water Supply and Recreation.* Washington, DC: U.S. Department of Agriculture, 1971.

Vallentyne, John R. *The Algal Bowl: Lakes and Man.* Ottawa: Environment Canada; Fisheries and Marine Service, 1974.

Wolman, M. Gordon, and Chamberlin, C. E. "Nonpoint Sources." *Proceedings of the National Water Conference*, Philadelphia Academy of Sciences, 1982, pp. 87–100.

12

SOIL EROSION, LAND USE, AND STREAM SEDIMENTATION

12.1 INTRODUCTION

Soil erosion may be the most serious land management problem facing humanity today. Simple estimates based on sediment discharged to the sea indicate that since the origin of agriculture some 12,000 years ago, the annual sediment loss worldwide has more than doubled—from 9 billion tons to more than 20 billion tons. This dramatic increase is even more alarming when we consider that only a fraction of the soil eroded from the land actually ends up in the sea. Most of the sediment from soil erosion is left in terrestrial and freshwater environments such as woodlands, stream valleys, wetlands, and reservoirs where it is a major source of habitat and water quality degradation.

Virtually all land uses contribute directly or indirectly to soil erosion, but agriculture is decidedly the single greatest contributor. Crop farming is the main cause, but grazing (cattle, sheep, and goats) also contributes to soil loss by causing extensive damage to rangeland. Broadly speaking, we can say that soil loss from erosion has followed the trend of world population. Today it is estimated that 4.5 million square miles of land have been seriously degraded by crop farming, deforestation, overgrazing, and other factors; this is an area larger than all of Canada and soil erosion is the principal culprit.

The costs of erosion are enormous. Scientists estimate that soil erosion costs $44 billion a year in the United States and $400 billion worldwide. These estimates include direct costs represented by loss in soil fertility (productivity) and indirect costs represented by damage to waterways, infrastructure, and human health. Indirect costs include sediment clogging of channels, which increases flooding and damage to land use facilities, and the reduction of water quality. Direct costs are calculated according to the value of fertilizers needed to replace soil fertility losses. Neither the direct nor indirect costs calculated for these estimates include damage to ecosystems.

For most soils, vegetative cover is the most important control on erosion. When vegetation is disturbed or removed, no matter whether it is grass or trees, erosion inevitably increases. Other factors influencing erosion by runoff include rainfall magnitude and frequency, slope of the land, cropland management, and soil erodibility. The most erodible soils are loams and silts (among others) situated on sloping ground without a protective plant cover. If this scene is coupled with heavy rainfall and no runoff controls, such as berming, terracing, or contour plowing, then soil loss may be as high as 1 to 2 inches a year. This is equivalent to more than 100 tons of soil loss per acre. At this rate it is only a matter of a few years before the topsoil is entirely wiped out, leaving only the underlying, and much less fertile, mineral soil at the surface.

What is our particular interest in soil erosion? Although nonagricultural land uses are minor contributors to soil loss worldwide, they are major sources of the problem locally, especially on the urban fringe where the environments most affected—streams, lakes, and wetlands—are among the most prized and heavily used. The agricultural problem has long held the attention of soil scientists and geomorphologists, and it is upon their research and management recommendations that we build our approach to soil erosion related to land development. In this chapter we are interested in understanding the process of soil erosion and sediment transport, methods for estimating soil loss rates, and techniques for controlling erosion and sedimentation.

12.2 SOIL EROSION, BIOCLIMATE, AND LAND USE

From field studies conducted in different parts of North America, we are able to sketch a brief picture of the general relationship between bioclimatic conditions and soil erosion. Barring the influence of land use, it appears that erosion is relatively low

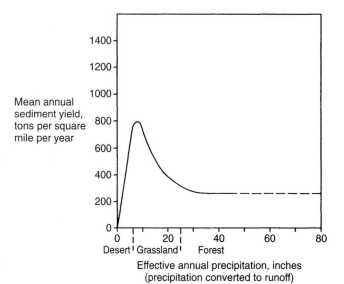

Mean annual sediment yield, tons per square mile per year

Effective annual precipitation, inches
(precipitation converted to runoff)

Fig. 12.1 Sediment yield related to bioclimatic conditions less the influences of land use. Rates are highest in the semiarid zone where short grasses provide poor protection against occasionally heavy rainstorms and runoff.

Semiarid grasslands

in the bioclimatically extreme environments, namely, arid and humid regions, but relatively high in semiarid regions, especially in their drier zones (Fig. 12.1). The explanation for this is related to the effects of precipitation and plant cover. In the semiarid zones precipitation is modest, averaging 10 to 20 inches a year in the drier sections, but is capable of producing substantial runoff events in most years. But evapotranspiration is also high and soil moisture is generally insufficient to support more than a weak and discontinuous cover of short grasses, which provides poor protection against erosion by runoff. In addition, rainfall is quite variable in these areas, which further promotes erosion because drought years weaken plant covers and wet years produce much higher than average runoff.

Humid and arid lands

In humid regions there is significantly more rainfall, and thus the greater *potential* for erosion. However, forest is the predominant vegetative cover and provides strong defense against the forces of erosion. This is reflected in the relatively low sediment loads in streams in the eastern and northwestern sections of the United States (Fig. 12.2). At the other bioclimatic extreme, in arid lands receiving annual runoff-producing precipitation of less than 7 to 8 inches, runoff rates are so low that, despite the erodibility of poorly protected desert soils, erosion is very limited.

Land clearing and farming

When land use is added to the picture, erosion rates change appreciably. In the semiarid grasslands, erosion by both wind and runoff usually rises dramatically with plowing, cropping, and grazing. Historically, droughts have been especially damaging to farmland, but today's widespread use of irrigation reduces the risk of massive soil loss from drought-related events such as the Dust Bowl of the 1930s. Nevertheless, wherever prairie grasses are removed for farming, with or without irrigation, erosion by runoff invariably rises, especially on sloping ground. The same situation holds in humid lands where forests are cleared for farming.

Land clearing breaks down the protective vegetative cover and, if not replaced by a permanent substitute cover, topsoil is subject to rapid erosion. On most cropland, fields are left barren for the winter months when soil moisture levels and the potential for runoff are high. In Canada and much of the northern United States, fields are often cropless for six months a year. Today, improved soil protection methods, such as no-till farming, call for leaving crop stubble on fields over winter. As a defense against soil erosion it is a poor substitute for the original plant cover however.

In the past century or so, urbanization has also become important in the soil erosion issue, especially as it is practiced in North America. The general sequence of

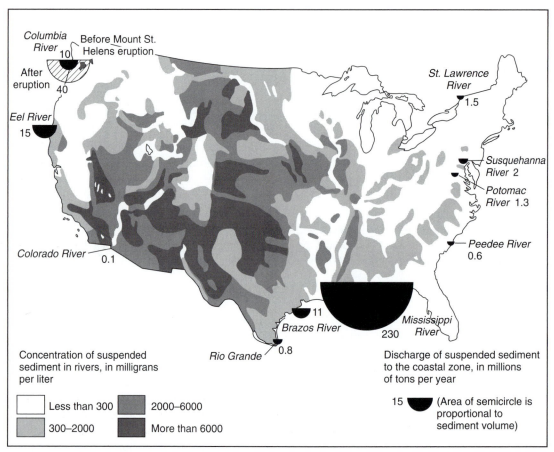

Fig. 12.2 Average concentration of suspended sediment in streams in the contiguous United States with figures on the average discharge of suspended sediment by major rivers. Suspended sediment is made up of fine soil particles, mainly clay, eroded by runoff and discharged into streams. It is a key indicator of soil erosion rates.

The road to urban development

land use change that ends in urban development and the corresponding rates of soil erosion are given in Figure 12.3. A description of the process begins with forest clearing and agricultural development sometime in the 1800s for most areas. Erosion rises with the destruction of natural vegetation and the establishment of cropland and pasture. This trend continues until the first half of the twentieth century when agriculture declines as rural population shifts to urban areas. As abandoned farmland is taken over by volunteer vegetation, soil erosion probably declines.

With the growth of cities in the second half of the twentieth century comes urban sprawl and the development of suburbia. Abandoned farmland and woodland, as well as active farmland on the urban fringe and beyond, are cleared for development. Wholesale exposure of soil during the construction phase of development often gives rise to massive erosion rates, particularly for large construction sites left open for many months. Although this phase is short-lived, its effects in terms of sedimentation of streams and wetlands can be profound and lasting. With completion of development the soil is more or less secured under buildings, roads, and landscaped surfaces, causing erosion rates to decline. With full urbanization and impervious cover approaching 75 percent or more, erosion rates appear to drop to levels approaching early agriculture. At this stage, the sediment appearing in stream discharge is less a product of soil erosion and more a product of (1) channel scouring caused by massive stormwater

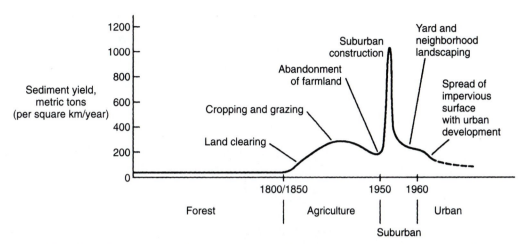

Fig. 12.3 The trend of soil erosion related to land use/cover change beginning with presettlement forest and ending with 20th century urban development.

discharges, and (2) pollution sources such as atmospheric fallout, automobile wastes, and debris from the wear and tear of streets and buildings (Fig. 12.3).

12.3 THE SOIL EROSION-SEDIMENT TRANSPORT SYSTEM

The system As with any environmental problem, the path to effective management begins with understanding the nature of the system. For our purposes, a **soil erosion–sediment transport system** is made up of four essential components: erosion, transport, storage, and export. These components are interconnected, but not necessarily sequential, especially the sediment storage and export components. In addition, the system operates within a relatively small spatial framework, the local drainage basin, with an area ranging from several square miles to a 100 square miles or more (Fig. 12.4).

System components *Erosion* is the process by which particles are dislodged from the soil. *Transport* involves the transfer of particles by runoff from the point of erosion to a storage site or point of export. For most sediment, transport distances are short. Small sediments (clays and silts) may go for long rides but only if they become entrained in a stream's

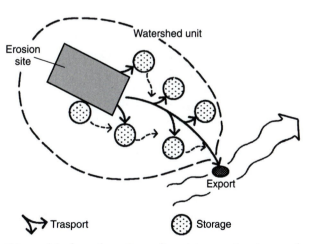

Fig. 12.4 A graphic model of a soil erosion-sediment transport system made up of four components: erosion, transport, storage, and export.

Fig. 12.5 Photograph of gullying in a Mississippi farm field. In erodible soils, gullies such as these can be cut in only several years.

discharge. *Storage* is merely deposition of sediment. This process begins with larger particles on and around an erosion site and continues as sediment is trapped in various places as it moves down the drainage basin. Storage may be brief, lasting only until the next runoff event, or it may be long term if it is immobilized by burial or by plant overgrowth. The sediment discharged after losses to storage within the basin constitutes the *export* component of the system.

Erosional processes   The erosional process itself begins with *rainsplash* from the impact of raindrops. When the drops hit the ground, they explode, driving small soil particles downslope. If the rainfall rate (intensity) exceeds the soil's infiltration capacity, then runoff in the form of overland flow results. Overland flow moves in thin sheets called *sheetwash* and tiny threads called *rivulets*. Both processes are capable of moving sediment and rivulets can cut micro-channels, termed *rills*, into exposed soil. The soil erosion from sheet-wash and rivulet runoff is also known as **rainwash**. At some point in the midsection of a slope, this runoff begins to concentrate in small channels, and as its volume and velocity increase, its energy rises dramatically giving it the power to erode gullies into the slope face. Known as **gullying**, this process is one of the most severe forms of erosion. In erodible soils such as silt and loose sand in exposed farm fields or construction sites, gullying may carve channels several feet deep and tens of yards long in a single year. Entire hillsides may be gutted in a matter of several years (Fig. 12.5).

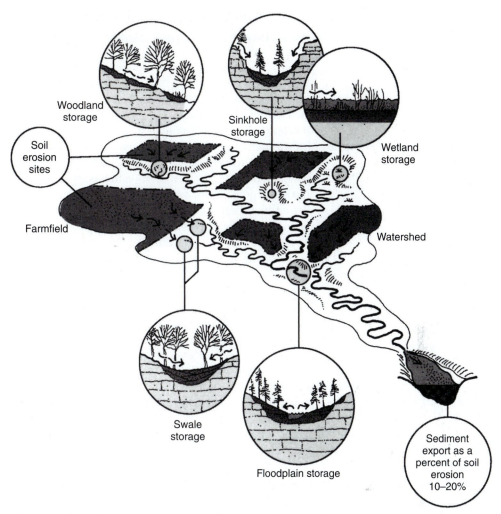

Woodland
storage

Soil
erosion
sites

Sinkhole
storage

Wetland
storage

Farmfield

Watershed

Swale
storage

Floodplain storage

Sediment
export as a
percent of soil
erosion
10–20%

Fig. 12.6 Sediment sinks in a small drainage basin. Only 10 to 20 percent of sediment released in soil erosion may be exported from the basin.

Sediment storage and export

Studies show that the vast majority of the sediment released from an erosion site is stored within the local drainage basin relatively close to its place of origin. Most of this sediment is deposited in various places, called **sinks**, near the erosion site such as wooded hillslopes, swales, stream channels, wetlands, and floodplains (Fig. 12.6). Based on a limited number of studies, it appears that only 10 to 20 percent or so of the sediment released through soil erosion is actually exported from small rural watersheds on an annual basis.

Scale and linkage

There is an important lesson here. Beyond the loss of soil itself, the greatest impacts of soil erosion occur from sedimentation close to the erosion site and within the local watershed. A key management consideration, then, is controlling the *linkage* between sediment sources and transport processes, mainly channel flows in swales, ditches, and streams that can distribute the sediment down the system and over larger areas. This explains why cropland erosion has such great impacts on aquatic habitats in humid regions. To expedite water removal for plowing and planting soil must be drained and fields are purposely tied directly to local streams by tile and ditch systems. Efficient movement of runoff means efficient movement of sediment. This arrangement serves to expand the zone of impact from sedimentation rather than contain it on or near cultivated fields.

12.4 FACTORS INFLUENCING SOIL EROSION

Rainfall erosion index

Any attempt to forecast soil erosion for a site must take into account four basic factors: vegetation, soil type, slope size and inclination, and the frequency and intensity of rainfall. Tests show that in general heavy rainfalls such as those produced by thunderstorms promote the highest rates of erosion. Accordingly, the incidence of such storms together with the total annual rainfall can be taken as a reliable measure of the effectiveness of rainfall in promoting soil erosion. The U.S. Natural Resources and Conservation Service has translated this into a **rainfall erosion index** that represents the erosive energy delivered to the soil surface annually by rainfall in an average year (Fig. 12.7). The index values vary appreciably over the United States, and in some regions, from one side of a state to another. For instance, values decline from 250 to less than 100 from the southeastern to the northwestern corner of Kansas meaning that, on the average, erosion on comparable sites should be more than 2.5 times greater in the southeast.

Vegetation controls

On most surfaces, vegetation appears to be the single most important factor regulating soil erosion. Foliage intercepts raindrops, reducing the force at which they strike the soil surface. Organic litter on the ground further reduces the impact of raindrops, and plant roots bind together aggregates of soil particles, increasing the soil's resistance to the force of running water. The one feature of vegetation that appears to have the greatest influence on erosion is **cover density**; the heavier the cover, either in the form of groundcover or tree canopy, the lower the soil loss to runoff.

Soil erodibility

If running water is applied to soils of different textures, sand will usually yield (erode) first. In order to erode clay, the velocity of the runoff would have to be increased to create sufficient stress to overcome cohesive forces that bind the particles together. Similarly, high velocities would also be needed to move pebbles and larger particles because their masses are so much greater than those of sand particles. Thus, in considering the role of soil type in erosion problems, it appears that

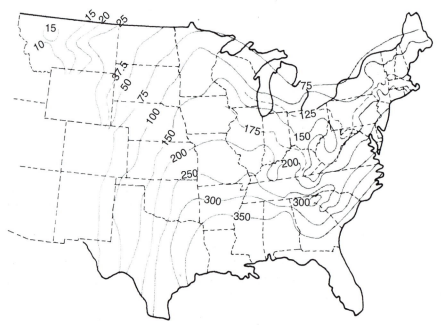

Fig. 12.7 Rainfall erosion index for the United States east of the Rocky Mountains. In the West, index values are less reliable and best calculated from local rainfall records.

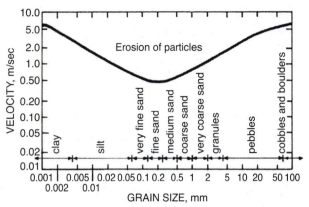

Fig. 12.8 Erosion thresholds for soil particles ranging from clay to cobbles and boulders. Sand has the lowest threshold or highest erodibility. Conditions in nature may vary with compaction, cementing, and other factors.

erodibility is greatest with intermediate textures, whereas clay and particles coarser than sand are measurably less erodible (Fig. 12.8). Other soil characteristics, such as compactness and structure, also influence erodibility, but in general *texture* can be taken as the leading soil parameter in assessing the potential for soil erosion.

The role of slope   The velocity that runoff is able to attain is closely related to the **slope** of the ground over which it flows. Slope also influences the quantity of runoff inasmuch as long slopes collect more rainfall and thus generate a larger volume of runoff, other things being equal. In general, then, slopes that are both steep and long tend to produce the greatest erosion because they generate runoff that is high in both velocity and mass. This is true only for slopes up to about 50 degrees, because at steeper angles the exposure of the slope face to rainfall grows rapidly smaller, vanishing altogether for vertical cliffs. In land use problems, however, greatest consideration is usually given to slopes much less than 50 degrees, particularly insofar as urban development, residential development, and agriculture are concerned. Table 12.1 gives relative values for soil erosion, or the potential for it, based on slope steepness and length for slopes up to 50 percent inclination and 1000 feet length. In using this table, note that the rate of change with slope steepness is not linear, whereas it is with slope length. For example, for a 500-foot slope of 10 percent the value is 3.1, whereas for the same length at 20 percent, the value is 9.3.

12.5 COMPUTING SOIL EROSION FROM RUNOFF

An estimate of soil loss to runoff can be computed by combining all four of the major factors influencing soil erosion. For this we use a simple formula called the **universal soil loss equation** (USLE), which gives us soil erosion in tons per acre per year:

Forecasting soil erosion

$$A = R \bullet K \bullet S \bullet C$$

where

A = soil loss, tons per acre per year
R = rainfall erosion index
K = soil erodibility factor
S = slope factor, steepness and length
C = plant cover factor

Table 12.1   Slope Geometry Factor Based on Steepness and Length

Slope Length in feet	Slope Steepness in Percent														
	4	*6*	*8*	*10*	*12*	*14*	*16*	*18*	*20*	*25*	*30*	*35*	*40*	*45*	*50*
50	0.3	0.5	0.7	1.0	1.3	1.6	2.0	2.4	3.0	4.3	6.0	7.9	10.1	12.6	15.4
100	0.4	0.7	1.0	1.4	1.8	2.3	2.8	3.4	4.2	6.1	8.5	11.2	14.4	17.9	21.7
150	0.5	0.8	1.2	1.6	2.2	2.8	3.5	4.2	5.1	7.5	10.4	13.8	17.6	21.9	26.6
200	0.6	0.9	1.4	1.9	2.6	3.3	4.1	4.8	5.9	8.7	12.0	15.9	20.3	25.2	30.7
250	0.7	1.0	1.6	2.2	2.9	3.7	4.5	5.4	6.6	9.7	13.4	17.8	22.7	28.2	34.4
300	0.7	1.2	1.7	2.4	3.1	4.0	5.0	5.9	7.2	10.7	14.7	19.5	24.9	30.9	37.6
350	0.8	1.2	1.8	2.6	3.4	4.3	5.4	6.4	7.8	11.5	15.9	21.0	26.9	33.4	40.6
400	0.8	1.3	2.0	2.7	3.6	4.6	5.7	6.8	8.3	12.3	17.0	22.5	28.7	35.7	43.5
450	0.9	1.4	2.1	2.9	3.8	4.9	6.1	7.2	8.9	13.1	18.0	23.8	30.5	37.9	46.1
500	0.9	1.5	2.2	3.1	4.0	5.2	6.4	7.6	9.3	13.7	19.0	25.1	32.1	39.9	48.6
550	1.0	1.6	2.3	3.2	4.2	5.4	6.7	8.0	9.8	14.4	19.9	26.4	33.7	41.9	50.9
600	1.0	1.6	2.4	3.3	4.4	5.7	7.0	8.3	10.2	15.1	20.8	27.5	35.2	43.7	53.2
650	1.1	1.7	2.5	3.5	4.6	5.9	7.3	8.7	10.6	15.7	21.7	28.7	36.6	45.5	55.4
700	1.1	1.8	2.6	3.6	4.8	6.1	7.6	9.0	11.1	16.3	22.5	29.7	38.0	47.2	57.5
750	1.1	1.8	2.7	3.7	4.9	6.3	7.9	9.3	11.4	16.8	23.3	30.8	39.3	48.9	59.5
800	1.2	1.9	2.8	3.8	5.1	6.5	8.1	9.6	11.8	17.4	24.1	31.8	40.6	50.5	61.4
900	1.2	2.0	3.0	4.1	5.4	6.9	8.6	10.2	12.5	18.5	25.5	33.7	43.1	53.5	65.2
1000	1.3	2.1	3.1	4.3	5.7	7.3	9.1	10.8	13.2	19.5	26.9	35.5	45.4	56.4	68.7

In problems involving agricultural land, a fifth factor, *cropping management* (P), is also included, but for problems involving nonagricultural land, abandoned farmland, and urban land types, it is not generally applicable.

Conditions of the USLE

In order to interpret the universal soil loss equation reliably, it is necessary to understand three points: (1) the computed quantity of soil erosion represents only the displacement of particles from their original positions; (2) only sheetwash and rill erosion are covered by the equation, whereas gullying is not; and (3) the soil loss equation is designed for agricultural land, and applications to other situations may or may not be appropriate. Cleared land, former cropland, and construction sites can generally be considered appropriate for application of this method.

Soil loss versus soil erosion

What constitutes soil loss depends on the study objectives involved. Strictly speaking, soil erosion takes place when particles are displaced from their place of origin no matter how far they are moved. But soil erosion may be different than soil loss. For example, much erosion can take place on a construction site, but if the resultant sediment is contained on site, it can be argued that no soil loss has taken place even though the soil has been degraded by erosion. For a site such as the one shown in Figure 12.9, for example, much of the soil lost from the upper surface will be deposited near the foot of the slope. Therefore, in computing the soil loss for the lower surface, we should take into account sediment added from upslope and stored on site and then solve for the net change in the soil mass. For purposes of environmental planning and management, however, the key consideration is often soil loss from the site and its impacts on neighboring environments.

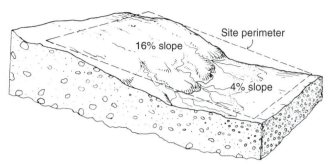

Fig. 12.9 A site comprised of two slope classes, one that yields sediment and the other that accumulates it. Erosion is significant on the upper slope, yet there may be little or no net loss of sediment from the site.

Data sources   The data needed to make a soil erosion computation can usually be obtained from topographic sheets, soil reports, and a few additional maps and tables. Field inspection of the site is recommended to determine plant cover and erosion and deposition patterns. If that is not possible, then ground and aerial photographs may be used instead, but the resultant data are approximations at best. Forecasting where soil lost to erosion goes beyond the site requires delineation of drainage lines (gullies, swales, ditches, stream channels) in the field with the aid of topographic maps and aerial photographs.

K factor   Soil type is expressed in terms of an erodibility factor, or *K factor*, which is a measure of a soil's susceptibility to erosion by runoff. This factor was derived from tests conducted by soil scientists on field plots for each soil series in a state and is given on a dimensionless scale between 0 and 1.0, though actual values range between 0.02 and 0.7. The higher the number, the greater the susceptibility to erosion. In some cases, two numbers are given for a soil, one for disturbed and one for undisturbed ground. Disturbed includes filled and rough graded ground. *K* factors are usually available from county or state offices of the U.S. Natural Resources and Conservation Service. If *K* values are not available, then an alternative method can be used based on laboratory analysis of a soil's particle size composition.

R, S, and C values   The rainfall erosion (R) values for most states may be read from the map in Figure 12.7. Reliable R values for the western states have not been generated because rainfall is highly variable owing to the diverse topography. For locations in the West, R values can be calculated from local rainfall records with best results using the 2-year, 6-hour rainfall figures. The slope factor (S) can be read from Table 12.1, and the plant cover factor (C) can be approximated from Table 12.2 based on ground-cover and canopy density.

Assuring reliability   The reliability of soil erosion computations based on the universal soil loss equation depends not only on the accuracy of the data used, but also on the way in which the problem is set up. For sites in areas of diverse terrain, the parcel should be divided into subareas within which soil, slope, and vegetation are reasonably uniform. Soil loss should be computed for each subarea and then adjusted for any deposition that may be received from adjacent subareas. In making the adjustment for deposition, it is important to consider not only the potential contribution from upslope areas, but also the patterns of gullies, streams, and swales through which runoff and sediment can be funneled across flat areas and discharged onto low ground or into streams, lakes, and wetlands.

To summarize, the procedure for computing soil loss from a site may be described in seven steps:

Computational procedure   **1.** Define the site (problem area) boundaries on a large-scale base map, preferably a topographic sheet.

Table 12.2   Plant Cover Factors

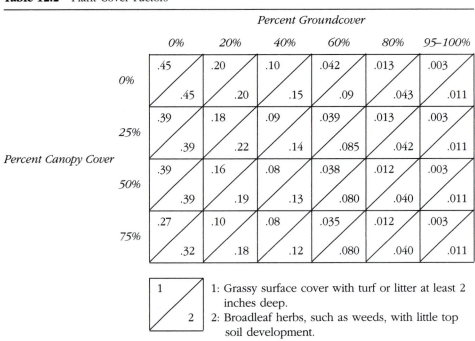

2. Using aerial photographs, soil maps, topographic maps, field inspection, and whatever other sources are available, divide the site into subareas. If the site is essentially uniform throughout, this is not necessary.

3. For each subarea, assign a *soil erodibility factor* (based on NRCS soil series *K*-factor designation), a *slope factor* (read from Table 12.1 based on steepness and length data taken from topographic map), and a *plant cover factor* (read from Table 12.2 based on aerial photographs or field observation).

4. Multiply the appropriate value from the *rainfall erosion* map in Figure 12.7 times the three factors assigned to the subarea in step 3. The answer is in tons of soil erosion per acre per year. If agricultural land is involved and one of the cropping management practices given in Table 12.3 is used, apply the appropriate P factor.

5. Examine the relations between slopes and drainage patterns in each subarea, identify where sediment would be expected to accumulate, and, if possible, adjust the quantity of soil erosion computed in step 4 and derive an estimate of soil loss.

6. Determine the total soil loss for the entire site by summing the net amounts of soil loss for each subarea. If this is not possible because of uncertainty over the

Table 12.3   Cropping Management (P) Factor

Land Slope, Percentage	P Values		
	Contouring	Contour Stripcropping	Contour Irrigated Furrows
2.0 to 7	0.50	0.25	0.25
8.0–12	0.60	0.30	0.30
13.0–18	0.80	0.40	0.40
19.0–24	0.90	0.45	0.45

relationship among various subareas, then the gross soil loss from the subareas should be summed for a gross site total.

7. Based on slope and runoff patterns both within the site and on neighboring land, identify the points and areas to which sediment is transported and where sedimentation (storage) will likely take place.

12.6 APPLICATIONS TO LAND PLANNING AND ENVIRONMENTAL MANAGEMENT

An important task of most local planning agencies is the review and evaluation of land development proposals for housing, industrial, and commercial projects. Proposals are evaluated according to a host of criteria including soil erosion, or the potential for it. Consideration of soil erosion stems not only from a concern over the loss of topsoil and depletion of the soil resource in general, but also from the impact of sedimentation on terrestrial vegetation, wetlands, river channels, and drainage facilities such as stormsewers. To gain the necessary information, planners are often asked to prepare a site plan and respond to critical questions such as these:

Management questions

■ What percentage of site lies in slopes of 15 percent or greater, and of this area how much (*a*) is proposed for development, and (*b*) if developed, will be affected by construction?

■ What percentage of this site is forested, grass covered, and shrub covered? How much of each of these covers will be destroyed as a result of development?

■ What are the minimum distances between the proposed development zone and (*a*) water features (streams, ponds, reservoirs, and wetlands), and (*b*) existing drainage facilities (stormsewers, stormwater ponds, and ditches)?

■ What erosion and sedimentation control measures are proposed during (*a*) the construction phase, and (*b*) the operational phase of the proposed project?

■ What is the anticipated length of the construction period, and which months in the year are proposed for (*a*) land clearing, (*b*) excavation and grading, (*c*) construction of facilities, and (*d*) regrading and landscaping? How does the proposed construction period relate to the seasonal pattern of rainfall, especially the months of heaviest rainstorms? Finally, how will these activities be phased so that not all the site will be opened at once (Fig. 12.10)?

Mitigation approaches and measures

Control of erosion and sedimentation during the construction phase of development is viewed very seriously in many communities—so much so that many have legislated erosion control ordinances. At the heart of such ordinances are the control or **mitigation measures**. These include a wide variety of techniques and approaches, and there is a very active dialogue among permitting agencies, scientists, landscape architects, civil engineers, construction management professionals, and others on this topic. The approaches employed vary widely, depending mainly on local requirements and the level of monitoring and enforcement employed by public agencies. Generally, however, the mitigation measures used are simple and small-scale, and most are aimed at (1) holding an exposed soil in place; (2) blocking sediment-laden runoff from draining into water features such as streams, ponds, or wetlands; or (3) filtering muddy water through fabric or straw. Measures include the placement of silt (cloth) fences, straw bales, and berms around construction zones, the use of fiber nets and fabric such as burlap on slopes, and the use of sedimentation (sump) basins to collect and isolate muddy water. (For additional comments on water quality mitigation techniques, refer to Chapter 11, Section 11.4.)

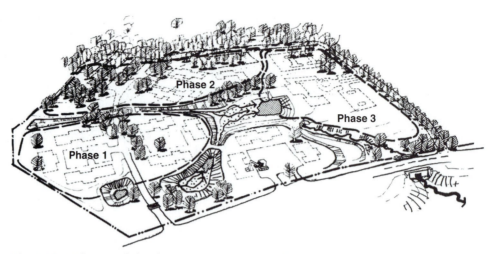

Fig. 12.10 Phasing of development is necessary in large projects in order to minimize soil erosion and sedimentation of water features near the site. In this project, only 25 to 30 percent of the site is open at any time.

Silt fence limitations

Perimeter berms

Despite the requirements of ordinances and related efforts to control erosion and sedimentation, mitigation programs on many, if not most, construction sites are inadequate. There are several reasons for this, including too great a reliance on silt fences, inadequate attention to drainage patterns and runoff volumes, and poor understanding of rainstorm magnitudes and frequencies. Especially critical is the use of silt fences around construction sites where potentially large volumes of runoff are involved. In general, silt fences can retain only small amounts of runoff and sediment such as that from exposed strips along roadways or embankments. For open areas larger than 5 acres or so, it is advisable to berm the perimeter, particularly on the downslope side, to contain both runoff and sediment. The height of the berm can be set according to the 10- or 25-year storm magnitudes and the site acreage. In this sit-

Silt fence overloaded by fill and sediment. A typical scene on construction sites.

(*a*)   (*b*)

Fig. 12.11 (*a*) Use of simple soil terraces to protect a steep slope; (*b*) a sediment-glutted stream channel, the result of heavy soil erosion upstream.

uation, the proper place for a silt fence is on the outer edge of the berm. For steeply sloping ground, perimeter berms may have to be supplemented by terraces to slow runoff and capture sediment before it reaches the edge of the site (Fig. 12.11*a*).

12.7 CONSIDERATIONS IN WATERSHED MANAGEMENT

Our perspective on soil erosion and stream sedimentation must extend beyond the site scale of observation. As we discussed earlier in this chapter, every site is situated in a drainage basin where most of the sediment released to runoff with soil erosion is deposited in various types of sinks. But we must also recognize that each drainage basin is linked together by a network of drainage channels that are the principal conduits of the sediment transport system. In most natural drainage systems, the size of an individual channel is adjusted to the size of flows it conducts, in particular to certain flows of larger than average magnitudes. We reason, therefore, that when two streams join in the network, the size of the resultant channel should approximate the

Channel capacity   sum of the two tributary channels. Actually it is the *capacity* of the channel, defined by the magnitude of discharge that it can accommodate, that increases with the merger of tributaries.

Sediment transport capacity also increases with the magnitude of stream discharge. As magnitude rises, sediment transport increases exponentially, and since very large events occur infrequently, streams tend to move most sediment in infrequent surges. Therefore, in assessing the disposition of sediment released to a stream from an erosion site, it is important to consider first its discharge regime, and then the size of the receiving channel and the relative position of the channel in the drainage network. Massive loading of small streams located in the upper parts of a watershed often results in channel glutting, which reduces channel capacities to carry large flows (Fig. 12.11*b*). Flooding increases, which carries additional sediment onto

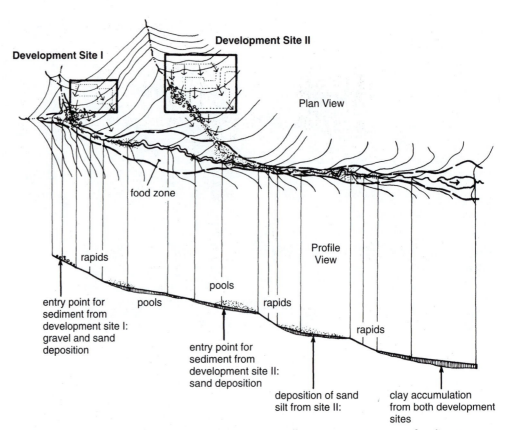

Fig. 12.12 Location of sedimentation sites in a small stream system. Sites of sediment accumulation vary with sediment size, loading rate, proximity to sediment sources, and channel size, shape, and gradient.

the floodplain, into wetlands, and other areas. Much of the sediment deposited in the channel and on the valley floor is relegated to storage, especially after vegetation has become established on it, and then not even large floodflows can displace it.

Sediment behavior in channel systems
    In a watershed where several sites are releasing sediment to a stream network at the same time, sediment will become differentiated into bodies or waves that move through the stream system at different rates, making it difficult to assess the impact on the system as a whole. Clays and silts that are carried in suspension in stream water will move through the entire system rapidly, whereas sand and coarser materials will not only move more slowly but also build up at selected points (Fig. 12.12). Often we can estimate where in the drainage system sediment is likely to accumulate based on sediment size and channel size, shape, and gradient. Broad reaches of slow-moving water are the favored deposition sites for sands, pebbles, and other large particles. Among these, wetlands and impoundments such as reservoirs and lakes are generally considered the most critical from a watershed management standpoint, and it is often necessary to estimate the amount of sediment that would be added to them based on land use practices in the drainage basin. On the other hand, knowledge of where and why sediment accumulates in a drainage system gives us clues about how to design facilities to capture fugative sediment, as is illustrated in Case Study 12.8.

Sediment loading estimates
    For purposes of watershed management, *sediment loading* of a drainage basin requires a yearly calculation of sediment production on a site-by-site basis for the entire watershed. The locations represented by each of these quantities must be pin-

pointed in the watershed and the drainage network, taking special care to determine where sediment would actually enter the network. In most cases where a site falls on a drainage divide within the watershed, the site may contribute sediment to two or more channels. Once these relationships are known, the receiving waterbodies downstream can be identified and a loading rate can be calculated.

Finally, the sediment loads must be converted to volumetric units to determine how much space in the channel or impoundment the sediments will actually take up. The following sediment densities are recommended:

- clay = 60–80 lb/ft³
- clay/silt/sand mixture = 80–100 lb/ft³
- sand and gravel = 95–130 lb/ft³

Using an intermediate value of 90 pounds per cubic foot, we find that a reservoir receiving 10,000 tons of sediment per year would lose 222,000 ft³ (8200 yds³) of volume each year. Given a reservoir capacity of 2,000,000 ft³ (one, for example, with dimensions of 200 feet wide, 1000 feet long, and 10 feet deep) the life of the reservoir would be only nine years. Thousands of reservoirs in North America were lost to sediment in the twentieth century.

For impoundments that have low resident times, that is, those that exchange water in only a matter of hours or several days rather than months or years, most of the clays will not settle out. Owing to their small sizes, clay particles remain in suspension, allowing most to pass through small impoundments. Impoundments with long residence times, which may be as great as 10 to 12 years in some lakes, retain fine sediments and deposit them over the lake floor, burying or coating bottom organisms.

12.8 CASE STUDY

Erosion and Sediment Control on a Creek Restoration Project, South Lake Tahoe, California

Steve Goldman

Erosion and sediment control methods must be tailored to the site conditions. No one plan will work for every site. In this case study—a creek restoration in a meadow near Lake Tahoe—conventional erosion and sediment control measures did not and could not do the job. The solution came from understanding the nature of the problem, and taking advantage of the natural filtering system that existed on this site.

This project involved the construction of 6000 feet of meandering stream channel. The stream, known as Cold Creek, is located in El Dorado County, California, a few miles south of Lake Tahoe, on land owned by the City of South Lake Tahoe. Cold Creek has a watershed of 13 square miles and is tributary to Trout Creek, which drains into Lake Tahoe. Above Cold Creek's confluence with Trout Creek is a 3000-foot-long meadow, through which Cold Creek originally meandered. During the 1950s, a dam was placed across the narrow part of the meadow, the creek was rerouted into a ditch along the northeast side of the meadow, and a small feeder channel was constructed from the creek to the meadow to create a lake there. The meadow-turned-lake, though originally constructed for agricultural purposes, became known as "Lake Christopher," an enhancement to a subdivision constructed on the plateau along the southwest side of the meadow. The creek was restored in 1994 by the City of South Lake Tahoe, with a grant from the California Tahoe Conservancy.

The objectives of the project were to restore a naturally functioning creek in the meadow, to reduce erosion, to protect Lake Tahoe water quality, and to enhance wildlife and fish habitat. The restored creek was connected, via a small side channel, to two waterfowl ponds, which were constructed in the meadow as part of an

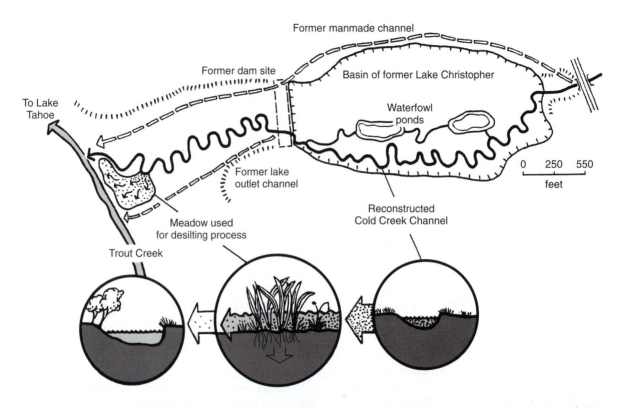

A view of the meadow near Trout Creek during desilting.

earlier project. A major challenge of the project was the desilting of the 6000 linear feet of newly constructed stream channel without allowing sediment or other pollutants to be discharged into Trout Creek.

The contractor hired to construct the project was also responsible for erosion and sediment control during construction. Because there are no standard methods for erosion control for stream restoration projects, the contractor chose to use conventional erosion control devices for construction sites, namely, silt fences and straw bales. Unfortunately, these devices proved to be ineffective at trapping sediment.

Because the flow greatly exceeded the ability of the fences to pass water, and the area behind each fence quickly filled with sediment, most of the water poured over the top of each, carrying its sediment load downstream. Turbid water was soon observed entering Trout Creek at levels significantly exceeding state standards.

State water quality officials then convened a meeting of the city, the contractor, and the project team, and it was agreed that the city would submit an erosion control plan for the final phase of the project, with the intent that this plan would become a model for future stream restoration projects at Lake Tahoe. A plan was developed that mimicked nature's way of trapping sediment. The plan involved using overbank flows into the meadow—similar to floodflows over a floodplain—to desilt the creek. First, a small wooden dam was constructed at the lower end of the newly-constructed creek channel, adjacent to a low beaver dam. The inflow to the new creek channel was controlled by placing sandbags at the junction point of the new creek channel and the old outlet channel from Lake Christopher. The sandbags allowed the creek flow to be routed in either channel at variable rates. Water was allowed to slowly fill the new channel, then gradually overflow into the meadow. Workers were stationed at various points to observe where the water was spreading and to make adjustment in the flow released to the new channel.

Based on estimates from photographs of the meadow, the spreading area covered 72,000 square feet. Using this figure, the inflow rate, and standard particle-settling velocities, the sediment trapping efficiency of the treatment area was calculated, and it was revealed that all sand- and silt-sized particles would be trapped in the meadow. Only clay-sized particles could possibly escape the meadow trap. However, the equation for this calculation does not include the effect of vegetation, and since the meadow was densely covered with sedges, the actual trap efficiency was probably greater than the calculation indicated. Because the silty water had to flow through thousands of plant stems on the way to Trout Creek, much of the sediment load was caught along the way. Resuspension of trapped fine sediments was unlikely, because of the low velocities across the meadow and because much of the settled soil tended to fall into the myriad crevices among the plant stems and roots. Thus, the water entering Trout Creek from the flooded meadow was clean, a sharp contrast to the highly turbid water that poured over the silt fences during the first phase of construction.

After spreading the water across the meadow for a week, the boards in the dam at the bottom end of the project were then removed one at a time over a five-day period. Each time a board was removed, a small amount of turbidity was observed, which typically lasted less than 30 minutes. When the last board was removed, the turbidity lasted about two hours, but it was at a much lower level than the turbidity that occurred during the first phase of the project. This method of controlling sediment proved to be a success, and no further regulatory actions were necessary.

We learned a lot from this project. As with so many aspects of environmental planning, if we model our approaches after natural processes we can often achieve highly satisfactory results at far lower costs with less environmental disturbance than would be possible through more conventional technical approaches.

Steve Goldman is an erosion control and stream restoration specialist with the California Tahoe Conservancy, South Lake Tahoe, California. He is senior author of Erosion and Sediment Control Handbook *(McGraw-Hill).* ∎

12.9 SELECTED REFERENCES FOR FURTHER READING

Environmental Protection Agency. *Erosion and Sediment Control: Surface Mining in the Eastern United States.* U.S. EPA Technology Transfer Seminar Publication, 1976.

Ferguson, Bruce K. "Erosion and Sedimentation in Regional and Site Planning." *Journal of Soil and Water Conservation* 36:4, 1981, pp. 199–204.

Glanz, James. "New Soil Erosion Model Erodes Farmers' Patience." *Science* vol. 264, 1994, pp. 1661–1662.

Goldman, S. J., et al. *Erosion and Sediment Control Handbook.* New York: McGraw–Hill, 1986.

Heede, B. H. "Designing Gully Control Systems for Eroding Watersheds." *Environmental Management* 2:6, 1978, pp. 509–522.

Hjulström, F. "Transport of Detritus by Moving Water." In *Recent Marine Sediments: A Symposium* (P. Trask ed.). Tulsa, OK; American Association of Petroleum Geologists, 1939.

Meade, R. H. and Parker, R. S. "Sediment in Rivers of the United States." In *National Water Summary 1984.* U.S. Geological Survey Water-Supply Paper 2275, 1985, pp. 49–60.

Patterson, R. G., et al. "Costs and Benefits of Urban Erosion and Sediment Control: The North Carolina Experience." *Environmental Management* 17:2, 1993, pp. 167–178.

Pimentel, David, et al. "Environmental and Economic Costs of Soil Erosion and Conservation Benefits." *Science,* 267, 1995.

Trimble, S. W. "A Sediment Budget for Coon Creek Basin in the Driftless Area, Wisconsin, 1853–1977." *American Journal of Science* 283, 1983, pp. 454–474.

Wischmeier, W. H., et al. *Procedure for Computing Sheet and Rill Erosion on Project Areas.* U.S. Natural Resources Conservation Service, Technical Release No. 51, 1975.

Wolman, M. G. "A Cycle of Sedimentation and Erosion in Urban River Channels." *Géografiska Annaler* 49A, 1967, pp. 385–395.

13

BEST MANAGEMENT PRACTICES, WATERSHEDS, AND DEVELOPMENT SITES

13.1 INTRODUCTION

In previous chapters we examined loading of the runoff system with stormwater, sediment, and nutrients and other contaminants. In each of those chapters (8, 9, 11 and 12) we highlighted and briefly discussed various management concepts and measures. In this chapter we focus exclusively on best management practices, or BMPs, and present a general approach or strategy that can help build BMP plans to mitigate the effects of land use on the runoff system. Through this approach we hope to broaden the conventional concept of BMPs and introduce a multitiered concept of BMP planning.

BMP definition   **BMPs** can be defined as measures taken to prevent or reduce the detrimental impacts of land use development and practices on the environment. They may address any part of the environment, but their most common use is associated with runoff systems, especially stormwater. In this context, BMPs usually take the form of structural devices such as detention ponds, designed to regulate stormwater discharge and help reduce soil loss and water pollution. But BMPs can also include other measures such as policy guidelines, planning initiatives, landscape design, information programs, and ecological restoration.

Our approach   Our approach to BMP planning begins with the watershed and calls for measures best described as preventative, that is, measures that are introduced before development as part of the planning process. This is followed by site-scale BMP planning that deals with corrective measures in the postdevelopment landscape. At the watershed scale, we are interested in understanding how the runoff system functions and how land uses can be designed to conform with these functions and capitalize on mitigation opportunities that already exist in the watershed. At the site-scale, stormwater and its contaminant load are addressed in terms of the three-part system outlined in Chapter 11 and illustrated in Figure 11.1: (1) production, (2) removal or release from the site, and (3) delivery to a receiving waterbody. The overall goal of the BMP package at both the watershed and site scales is to mimic the predevelopment performance of areas subject to development and to reduce the need for large, structural mitigation measures.

13.2 THE WATERSHED RUNOFF SYSTEM AND BMP OPPORTUNITIES

BMPs have traditionally been the domain of civil engineering, and their application has been mainly corrective rather than preventative. Where development is already in place, engineered BMPs remain an appropriate tool, but for undeveloped lands

Proactive BMP planning   they should not be our first choice. Instead, **BMP planning** should take place before development, as part of land use planning and design, and it should focus on finding where the watershed has the capacity to take on development without the assistance of structural devices to mitigate stormwater. In other words, BMP planning should be proactive rather than reactive in the planning and development process.

If we were able to plan and design land uses to properly fit the land and its carrying capacity, new development would not need to be retrofitted with structural BMPs. Unfortunately, we have gotten into the habit of planning less for land and more for real estate and zoning interests, therefore, we commonly miss opportunities to reduce or even eliminate elaborate and costly structural measures in attempting to mitigate stormwater runoff. In short, if we were better problem solvers in landscape planning and design, using land use densities, mixes, and configurations appropriate for various terrains and drainage systems, there would probably be little need for pipes and ponds to take care of stormwater.

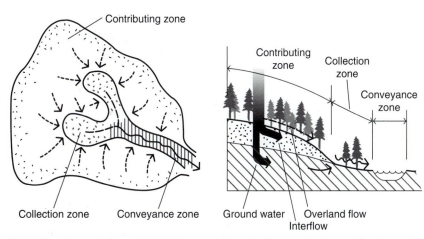

Fig. 13.1 The three main hydrological zones of a small watershed: contributing, collection, and convenience. The contributing zone represents the large part of the watershed and offers the greatest opportunity for stormwater mitigation.

Site as part of a watershed

Every development site, no matter where it is located, belongs to a watershed and is inherently part of a drainage system. For undeveloped lands, the first step in the BMP process is to define where you are in the drainage system, what runoff processes are operating there, and what conditions are associated with that location and its processes. In small watersheds (where most development takes place), a site will fall within one of the *three hydrologic zones* we discussed in Chapter 9 and are named above (Fig. 13.1).

Each zone operates differently in the drainage system in terms of how it handles water, and it is necessary to understand these functions to make BMP decisions at the watershed scale. Once the essential processes and related conditions have been identified, it may be helpful to model the subject area to estimate predevelopment hydrologic performance. This should include not only estimating how much stormwater will be generated but also how that stormwater moves over the area, that is, the patterns and processes involved in its movement downslope.

Watersheds as partial area systems

Most watersheds function as *partial areas systems*; that is, only part of the basin actually contributes surface runoff to streamflow in response to a rainstorm (see Fig. 8.6). The total area contributing stormwater typically ranges from 20 to 40 percent, depending on the magnitude of the rainstorm and prestorm basin conditions such as soil moisture, ground frost, and snow cover. It is extremely important to define the distribution of noncontributing areas in order to figure out, first, how land uses should be assigned to the watershed and, second, where there are opportunities to take advantage of "built-in" BMP features such as wetlands and sandy soils.

Finding noncontributing areas

Finding *noncontributing areas* begins with basic mapping of slopes, soils, vegetation, water features, and land uses. Areas with permeable soils and substantial vegetation are good candidates, as are wetlands, especially closed or partially closed ones (Fig. 13.2). Once candidate areas have been defined, a field examination should be conducted to verify the absence or paucity of surface runoff. The main features to look for are small channels such as swales, gullies, and rills. Where there is no evidence that the ground is serviced by ephemeral channels, even microchannels such as rills, then it is doubtful that the site generates stormwater runoff in the conventional sense of the term, that is, as overland flow, or that surface runoff is so weak and/or infrequent that it is incapable of mounting flows large enough to etch the surface.

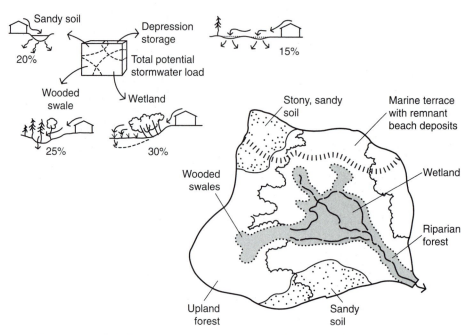

Fig. 13.2 Potential BMP opportunities in a watershed. To capitalize on these opportunities requires both sensitive site analysis and thoughtful land use planning and site design. The inset suggests how stormwater might be allocated in a diverse watershed.

In such areas runoff is moved in other ways, as interflow and/or groundwater after infiltration has taken place, or it is taken up in interception and depression storage or as soil moisture. This is often the case in forested watersheds where previous land uses have not erased the microtopography and topsoil and have not extended swales and built roads, ditches, field drains, and the like (see Figure 8.11). Because of tree throws, rotting stumps, burrowing animals, and other disturbances, the forest floor is typically rough and porous and unless it has been graded in some fashion at an earlier time, it is usually difficult for it to generate overland flow. Identifying such conditions is important in BMP planning because it is a clue that an opportunity exists to incorporate natural mitigation measures into the land use plan, thereby reducing or even eliminating the need to construct BMP facilities with or after development. And where there are ample opportunities for natural mitigation, it is important to take advantage of such sites by clustering development there.

Green infrastructure and source control

This approach is fundamental to the application of **green infrastructure** concepts because the green infrastructure approach to stormwater management relies on **source control** of runoff. To employ source controls, you either use existing means to hold and infiltrate stormwater or you must create them. The most successful source control measures, such as infiltration beds or grass-lined swales combine existing opportunities such as permeable soils with grading, soil treatments, and planting schemes as a part of site planning and design.

Meaning of drainage density

Another attribute of watersheds that is important in making assessments for BMP planning at the watershed scale is drainage density. Low-drainage densities indicate that precipitation is being taken up and stored, being returned to the atmosphere via evapotranspiration, and/or is running off by means other than surface flow, that is, by interflow and groundwater. High-drainage densities, on the other hand, indicate that stormwater and channel flows are abundant and that few, if any, sites in the watershed are not serviced by a stormwater channel, either natural or man-made (see Figure 8.10). Where a watershed is laced with stormdrains, the opportunities to use

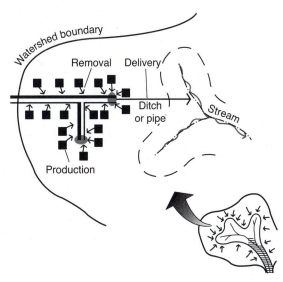

Fig. 13.3   The site stormwater system consists of three components: production, removal, and delivery. This diagram illustrates the arrangement of these three components in a conventional stormwater system.

natural opportunities for BMP planning are greatly diminished, especially at the source control level, because the stormdrain system taps into most sites and in many instances overrides the watershed's built-in opportunities. In some cases it is possible to detach a site from this network of stormdrains, but short of ripping out the stormdrain system or abandoning it, it is advisable under such conditions to look to the site itself as the place to apply BMPs

13.3   THE SITE STORMWATER SYSTEM

Within the watershed, each development site has its own stormwater system. This system consists of three components: (1) the on-site production of stormwater; (2) the removal of stormwater from the production site; and (3) the delivery of stormwater and its contaminant load to a receiving waterbody. Production is defined as surface water (and its contaminants) generated from cleared and developed ground that is available for stormwater runoff if it is released from the site. Removal is the means by which stormwater is released from the site and discharged into a delivery system, and delivery is the means by which stormwater is conducted to a receiving water body (Fig. 13.3).

System overview   The production system consists of all the facilities and land use practices added to a site that result in increased amounts of stormwater and contaminants. Production begins with land clearing and commonly includes soil compaction, construction of impervious cover, lawn fertilization, and garbage burning. The removal system is usually made up of gutters, downspouts, yard drains, and field tiles, whereas the delivery system is usually made of curbs, gutters, ditches, and stormsewers. Once in place, the removal and delivery systems function as bypass devices that override the site's natural hydrological system, making it "more efficient" as a flow system. In conventional land use development, stormwater is produced quickly, removed from the site quickly, dumped quickly into a delivery system, and quickly discharged into nearby streams.

The choice of BMPs for the site stormwater system is based on how the development and associated drainage system functions to produce, convey, and deliver

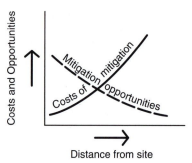

Fig. 13.4   The general trends in mitigation (BMP) opportunities and costs with distance from the site.

stormwater from the site to the receiving waterbody. Three categories of BMPs are defined, corresponding to each phase of this three-part system: planning, design, and engineering. Various combinations of BMPs representing these three levels of control may be selected for different settings and problems. Once stormwater passes through the system and is delivered to the receiving waterbody, BMP opportunities are greatly diminished or lost altogether. As a rule of thumb, we expect BMP opportunities not only to decline with distance from the site, but to rise in cost (Fig. 13.4)

13.4 PRODUCTION BMPS

Production BMPs are used to decrease the on-site production of stormwater runoff and contaminants. This class of BMPs is typically ignored in conventional development in preference to "catch-all" BMPs, such as stormdrains and detention basins, designed to capture large amounts of stormwater runoff from one large site or many small ones. Production BMPs utilize planning strategies, planning and development policies, and educational programs to help reduce runoff and pollution production on individual sites. This includes not only siting facilities in appropriate locations, but evaluating density and assessing site "carrying capacity," as well as designing facility configurations that optimize site conditions and advance sustainable design principles at both the watershed and site scales (Table 13.1).

Classes of production BMPs   In general, production BMPs fall under three major headings: planning policy; architectural and landscape design; and site management practices. Planning policies that affect the production of stormwater and related pollutant loading include zoning regulations addressing land use type and density, lot coverage restrictions, wetland protection regulations, and slope limitations on street, road, and building placement. Architectural and landscape design BMPs include limits on roof area to floor area ratios; limits on site coverage by outbuildings, walks, and drives; rules on foundation and basement design; and guidelines on grading and the use of plant and soil materials in landscaping. Site management practices address BMPs related, among other things, to waste disposal, fertilizer and pesticide application, irrigation and water recycling, and the number and type of domestic animals housed on a site.

Implementation   Production BMPs can be implemented in various ways. At the planning policy level they can be implemented through community planning (e.g., designating land use types), through zoning bylaws (e.g., restricting lot coverage and density), during subdivision review (e.g., directing road location and size), through development permits (e.g., designating environmentally sensitive areas as development permit areas), and through the building permit process (e.g., requiring onsite stormwater measures). The same implementation recommendations hold for architectural and land-

Table 13.1 Planning Options and Production BMP Measures at the Watershed and Site Scales

Planning options that affect stormwater and contaminant production	Production							
	Watershed Scale			Site Scale				
	Preservation of natural wetlands	Preservation of permeable soils	Preservation of natural depression storage areas	Density-site carrying capacity	Vegetation conservation and riparian buffers	Road configuration and design	Production-based policies (e.g., hazardous waste disposal)	Education Programs
Land cover	✓	✓	✓		✓	✓		✓
Impervious area				✓		✓		✓
Runoff storage	✓		✓					✓
Runoff treatment	✓	✓			✓			✓
Runoff infiltration	✓	✓			✓			✓
Pollution production				✓		✓	✓	✓

scape design BMPs. At the site management level, however, implementation demands an approach that combines public policy enforcement (e.g., garbage disposal), incentives (e.g., tax abatement for stormwater recycling), and public information and education (e.g., fertilizer applications and car washing).

Production BMPs include the following:

Key measures

- Limits on impervious cover including decreased road widths, smaller driveways, and reduced roof areas.
- Site management guidelines that encourage retaining woodland and wetland areas.
- Road and neighborhood configurations designed to reduce automobile travel distances and encourage sustainable modes of transportation.
- Roof water storage and recycling through the use of cisterns coupled with garden and lawn applications.
- Backyard burning, pet feces cleanup, and household and industrial hazard waste disposal practices to reduce onsite pollution production.
- Education and information programs to encourage home and business owners to adopt more sustainable waste disposal and landscape maintenance practices.
- Incentive programs including property tax abatements for installing stormwater reduction measures such as porous pavers, dry wells, and woodlands.

Retrofitting the developed site

Production BMPs can be used for both undeveloped and developed areas, but we usually have less flexibility and fewer options in developed areas. Not only are buildings already in place but stormwater facilities often also exist. Nevertheless, mitigation opportunities may be available in neighborhood parks, greenbelts, street

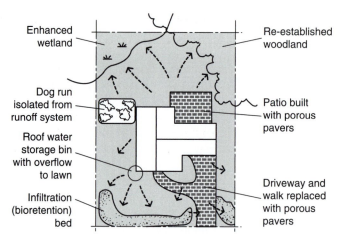

Figure 13.5 Measures that can be used to retrofit a residential parcel to reduce onsite production of stormwater and contaminants.

parkways, and remnant wetlands. These areas need to be evaluated to determine whether they have stormwater management potential and can be integrated into a larger management plan.

Developed sites can also be retrofitted in various ways to reduce stormwater and contaminant production. A simple example is replacing worn out driveways and side walks with porous pavers. Another is capturing part of roof drainage in cisterns with overflow quantities going to dry wells or infiltration beds on the yard perimeter (Fig. 13.5). The resulting reduction in stormwater production will not only lower environmental impacts but reduce wear and tear on stormdrains, thereby increasing their life cycle periods.

Plans and designs must maximize opportunities to reduce stormwater production and maximize onsite infiltration. To test the efficacy of a proposed plan, runoff simulation models can be applied to determine whether it meets predevelopment performance standards. In certain situations, management programs that utilize only production-level BMPs will not be sufficient to effectively treat and manage stormwater production from a developed site.

13.5 SITE REMOVAL (RELEASE) BMPS

The next level of BMP application is intended to prevent excess water from leaving the site and from contributing to stormwater surcharging of streams. Site removal, or release, BMPs are designed to disconnect the site as a source of stormwater from the larger watershed drainage system. BMPs such as lawn and woodland infiltration, porous pavers, and depression storage are important lines of defense at this level because they are designed to capture and treat excess stormwater before it leaves the site. Site release BMP measures are designed mainly to handle low- to moderate-magnitude runoff events because for many regions these events produce the most stormwater and contaminants in total. On the other hand, stormwater runoff from larger events can usually be handled by the delivery BMPs, which are taken up below.

Isolating and disconnecting the site

In conventional designs, stormwater runoff produced onsite is quickly released from the site and dumped into stormdrain systems. Removal BMPs, however, intercept this runoff and direct it to spots such as infiltration beds that are designed to detain and absorb it. The objective of these BMPs is to isolate the site as a source of stormwater. Treatment and infiltration of runoff from rainfall events producing less than an inch of water, for example, will go a long way toward improving the hydrological performance of the entire watershed. Slowing the release of runoff and disconnecting the site from the watershed can be achieved by: (1) increasing travel time through the use of longer and slower routing schemes (for example, see Fig. 9.11c); (2) increasing surface roughness to slow down overland flow and delay the formation and advance of channel flows; (3) disconnecting impervious surfaces from drainage systems to reduce flow continuity and stormwater drainage area; and (4) utilizing grading and planting designs to diffuse runoff and promote infiltration (for example see Figure 8.12 b). The key principle of the site removal BMPs is to treat stormwater runoff at the site scale, that is, at its source, and not at the neighborhood, community, or watershed scale. Studies have found that site-level treatment is more effective and considerably cheaper than broadly integrative stormwater BMPS.

Site removal BMPs include:

Key measures
- Disconnection of downspouts and yard drains from the stormwater system.
- Bioretention facilities such as stormwater gardens and depression storage areas that take up stormwater runoff and treat pollutants.
- Infiltration facilities such as permeable swales, trenches, and dry wells that enhance soil intake of surface water.
- Diversion channels that direct stormwater away from stormwater delivery systems.
- Depression storage cells and other grading features that slow the movement of runoff.
- Infiltration beds, trenches, and swales (grass and cobble lined).
- Special planting schemes such as "stormwater gardens" designed to enhance infiltration, to reduce soil erosion, and to intercept and treat stormwater pollutants.

13.6 DELIVERY BMPS

Rationale
Delivery BMPs are the final set of practices that can be used to manage stormwater runoff and pollution before it enters the receiving waterbody. Although site-scale BMPs can manage the majority of runoff events, they may not be able to accommodate larger, infrequent ones. Delivery BMPs are designed to reduce the risk of flooding and property damage associated with larger rainfall events, especially those that yield heavy downpours over extended periods. Delivery BMPs, like the site export BMPs, should slow stormwater delivery rates by increasing travel time and channel roughness. These BMPs are usually within the engineer's domain and favor structural measures. Engineering BMPs are related to roads and drainage structures, including stormdrains and detention ponds, and are decidedly the most widely used BMPs in stormwater management (Figure 13.6).

Delivery BMPs include:

Key measures
- Open (swale) drains without curbs and gutters.
- Diversion channels that direct stormwater away from valued habitat and water features.
- Rerouted flow patterns that lengthen travel times and slow delivery rates.
- Storage basins, both retention and detention classes.
- Very low gradient delivery systems such as wide swales with roughened beds.
- Constructed wetlands.
- Infiltration trenches.

(*a*)

(*b*)

(*c*)

Fig. 13.6 Not all open swales are appropriate BMPs in delivery systems: (*a*) is an example of one that probably performs poorer than a pipe; (*b*) is grass-lined and much better; and (*c*) is better yet.

13.7 STEPS TO BMP PLANNING

The recommended steps in formulating a BMP plan according to the methodology we have described are as follows.

Toward a BMP plan

1. Delineate the watershed and its subbasins and locate existing and proposed development within this area.

2. Define the watershed's drainage system, showing the network of channels and other drainage features such as lakes, wetlands, floodplains, and seepage zones, and delineate the flow system. Highlight key habitats, especially fish-bearing streams and connecting features.

3. Build an inventory of biophysical features that have stormwater management potential for the watershed as a whole and/or around the proposal development area, and highlight critical environmental features, such as streams.

4. Formulate a watershed plan and/or a site design model that:
   a. defines buildable land units;
   b. relates land units to watershed BMP opportunities;
   c. minimizes impervious cover;
   d. segregates impervious cover from the existing drainage system;
   e. provides for stormwater retention/infiltration onsite; and
   f. includes covenants that minimize onsite stormwater (and contaminant) production and release from the site.

5. Calculate predevelopment and postdevelopment discharges for a modest design storm and allocate excess stormwater to existing biophysical features on and near the site, taking care not to exceed their loading capacities.

For any stormwater that remains after biophysical uptake, reexamine the site design model (Step 4), and modify it to accommodate excess stormwater. If a sizable surplus still exists, consider delivery-based BMPs.

13.8 SUMMARY AND RECOMMENDATIONS

BMP planning requires building multiple lines of defense in the control of stormwater (Fig. 13.7). It begins with BMPs at the watershed scale, taking into consideration that many BMP opportunities already exist in the watershed. The first step, therefore, is to survey the watershed, recording its various hydrologic systems and essential biophysical features. Land use planning and design must not only honor these watershed attributes but integrate functionally with them, seeking balance with predevelopment performance levels. In other words, land use must be an integral, working part of the watershed, not an alien appendage superimposed upon it.

Focus on source controls

BMP efforts should focus on source control because the site offers the greatest number and the least expensive opportunities for stormwater management (Table 13.2). In the production–removal–delivery scheme outlined here, the emphasis should be on production and removal, not on delivery. Offsite BMPs such as those built into delivery systems should be seen as backup measures to be used selectively rather than habitually. The argument that large detention basins are everywhere needed to curtail large flows is untenable for many small watersheds in the face of evidence that (1) in regions such as the Pacific Northwest small streams produce most runoff; (2) many, perhaps most, small watersheds function as partial area systems, producing much less stormwater than rainfall produces; (3) stream systems should not be deprived of large flows because these flows are essential to the movement of

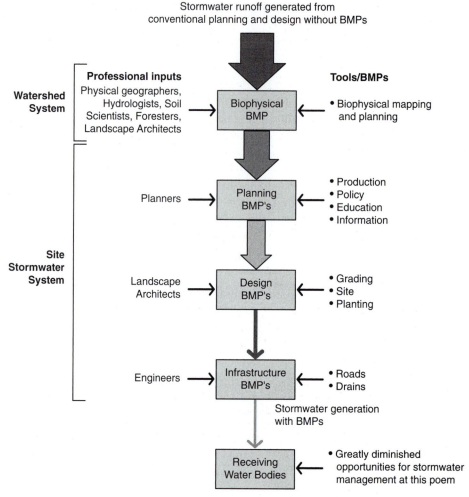

Fig. 13.7   Best management practices planning and design process utilizes a variety of planning, landscape design, and engineering tools, with different professions in the lead at each level.

sediment, shaping of channels, and maintenance of habitats; and (4) with proper planning large, infrequent flows can be handled as risk management events.

Although land use should be integrated into the watershed, it is also important to avoid transforming watersheds into systems that are more hydrologically efficient than they were before development. Indeed, advances in the efficiency of stormwater drainage have brought us to the depressed state of affairs we face today in watershed management. Therefore, as a general rule it is advisable to think in terms of disconnecting the site stormwater system from the system of natural channels and minimizing effective impervious cover rather than vice versa. This means forcing stormwater generated from land use to find slower, more natural means of getting to a channel. All this, of course, points to green infrastructure as the preferred approach to stormwater management, an approach that endeavors to integrate low-impact architecture, landscape design, and engineering into a balanced whole. In this way we may be able to build watershed arrangements that include people, their facilities, and natural features functioning together in a sustainable manner that is not measurably different from nature operating alone.

Guarding against efficiency

Finally, BMP programs should not be narrowly focused and limited to the perspectives and services of one professional discipline, but rather should include a

Table 13.2 Planning, design, engineering BMPs for undeveloped and developed sies.

BMP	Undeveloped Site	Developed Site
Planning		
Location and Siting	●	○
Housing Density/Impervious Area	●	○
Urban Design and Layout	●	○
Drainage Features	●	◓
Infiltrating Soils	●	◓
Vegetation Conservation	●	◓
Fertilizer/Pesticide Application	●	●
Household Chemical Disposal	●	●
Pet Feces Cleanup	●	●
Backyard Burning	●	●
Education	●	●
Design		
Depression Storage/Grading	●	◓
Filter/Buffers	●	◓
Bioretention Gardens	●	◓
Vegetated Swales	●	◓
Planting Design	●	●
Infiltration Zones/Dry Wells	●	●
Engineering		
Ditch and Stormdrain Diversions	●	●
Constructed Wetlands	●	●
Infiltration Trenches	●	●
Retention and Detention Ponds	●	●
Low Grade Delivery Systems	●	●

● Most appropriate   ◓ May be used if opportunities exist
○ Not Applicable

wide range of measures and multiple professions. At a minimum, the balanced BMP program should include planning, landscape design, and engineering components, and draw on the services of scientists, landscape architects, planners, and engineers, as Figure 13.7 suggests.

13.9 SELECTED REFERENCES FOR FURTHER READING

Department of Environmental Resources, *Low-Impact Development: An Integrated Design Approach*. Prince George's County, MD, 1999.

Marsh, William M., "Toward a Management Plan for Lazo Watershed and Queen's Ditch," Regional District of Comox-Strathcona, Courtenay, British Columbia, 2002.

Marsh, William M., and Hill-Rowley, Richard. "Water Quality, Stormwater Management, and Development Planning on the Urban Fringe." *Journal of Urban and Contemporary Law*. 35, 1989, pp. 3–36.

Murdoch, Scott P. *The End of the Pipe: Integrated Stormwater Management and Urban Design in the Queen's Ditch*. Thesis: Master of Landscape Architecture. University of British Columbia, Vancouver, 2001.

Richman, Tom, and Associates. *Start of the Source: Residential Site Planning and Design Guidance Manual for Stormwater Quality Protection*. Bay Area Stormwater Management Agencies Association. Palo Alto, CA 1997.

Sabourin, J. F. and Associates. *Evalution of Roadside Ditches and Other Related Stormwater Management Practices*. Toronto: Metropolitan Toronto and Region Conservation Authority, 1997.

U.S. Environmental Protection Agency. *Urbanization and Streams: Studies of Hydrologic Impacts*. Office of Water, Washington, DC, 1997.

14

STREAMS, CHANNEL FORMS, AND THE RIPARIAN LANDSCAPE

14.1 INTRODUCTION

Stream channels are among the truly spectacular environments on the planet—wonderfully diverse, aesthetically pleasing, scientifically challenging, economically valuable, and above all, ecologically rich. Yet for all these qualities, streams are one of our most maligned, degraded, and mismanaged natural phenomena. In North American and Eurasia, only 20 percent of the major rivers remain without structural barriers such as dams and locks. The great channel of the Missouri between Kansas and Montana, which Lewis and Clark traversed in 1804 and 1805, today is little more than a long string of dams and reservoirs.

The channels of 70 percent of the streams in the United States have been altered by land use activity. In agricultural and urban areas, countless streams are routinely deepened, widened, straightened, piped, and diverted; in other words, effectively destroyed as natural entities and converted into conduits to carry water and sediment. Although no recent data are available, it appears that more than 300,000 miles of streams in the United States and Canada have been treated in this manner. As land use pushes farther into watersheds and marginal and submarginal lands are taken up by settlements, roads, and farms, stream systems will suffer further manipulation as a part of drainage schemes, road construction, irrigation projects, flood control programs, and navigation projects. At the other end of the spectrum are efforts to recover stream channels and restore their potential as habitats, sources of clean water, and scenic environments. Driven by a variety of concerns including erosion and sediment control, aquatic habitat conservation, and riparian corridor ecology and restoration, these activities all require analytical understanding of the processes, forms, and features of stream channels.

14.2 HYDRAULIC BEHAVIOR OF STREAMS

Stream velocity

In order to understand how streams erode, deposit material, and shape channels, let us first examine a few principles about running water beginning with velocity. **Flow velocity** in stream channels is apparently related to three factors: the slope of the channel, the depth of the water, and the roughness of the channel. These variables form the basis of the *Manning equation*, an empirical formula commonly used to calculate the mean velocity of a stream:

$$v = 1.49 \frac{R^{2/3}s^{1/2}}{n}$$

where:

v = velocity in feet or meters/sec
R = hydraulic radius, which represents depth and is equal to the wetted perimeter (P) of the channel divided by cross-sectional area (A) of the channel
s = slope or gradient of the channel
n = roughness coefficient:

n value	Channel Material
0.020–0.025	Soil material (e.g., loam)
0.040–0.050	Gravel with cobbles and boulders
0.050–0.070	Cobbles and boulders
0.100–0.150	Brush, stumps, boulders

Controls on velocity

Increases in slope or depth (or hydraulic radius) cause increases in velocity, whereas an increase in the channel roughness causes a decrease in velocity. The effect of an increase in depth is slightly greater than that of an increase in slope of equal proportion. Hence the fastest velocities in stream systems may be found not in the mountainous headwaters, but farther downstream, where the depth and discharge are greater. For example, average flow velocity near the mouth of the Amazon, where slope is only a few inches per mile and the depth is 150 to 200 feet, is about 8 feet per second, whereas average velocity in the Grand Canyon River of the Yellowstone National Park, where slope is 200 feet per mile and the depth is 3 to 6 feet, is only about 3 feet per second.

Velocity distribution   Within a stream channel the velocity is lowest near the bottom and the sides of the channel and highest near the top in the middle. The highest velocity is generally slightly below the surface because of friction between the water and the overlying air. At channel bends, the faster-moving water responds more to centrifugal force than slower water does and slides toward the outside of the bend (Fig. 14.1). In addition,

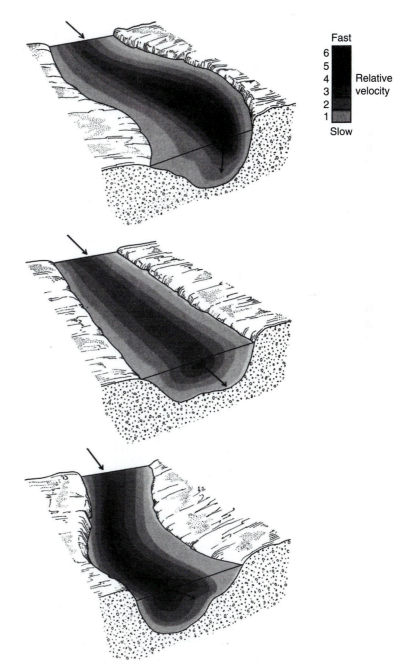

Fig. 14.1   The distribution of streamflow velocity in three different channel forms.

this may also produce superelevation of the stream, in which the water surface near the outside of the bend is higher than the water surface near the inside. Where channels meander, the belt of fast-moving water is thrown from one side of the channel to the other in successive bends much as a bobsled team does in negotiating its run.

Turbulent motion

Streamflow is always characterized by *turbulent motion*, that is, by the rolling and swirling action of eddies. This mixing motion is a primary source of flow resistance between the slower-moving water molecules on the streambed and the faster-moving water above the bed. Although some streams, especially those with smooth water surfaces, appear to be dominated by laminar flow, none in fact is. Laminar flow occurs only in a very thin sublayer immediately at the streambed where velocity is effectively zero. Above this sublayer, the intensity of turbulence in streams increases with flow velocity.

Any inferences we make about stream behavior based on average flow velocities are limited by an important observation: Flowing water very quickly establishes an equilibrium with the prevailing frictional environment of the channel. Unless there is a change in channel shape or roughness, or unless the discharge is increased or decreased, the water does not accelerate or decelerate. Because the downhill force of the moving water mass does not result in any acceleration of the stream, there must be an equal and opposite force holding the water back. This opposing force is

Bed shear stress

the **bed shear stress**, which is the amount of energy loss over the stream bed. Bed shear stress can generally be given as a quantity equal to water density times gravitational acceleration times water depth times channel slope, or:

$$\text{Bed shear stress} = \rho\, g\, D\, S.$$

The density (ρ) of water varies slightly with temperature, but for our purposes we can consider it constant, at 1000 kg/m³. Similarly, gravitational acceleration (g) varies slightly with latitude and altitude, but we can consider it constant, at 9.8 m/sec². Therefore, the only variables are depth (D) and slope (S), and the mean bed shear stress is proportional to the product of depth and slope. The product of bed shear

Energy loss

stress and discharge is the *rate* at which the stream is losing energy. Variations in energy, in turn, can be related to changes in either discharge or slope and water depth. These three variables are at the root of most problems in managing open channel flow and stabilizing and restoring channel environments. In the end most of a stream's energy is converted to heat as streamflow converts kinetic energy into thermal energy. This explains why streams do not freeze out in winter. But a small amount of energy is also devoted to the work of dislodging and moving sediment.

14.3 STREAM EROSION AND SEDIMENT TRANSPORT

Channel scouring

Channel erosion occurs mainly by **scouring**. This process involves heavy particles bumping and skidding along the bottom, loosening and freeing channel material. Scouring is highly effective in eroding unconsolidated materials such as the various soil and sediment materials that make up most stream channels. Surprisingly, it is also effective in eroding bedrock as evidenced by channel potholes, which are pits etched into the bedrock on the streambed by stones swished around and around by currents. Although streams can be important in the erosion of rock, especially in their upper reaches, most of their work involves the erosion or re-erosion of transported material already in the valley, especially deposits on the valley floor, or floodplain.

In addition to the sediments eroded from these deposits, much of the sediment transported by streams is brought to the channel by tributary streams and by hillslope processes on the valley margins. The total mass of material entrained and transported

Sediment loads　　by a stream is called **sediment load**. There are three types of loads carried by streams: *bed load, suspended load*, and *dissolved load*. Bed load consists of larger particles (sand, pebbles, cobbles, and boulders) that roll and bounce along in virtually continuous contact with the streambed. **Suspended load** consists of small-size particles, principally clay and silt, that are held aloft in the stream by turbulent flow. The amount of the suspended load carried by a stream is determined mainly by sediment supply rather than the force of the stream's flow, and most streams are capable of carrying a large suspended load if the sediment is available. On the other hand, the size and amount of bed load transport are determined by the level of bed shear stress.

Bed-load transport　　The mechanism of **bed-load** movement involves two forces: *drag*, a force along the bedstream, and *lift*, an upward force. The lift force is of primary importance in initiating motion; laboratory flume experiments have demonstrated that a lift force of about 70 percent of the submerged weight of a particle is sufficient to pivot it upward on its axis, whereupon the drag force can push it downstream. For any size class of particles there is a *threshold of force*, or critical level of bed shear stress, below which no movement takes place. Above this threshold value, the rate of sediment movement increases as some power of bed shear stress. In simplified form, this concept can be expressed in terms of threshold velocities required to dislodge and move particles of different sizes (Table 14.1). These data have practical value in channel management because they tell us the size particles that need to be placed on stream banks, in ditches, and in gullies to stabilize them against erosion under different flow velocities.

Particle erodibility　　Sand particles are the most erodible sediment in stream channels. Clays are less erodible because of particle cohesion (see Fig. 12.8). A relatively large bed shear stress is required to overcome this cohesive force, and the smaller the clay particle, the greater the force required, especially if the material is compacted. Clay particles are so small, however, that once dislodged, they will remain in suspension even under the slowest of stream flows. Accordingly, clay sediments are extremely mobile in stream systems, and once entrained, tend to be moved great distances (see Fig. 12.2).

Dissolved load　　**Dissolved load** is that material carried in solution. Ions of minerals produced in weathering are released into streams mainly through groundwater inflow. Total ionic concentrations in stream water are generally on the order of 200 to 300 milligrams per liter, but in areas with low relief and high rates of soil leaching, as in parts of the American East and South, concentrations may reach several thousand mg per liter. In dry areas dissolved load is often much lower, typically less than 10 percent of the total stream load. This is explained by the low rates of chemical weathering and soil leaching in arid climate watersheds.

Table 14.1　Relative Resistance of Channel Materials to Stream Erosion

Material	Relative Resistance[a]
Fine sand	1.0
Sandy loam	1.13
Grass-covered sandy loam	1.7–4.0
Fine gravel	1.7
Stiff clay	2.5
Grass-covered fine gravel	2.3–5.3
Coarse gravel (pebbles)	2.7
Cobbles	3.3
Shale	4.0

[a]Resistance relative to fine sand, the least resistant channel material.

14.4 CHANNEL DYNAMICS AND PATTERNS

Flow-channel interplay

In trying to interpret the features we see in and around stream channels, we must keep two facts in clear perspective. *First*, the types and magnitudes of stream processes that can be observed on most days may have little or nothing to do with the origins of most channel forms and valley features. Rather, it is the processes associated with a limited number of flows of particular magnitudes or selected combinations of flows that determine many features. *Second*, through erosion and deposition the stream is continually adjusting its channel and in turn its slope, depth, and bed shear stress. If, for example, slope and bed shear stress are low at one point in the stream profile, deposition will take place there. But this in turn will increase the slope and thus the bed shear stress, eventually leading to restoration of equilibrium, whereby the sediment brought to that point will continue to be transported downstream. In fact, most streams are in a constant state of flux, or dynamic equilibrium, as they trend toward balance in response to fluctuations in discharge and changes in channel forms, slope, and sediment load.

Riffle-pool sequence

One feature that illustrates the complexity of the processes involved in channel formation is the **riffle-pool sequence**. During low flow in some streams, the water is distributed in a sequence of quiescent segments (pools) linked together by rapid segments (riffles). Further investigation reveals that the pools contain fine sediment, whereas the riffles contain gravels and larger particles. Moreover, the spacing is surprisingly regular, generally from five to seven times the channel width from pool to pool (Fig. 14.2). As habitats, riffles with their shallow turbulent flows and pools with their deeper, less turbulent flows are distinctly different. Among other things, trout lay their eggs in the riffles where the newly hatched fish can escape through the grav-

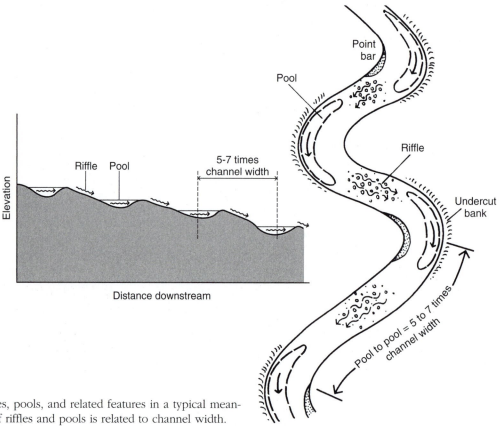

Fig. 14.2 The pattern of riffles, pools, and related features in a typical meandering channel. The spacing of riffles and pools is related to channel width.

els into the stream, but adult trout occupy the pools and feed where riffle water pours into pools.

Riffle-pool maintenance

How is the pool–riffle sequence maintained? The key to this question lies in the high flows rather than the low flows. During the low-flow period, we find, *first*, that depth is greatest over the pools, but the slope is so low there that the bed shear stress is lower than over the riffles. *Second*, we find the bed shear stress over the riffles is *not* great enough to dislodge the larger particles, but it is great enough to dislodge the fine particles and they are swept from the riffles and deposited in the pools. Taken to the extreme, the pools will ostensibly fill with sediment.

The role of big flows

If we return to the stream at high flow, we immediately notice that the pool–riffle sequence is drowned, and the distinction between pools and riffles is not apparent. Closer inspection reveals that the pools and riffles are still there, but the former difference in the slope of the water surface between pool and riffle is no longer present; in other words, the slope of the stream surface now is almost the same over both pool and riffles. But the water depth is still greater over the pools, therefore bed shear stresses are maximized over the pools. Hence, at high flow the pools are scoured and deepened and the larger particles are moved downstream to the next riffle area. Thus the modest flows we are apt to see during a visit to a stream appear to have little to do with the formation of the riffles and pools that produce the sequence of fast water and slow water segments that dominate stream channels.

Degradation and aggradation

Streams typically undergo substantial changes in channel geometry in response to changes in discharge and sediment supply. When discharge rises, both velocity and water depth increase, producing scouring of the streambed and increased sediment load (Fig. 14.3). Conversely, much of the channel bed fills in with sediment as the discharge falls. These channel changes, referred to as **degradation** and **aggradation**, respectively, are also associated with land use changes. Aggradation is initiated with early land clearing and cultivation and then peaks years later with massive sediment inputs brought on by suburban construction activity (Fig. 14.4). As urbanization continues and the land is secured under pavement and buildings, sediment loading declines while stormwater discharge rises, resulting in the degradation of stream channels. Such conditions are manifested not only by the absence of typical

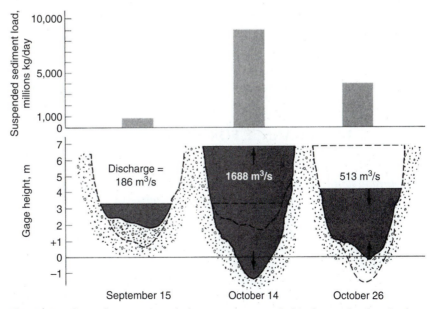

Fig. 14.3 Channel scouring and changes in suspended sediment load with changes in discharge; San Juan River near Bluff, Utah.

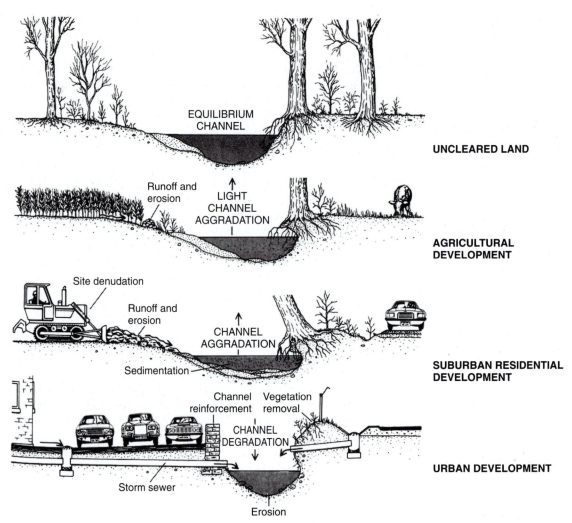

Fig. 14.4 Stream channel changes associated with land use: aggradation with agriculture and suburban development and degradation with urbanization.

amounts of channel sediments, but in severely eroded channels, which are both deeper and wider than before development.

14.5 CHANNEL FORMS, MEANDERS, AND RELATED PROCESSES

Single-thread channel   Two distinctive channel patterns occur in natural streams: braided and single-thread. In **single-thread** channels water is confined to one conduit. The channel form is relatively stable and characterized by one or two steep banks held in place by bedrock, soil materials, and plant roots. The main axis of the channel is marked by a zone of relatively deep, fast-moving water known as the *thalweg* (see Fig. 14.1). Most single-thread channels are sinuous or curving, within which the thalweg swings from middle channel to one bank and then the other. Many are also conditionally stable or metastable. This means that they owe their stability to a few key factors, such as the bank vegetation and the loss of which (due, for example, to excessive sediment loading or construction activity) can cause the channel form to break down and become braided.

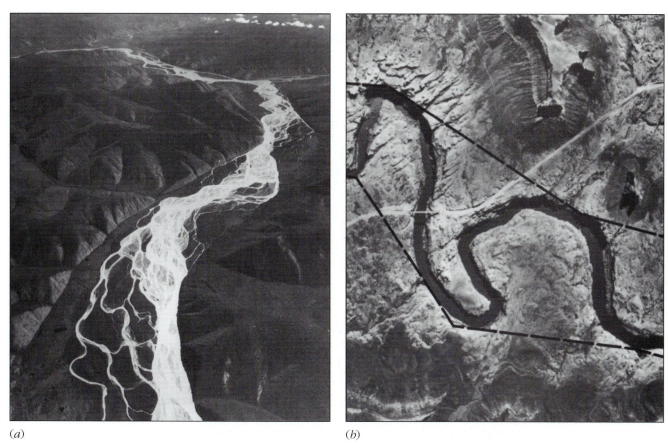

(a)   (b)

Fig. 14.5 (*a*) A braided stream channel in the Canadian Rockies. (*b*) A meandering, single-thread channel with the meander belt depicted (broken lines).

Braided channel

Braided channels are often found in geomorphically active environments, such as near the front of a glacier or at an unmanaged construction site, where massive amounts of sediment are being moved and stream banks are erodible and poorly defended by vegetation (Fig. 14.5*a*). Braided channels usually form in coarse, non-cohesive bed material under conditions of sharply fluctuating discharge. While the discharge is decreasing, coarse debris is deposited on the channel bed in the form of gravel bars. When the discharge declines further, the bar forms a barrier and splits the flow. Subsequent high discharges may erode the gravel bar, completely modifying the appearance of the channel. If a bar survives several seasons, plants may become established on it, adding to its stability and forming a small island.

Meander forms

Meandering channels occur in a wide variety of environments—in arid and humid regions and in all kinds of surface materials. They also occur in a wide variety of patterns, from slightly bending channels to those that are excessively sinuous called *tortuous* channels. Analysis of meanders usually begins with a description of the meander geometry, including meander belt width, wavelength, and sinuosity. The meander belt is the corridor containing the meander system as illustrated in Figure 14.5*b*. The width of the meander belt varies with the size of the stream and appears to be related to the river's mean annual discharge; the larger the discharge, the greater the width.

Sinuosity

A line drawn down the center of the meander belt is the *meander belt axis*. **Sinuosity** is the ratio of channel length along the curve of the meanders to the length of the meander belt axis. Geomorphologists class channels with sinuosities of less

Table 14.2   Stream Channel Types and Characteristics

Channel Type		Morphology	Sinuosity	Load Type	Behavior
	Meandering	Single-thread channel	>1.5	Suspended or mixed	Lateral erosion and point bar formation
	Straight	Single-thread channel	<1.5	Suspended, mixed, and/or bed load	Minor downcutting and widening
	Braided	Multiple, intertwined subchannels	<1.3	Bed load	Aggradation
	Anastomosing	Two or more single-thread channel with large islands	>2.0	Suspended load	Slow lateral migration

Source: Adapted from Selby, M. J., 1985.

than 1.5 as straight or relatively straight, and those greater than 1.5 as meandering. At sinuosities approaching 4.0 or more, a channel is tortuous. The most extreme channels, called *anastomosing*, are streams with multiple single tread channels all with meandering, even tortuous, patterns. Table 14.2 summarizes the forms and conditions of the various channel forms and patterns.

Management implications   Understanding the dynamics of meanders is essential to managing the channel environment, including facility planning and maintenance, habitat restoration, and sediment control. Where pools and riffles occur, we often find successive riffles favoring opposite sides of the stream. In a meander sequence, the pools are generally found at the bends, with the riffles between them. Hence there are two sets of riffle-pool sequences in a full meander (Fig. 14.2).

Detailed investigations of the distribution of bed shear stresses along meandering and straight reaches of the same rivers show that the meandering reaches have less variable energy distributions. Thus, as a river trends toward a meandering course, the distribution of energy along the channel grows less variable. This fact indicates that meandering flow represents a more stable state than does flow in a straight channel and suggests that ditches, straightened streams, and earthen stormdrains are fundamentally unstable channel forms that eventually will cut into banks, roads, field margins, and so on.

Meander development   Active meanders in most streams are continuously changing with lateral channel erosion and deposition. Erosion is concentrated on the outsides of bends and slightly downstream from them where it cuts back the bank, forming an *undercut bank*. Deposition occurs on the insides of bends, forming features called *points bars* (Fig 14.6*a*). Each year or so a new increment is added to the point bar while the river erodes away a comparable amount on the opposite undercut bank (Fig. 14.6*b*). In this way the river shifts laterally, gradually changing its location in the valley. But the river can also undergo sudden changes in location when it erodes new segments of channel and abandons old ones. This is especially commonplace where a meander forms a large loop and the river erodes toward itself from opposite sides of the loop, eventually breaching the meander. The old channel is abandoned because the new route is steeper and thus more efficient. The old channel forms a small lake, called an *oxbow*, but in time it fills with sediments and organic debris, becoming a wetland.

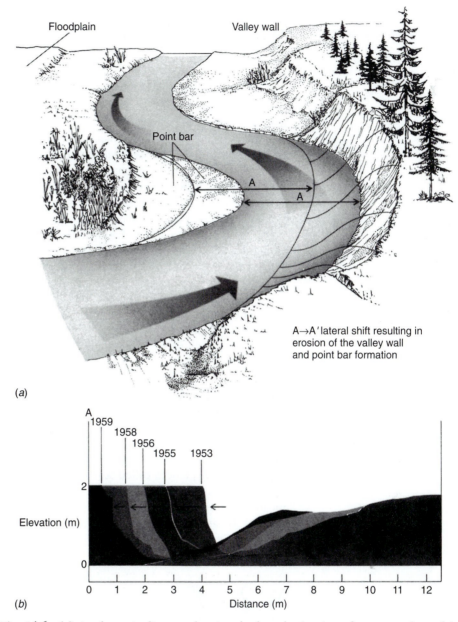

Fig. 14.6 (*a*) A schematic diagram showing the lateral migration of a stream channel in a meander bend. (*b*) The actual record of lateral movement of a small stream channel over a 6-year period.

On aerial photographs of floodplains, such features are easily identified on the basis of the swamp or marsh vegetation over them (see Fig. 3.3).

14.6 FLOODPLAIN FORMATION AND FEATURES

Floodplain formation

The flat low ground in a stream valley that contains the stream channel is the **floodplain**. Floodplains form mainly by channel processes related to the lateral shifting of streams on the valley floor. The process works as follows. When the river flows against the high ground at the edge of its valley, called the *valley wall*, it undercuts the wall, which fails and thereby retreats a short distance. At the same time, new ground is being formed on the opposite bank in the form of a point bar as described above regarding Figure 14.6*b*. The new ground forms at a low elevation, near that of the stream. In time the valley walls are cut back so far that a continuous ribbon of low ground is formed along the valley floor. This is the floodplain, and it is composed principally of channel deposits of sands, gravels and other materials that were part of point bars, riffles, pools, and other channel features (see Fig. 10.7 for examples of different valley and floodplain forms). When the river floods, the floodplain receives the overflow; hence the basis for the term. Although floods can alter the surface of the floodplain by eroding it and leaving deposits on it, they are clearly not the main cause of its formation, and in this regard the term *floodplain* is a little misleading.

Processes along the valley wall
In addition to lateral cutting by the channel in widening the valley and building the floodplain, other processes also contribute to valley formation and composition. These are various slope processes such as slumping, overland flow, and gullying that work on the valley walls, wearing them back and adding debris deposits to the floodplain. Especially significant are aprons of *talus* that develop along steep bedrock walls and *alluvial fans* and *colluvial deposits* where runoff processes are active. Colluvial deposits are mixes of alluvium and mass wasting debris (such as talus) and are very common on valley margins in mountain areas (Fig. 14.7).

Deposits and soils
The resultant composition of the floodplain is often very diverse indeed. Various types of channel deposits are laced together with deposits from floods, organic materials from forests and wetlands, and hillslope deposits from the valley walls. Floodplain soils are understandably diverse and require detailed investigation in planning and engineering projects. But it is also necessary to recognize that soil materials are not completely random in their distribution in the valley. At least three zones can be identified with different concentrations of floodplain, alluvial fan, and valley wall materials (Fig. 14.7).

Topographic features
Floodplains abound with distinctive topographic features. *Meander scars* and *oxbows* are the most salient features in many floodplains. As the term implies, meander scars are the imprints left by former channel locations. They often occur in regular series or sequences called *scrolls*, marking the progressive lateral shift of a channel (see Fig. 3.3). Oxbows are most abundant in the valleys of streams with tortuous meander patterns. Although meander breaching is a natural process in meandering streams, breaching is also performed by engineers to make streams straighter and more efficient as navigation channels.

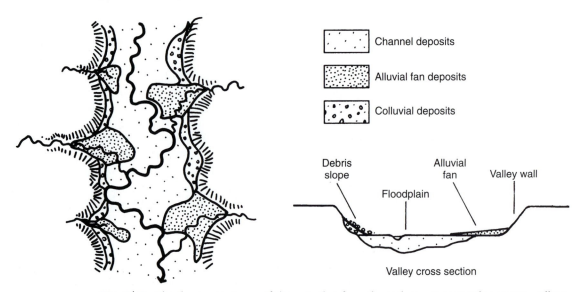

Fig. 14.7   The three main types of deposits that form the soil parent material in stream valleys.

Levees, backswamps, and terraces

Natural levees, scour channels, backswamps, and terraces are also common flood-plain features. *Levees* are mounds of sediment deposited along the river bank by flood-waters. They occur on the bank because this is where flow velocity declines sharply as the water leaves the channel, which causes it to drop part of its sediment load. In the low areas behind levees, water may pond for long periods, forming *backswamps*. *Scour channels* are shallow channels etched into the floodplain by floodwaters. They often form across the neck of a meander loop and carry flow only when floodwaters are available. *Terraces* are elevated parts of a floodplain that form when a river down-cuts and begins to establish a new floodplain elevation (See Fig. 10.6e).

Human influences

Changes in the elevation of stream channels do not necessarily require long (e.g., geologic) periods of time. Significant changes can take place in a matter of years and such changes are often great enough to pose serious management problems. Two examples of problems brought on by changes in sediment load are noteworthy. When a stream is dammed and its sediment load is captured in a reservoir, the chan-nel below the spillway generates a new sediment load by severely eroding its chan-nel. Under large spillway discharges, the channel degrades, thus rapidly reducing it to large boulders and exposed bedrock. At the opposite extreme are channels build-ing up from excessive sediment loading. The Yellow River of China suffers from such massive aggradation that the channel elevation is now above parts of the floodplain and its waters are held in only by constructed levees (see Fig. 10.7g). Heavy erosion in tributary basins upstream is the cause and, when the levees eventually break, the flooding and sediment burial of farmland and settlements will be disastrous.

14.7 RIPARIAN AND CHANNEL HABITATS

Stream channels and valleys are complex and dynamic environments. As habitats, they are both attractive and risky places. Their diverse forms, compositions, and scales offer opportunities for a huge number of organisms and their communities, while presenting constraints to others. Life is nurtured and sustained by the stream of water, energy, and nutrients and at the same time is threatened, rearranged, and even destroyed by the same flow system. Stream valleys are places of fresh starts for both humans and other organisms.

Adaptation

Survival in and around stream channels is improved through *adaptation*. Although humans have struggled with adaptive land use practice—preferring instead to manipulate the environment with dams, channels, levees, and the like—many organisms have shown great success in adjusting to the radical behavior of stream-flow and the related geomorphic changes. Floodplain and riparian vegetation is dis-tinctive, among other things, for both its resilience and sensitivity to variations in drainage processes and conditions on the valley floor.

Floodplain vegetation

Floodplains contain some of the most diverse vegetation in North America. In Canada and the United States more than 5000 species of wetland plants are found in floodplains. Most of these plants are adapted to a particular *hydrologic habitat* gov-erned by two factors: (1) the frequency and duration of flooding, and (2) ground-water conditions. For most floodplains the influence of flooding and groundwater can be approximated on the basis of elevation changes associated with landforms such as levees, oxbows, scour channels, and terraces. This is illustrated in Figure 14.8 for a typical floodplain in the American South where bald cypress, tupelo, and water elm occupy wetlands immediately along the channel and a different association led by willow oak and pin oak occupies slightly higher ground where flooding can be expected every one to two years. On the margin of the floodplain next to the valley wall where the flood frequency is greater than 10 years are the least water-tolerant species. A broadly similar pattern holds for stream valleys in the North and West, but the species are different and the biodiversity generally lower.

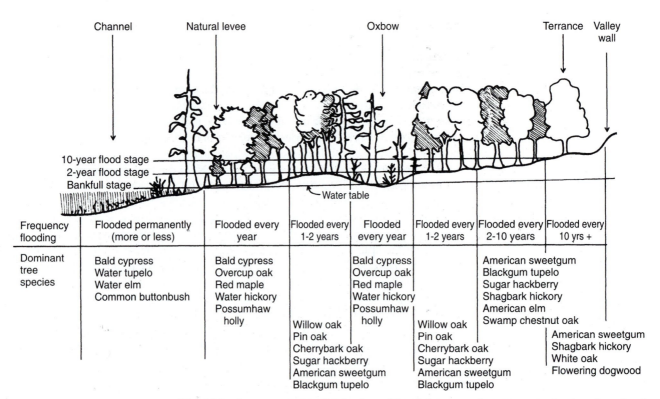

Fig. 14.8   A representative distribution of floodplain trees for streams in the American South and the related flood frequency of their habitat.

Point bar habitat   Another distinctive vegetation pattern is related to point bar deposits on migrating channel meanders. As point bars are formed on the insides of meanders, riparian plants such as cottonwood and poplar seedlings become established there. The channel subsequently shifts, creating a new point bar and new tree habitat. In time a belted pattern of vegetation is established within meander loops that is characterized by a sequence of tree stands trending from young to old along the meander axis.

Tolerance to floodflow   Another notable trait of riparian and floodplain vegetation is its **tolerance** to extreme levels of disturbance from powerful floodflows. Many shrub and tree species are able to regenerate after heavy damage from flooding, ice ramming, and sediment burial. Trees such as willow, poplar, cottonwood, and sycamore, for example, can generate new stems after severe disturbances that include complete uprooting and burial. The same species and others also have the capacity to raise their rooting level when new layers of sediment are added to the floodplain surface. Furthermore, they have the ability to develop dense root systems and spread by cloning, creating thickets of stems along banks, point bars, and levees. These species are true survivors and understandably are excellent ground stabilizers as well as important agents in shaping and sustaining riparian habitats.

Riparian forest   **Riparian forest** is also important to the aquatic habitat of the channel. Root masses and fallen trunks alter waterflow patterns that in turn diversify channel topography and improve fish habitat. Parenthetically, we should add that these are the very features removed from channels to decrease roughness and quicken flow as a part of traditional flood-control programs. For small streams, forest canopies shade the summer channel, thereby helping maintain cool water temperatures essential to trout habitat. And, riparian vegetation is the principal source of food for aquatic organisms. In forested watersheds it is estimated that well over 90 percent of the aquatic food supply originates in the riparian forest as insects and plant matter are dropped, blown, and washed into streams.

14.8 MANAGEMENT PROBLEMS AND GUIDELINES

Approaches to management

Stream environments are natural sinks in the landscape where the excess water, sediment, land use debris, sewage effluent, and many other things collect. At the same time, they are the environments on which we depend for water supply, navigation, farmland, habitat, recreation land, and open space. Fluvial environments are deeply imbedded in our history and mythology, so much so that streams and their valleys are central to our geographic ethos. Not surprisingly, we devote tremendous social, political, scientific, and economic energy learning how to use, manage, and reuse them. Much of the energy and ingenuity, however, has been misdirected over the years. Until relatively recently our approach to river management was overwhelmingly utilitarian, concentrating on structural manipulation of channels and floodplains in an effort to control flooding, water supplies, and navigational resources. After enormous capital expenditures, untold environmental damage, and disappointing results, especially in the area of flood control, the balance of things has now shifted somewhat to a more integrated approach that incorporates environmental, cultural, land use, and many other "soft" agendas.

Small stream abuse

But a great deal of damage has already been done that is not limited to large rivers and big projects, for local streams have also been abused. Here the responsibility and the jurisdictional circumstances are not so clear as they are with the major rivers. They often involve lists of private property owners as well as drain commissions, forestry operations, state and local highway departments, railroads, and many other parties operating in different ways at different times. To remediate the cumulative environmental impacts and protect the local stream resources we have remaining has become an issue in nearly every community.

Impacts on large streams

As for large streams, the list of damages is nearly overwhelming and the responsibility, of course, is much broader. Between 1940 and 1971, the U.S. Army Corps of Engineers assisted in projects that modified (e.g., channelized, diverted, dammed, or built levees) more than 10,000 miles of riparian environment. Between the 1940s and 1980, the U.S. Natural Resources Conservation Service was involved in alterations of nearly 11,000 miles of streams. In the Tennessee River Valley, nearly 5000 miles of streams are impounded by reservoirs or have flows regulated by structures such as locks. In the Sacramento River Valley of California where in 1850 there were 775,000 acres of riparian forest, today only 18,000 acres remain. This huge loss of forest—which is typical of many other California streams as well—was mainly the result of clearing for agriculture, irrigation projects, urban water supply, and flood control.

Currently, there are more than 75,000 dams higher than six feet in the United States and thousands more in Canada. In a given year, more than 60 percent of the entire streamflow of the United States can be stored in existing reservoirs. Dams on the Colorado River (see Fig. 2.16) can store an astounding four years of discharge when the stream is flowing at a typical rate. About 3 percent of the land area of the United States is covered by reservoirs. In terms of floodplain coverage, between 30 and 40 percent of the area classified as 100-year floodplain is under water in the United States.

Dams and reservoirs are very serious intrusions on riparian and related aquatic environments. A full discussion of their impacts is beyond our scope here, but several highlights are noteworthy:

Impacts related to dams and reservoirs

■ *Reservoir Siltation:* Dams interrupt the normal transport of sediment, trapping huge loads in their reservoirs. Eventually all or part of the reservoir is filled and its utility for water supply, flood control, or recreation is greatly reduced or lost. More than

3000 such dams in the United States have been retired (abandoned), but the original channels and valley floors are lost under massive layers of sediment, and in the main, are beyond restoration to a condition approximating their original states.

■ *Channel Erosion:* To make up for the sediment captured in reservoirs, especially bed load sediment, streams scour their channels downstream of dams. Not only is the channel deepened, but the streambed is often reduced to a belt of boulders. After construction of the Glen Canyon Dam on the Colorado, for example, the river cut its bed down by 15 feet, destroying much of the benthic (bottom) habitat for some distance downstream (Fig. 14.9*a*).

■ *Reduced Flow Variability:* One of the main objectives of dams is to regulate streamflow by reducing its natural variability (Fig. 14.9*b*). By decreasing the magnitudes and frequencies of peak discharges, particularly floodflows, one of the principal forces acting on the aquatic and riparian habitat is altered. In addition to the change in water supply and distribution, an important source of disturbance is reduced, thereby allowing less tolerant plant and animal species to invade aquatic and floodplain ecosystems.

■ *Habitat Fragmentation:* Dams and reservoirs obviously break up habitat corridors and thereby reduce the diversity, productivity, and various functions of riparian

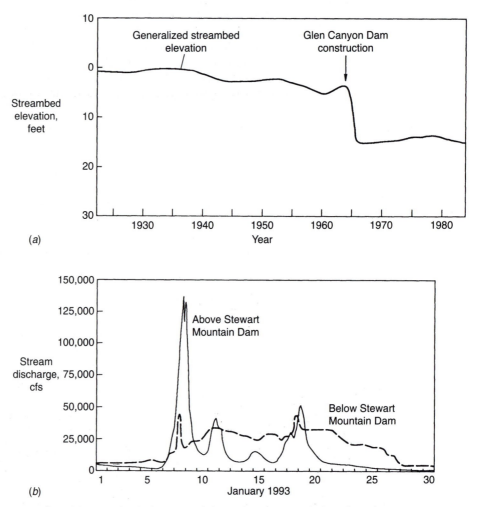

Fig. 14.9 (*a*) Streambed elevation of the Colorado River below the Glen Canyon Dam site before and after dam construction. (*b*) An example of the influence of dam construction on flow variability, Salt River, Arizona.

and aquatic ecosystems. The most glaring examples are provided by migratory fish such as salmon. On the Columbia River, between 5 and 14 percent of adult salmon are killed at each of the eight dam sites on their way upstream. The survival rate of young salmon migrating downstream through the reservoirs is even lower. Chinook salmon are now listed as a threatened species on the Columbia River.

Trends in stream management

The construction of dams in the United States and other developed nations has declined sharply in the past two decades (Fig. 14.10). Only a few potential sites for large dam projects are now under consideration in the United States and Canada. In addition to the several thousand dams in the United States already retired, the next 50 years will see the abandonment of thousands more. These environments, and related ones damaged by levee construction, channelization, and similar activities, will have to be repaired and reincorporated into the larger fluvial/riparian environment.

The stream restoration movement

A new wave of stream and river restoration programs has emerged with broadly based support from communities, environmental organizations, businesses, educational institutions, and state/provincial and federal agencies (Fig. 14.11).Among these are thousands of local nongovernmental organizations of watershed stewards and streamkeepers. Their programs call for a variety of efforts including repairing damaged habitat, planning open space systems, and managing riparian ecosystems. In one way or another, all these activities are rooted in hydrology, geomorphology, and ecology. Studies in these fields have taught us a lot about the complexities and dynamics of stream corridors. To restore and manage them effectively takes more than an understanding of the environment based on traditional measures of site planning, namely, slope, soil, drainage features, topography, vegetation, and habitat. Planners, designers, and managers must embrace certain governing principles based on a process-system perspective, and several are offered here for your consideration:

■ *Streams are parts of flow networks* that gain their water supply from a watershed, concentrate this water in channels, and direct it downstream. It is a one-way flow system; therefore, we must always be cognizant of location in the network and its meaning in terms of upstream/downstream land use relations. In managing stream systems, the first priority is to protect the headwater areas and the second is to protect the watershed as a whole against hydrologically abusive land uses.

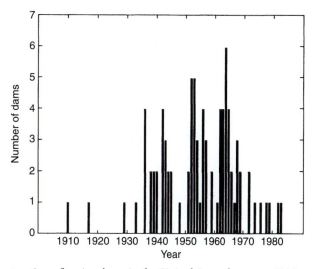

Fig. 14.10 Construction of major dams in the United States between 1910 and 1983. A major dam has a reservoir capacity one million acre-feet or more. In developing countries, such as China and Brazil, new dam construction shows no signs of abatement.

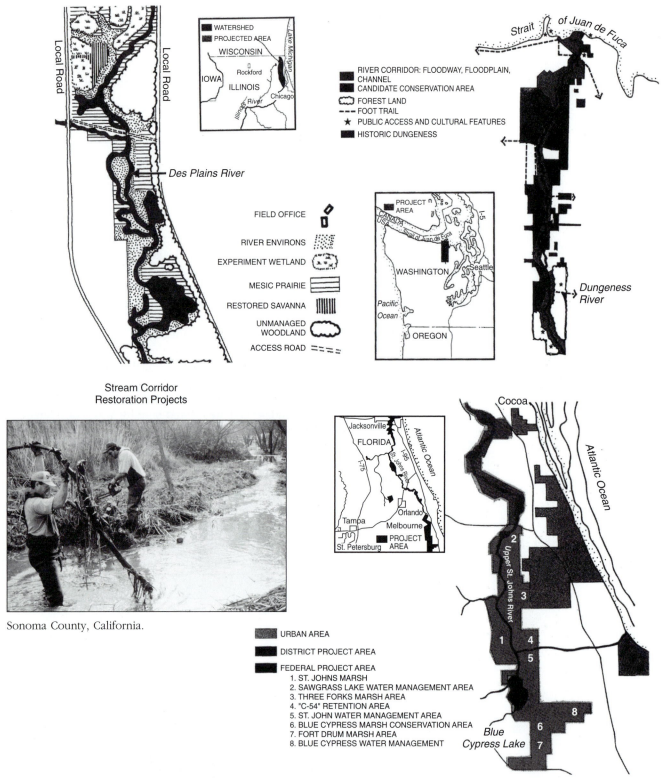

Fig. 14.11 Examples of stream corridor restoration projects carried out under the Rivers, Trails, and Conservation Program of the U.S. National Park Service. The program's objectives include streambank restoration, flood loss reduction, fisheries improvement, and improved realization of recreational resources. The photograph shows volunteer workers restoring a degraded stream in Sonoma County, California.

*Guiding principles
for stream restoration
and management*

■ *Fluvial systems are more than simple threads of water winding through a system of valleys.* They are actually hydrologic corridors where several water systems, notably, channel, flood, groundwater, wetland, and lake systems coalesce and interact to shape and change the riparian landscape. Water is decidedly the driving force of stream corridors both on and below the surface, and it follows that, no matter what their objectives may be, planning and management programs must not lose sight of the hydrologic imperative.

■ *Streamflow and related hydrologic systems fluctuate radically over time.* Average flow conditions may have little to do with the forms and features of channels and floodplains that we see on site visits and record in our field inventories. Therefore, in reading cause-and-effect relations into this landscape, we must develop a sense of the magnitude and frequency of significant discharge events based on multiple lines of evidence including discharge records, topographic features, vegetation patterns, old maps and aerial photographs, and local histories, both written and oral (Fig. 14.12).

■ *Erosion and deposition are natural processes in stream channels and floodplains.* But it is necessary to distinguish between natural, background levels of geomorphic change and those induced by urban runoff, deforestation, cropland runoff and sedimentation, and structural facilities such as bridges. Undercut banks and point bars are part of a stream's natural dynamics and most should not be earmarked for stabilization in landscape management programs.

■ *The principal biogeographical expression of stream systems is the habitat corridor.* In many respects the stream habitat corridor functions as a biological right-of-way through a watershed, providing linkage and continuity among plant and animal communities. Because they are narrow and continuous over great distances, corridors are prone to breaching from land use systems, which results in fragmentation and reduced biodiversity. The true impact of a breach, however, depends on the relative permeability a barrier such as a highway or bridge poses to plant and animal populations.

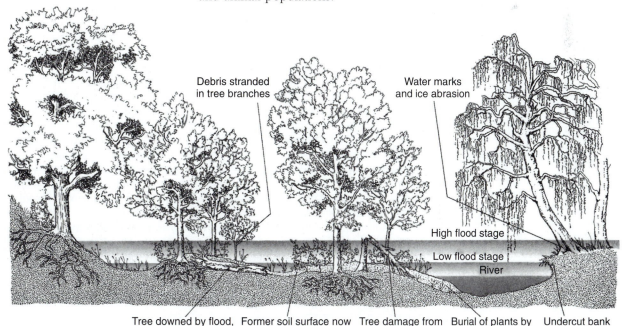

Fig. 14.12 Indicators of past floods, such as water marks, stranded debris, and damaged vegetation, are part of the record of a stream's flow regime.

■ *Many of the species and ecosystems inhabiting stream corridors are not only adjusted to the variable flow regime but are dependent on it* for nutrient supplies, propagule dispersal, seed propagation, and many other life processes and resources. The system tends to be episodic in its behavior and management concepts based on the notions of gradual ecological change such as illustrated by the community-succession model are not tenable for most stream valleys. Further, management programs, such as flood control projects, aimed at reducing variability, can seriously alter aquatic and riparian ecosystems. Among many other things, the absence of stress from low-and medium-magnitude flood events allows all sorts of flood-intolerant species to invade these habitats.

14.9 CASE STUDY

■ ### Stormwater Management and Channel Restoration in an Urban Watershed

Bruce K. Ferguson and P. Rexford Gonnsen

In the last few decades the southern Piedmont has become one of the fastest-urbanizing regions of the United States. The development of the 140 acre watershed draining Coggins Park, an industrial park in Athens, Georgia, began in the 1980s with the construction of roads, utilities, selected buildings, and parking areas. Following standard stormwater management practices for industrial sites, a "dry" detention basin was added to the watershed's major tributary. Only four years later, half the industrial lots were occupied. Impervious cover on the developed parts of the park exceeded 90 percent, and over the whole watershed, it totaled 30 percent.

The resultant increase in stormwater loading of the three tributary channels draining the park was severe. Stormflow peak rates had more than doubled over predevelopment levels. The larger and faster flows seriously eroded stream channels: banks sloughed in, bed material shifted, and the streams generated huge sediment loads as they adjusted to their new flow regime driven by stormwater. In some places, builders further destabilized the channels by digging trenches to drain old farm ponds and moving streams to make room for construction fill. The detention basin originally constructed as a part of the park's infrastructure proved to be ineffective at either suppressing peak flows or capturing mobile sediment. It neither reduced runoff nor changed the total volume of water released downstream.

Faced with stormflows that eroded away landscape plantings and sediment that turned the stream brown, smothered aquatic vegetation, and even filled a waterfall pool, downstream residents sued, demanding correction of the problem. Decisive countermeasures were called for and a stream and watershed rehabilitation project was conceived to capture mobile sediment, stabilize stream channels, suppress peak storm flows, augment baseflows, establish wetlands, and improve water quality.

To reduce bank erosion and stabilize the channel, numerous check dams were added to the stream. This established a series of intermediate base levels below which the stream could not erode while the pools behind the dams captured sediment. As the pools filled with sediment the channel gradients upstream of dams became gentler, thus reducing the stream's energy to erode and transport sediment. This led to further buildup of the streambed, and as the channel stabilized, wetland and riparian vegetation became established on the deposits of channel sediment. In addition, the sediment promoted streambed infiltration and recharge of the alluvial aquifer, which enhanced stream baseflows during dry periods.

Next, the detention basin was restored and enlarged. It now retains sediment from upstream and this improvement coupled with channel stabilization, has resulted in much clearer discharge water. Additional, but smaller, holding basins were placed in the disturbed streams where existing roads, utilities, and land use allowed, and more check dams were added. The goal was to induce such substantial channel filling that check dams and reservoirs would become buried, resulting in a more

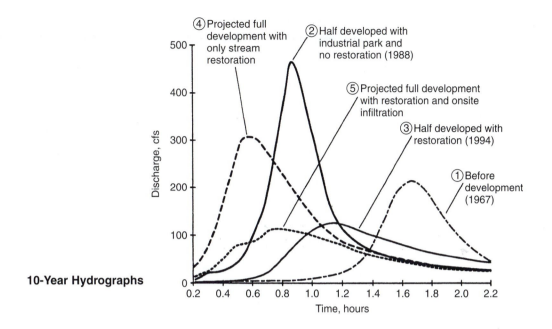

④ Projected full development with only stream restoration

② Half developed with industrial park and no restoration (1988)

⑤ Projected full development with restoration and onsite infiltration

③ Half developed with restoration (1994)

① Before development (1967)

10-Year Hydrographs

uniform channel gradient that would approach an equilibrium condition. At this point a stream channel would be able to accommodate the flows delivered to it without excessive channel erosion or sedimentation.

Hydrologic modeling of the overall drainage system indicated that the restoration effort consisting of check dams and reservoirs had indeed reduced the magnitude of peak flows to predevelopment levels (see the 10-year hydrograph). However, additional measures were needed to manage the runoff that would be generated by future development at full build out of the industrial park. Analysis showed that onsite reduction of runoff from future building sites was needed. In urban areas, such as Coggins Park, however, open space for surface basins on industrial plots was not

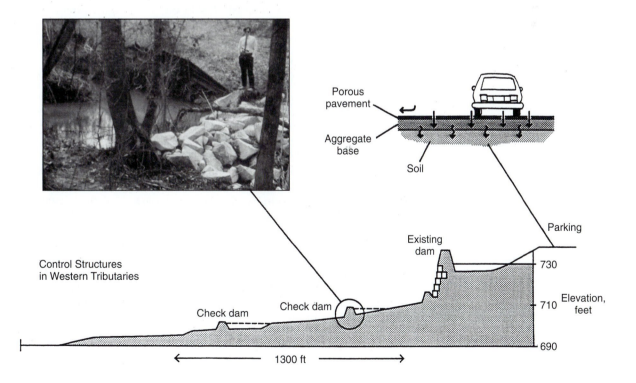

Control Structures in Western Tributaries

industrial plots was not available. As an alternative, we turned to porous pavements with aggregate bases as a means of increasing infiltration and reducing runoff. Infiltration attacks the problem of urban stormwater at its very source. Rainwater can pass through a porous asphalt surface and be stored in the void spaces of stone aggregate while infiltrating the underlying soil. In addition to reducing the volume of water, infiltration reduces stormwater contaminants such as suspended solids and automotive oils. The porous pavement infiltration systems were designed to work in concert with the check dams and reservoirs to help hold future peak flows to predevelopment levels. Our hydrologic models indicate that in combination these measures will produce acceptable levels of streamflow, bring environmental damage under control, and eliminate serious offsite impacts.

Recently, beaver tracks were seen along the channels of Coggins Park. Beavers may be taking over the business of flow regulation and check dam operation. Eventually, they may eliminate the need for perpetual human vigilance and maintenance of the drainage system. In any case they are a sign of recovery in a small stream system that had been devastated by urbanization and its fugitive runoff.

Bruce K. Ferguson is Professor of Landscape Architecture at the University of Georgia who specializes in stormwater hydrology and P. Rexford Gonnsen is a partner and a landscape planner in Beall Gonnsen & Company, Athens, Georgia. ■

14.10 SELECTED REFERENCES FOR FURTHER READING

Collier, Michael, et. al. "Dams and Rivers: A Primer on the Downstream Effects of Dams." *U.S. Geological Survey Circular* 1126, 1996.

Doppelt, Bob, et. al. *Entering the Watershed.* Washington, DC: Island Press, 1993.

Ferguson, B. K. "Urban Stream Reclamation." *Journal of Soil and Water Conservation* 46:5, 1991, pp. 324–328.

Gorman, O. T., and Karr, J. R. "Habitat Structure and Stream Fish Communities." *Ecology* 57, 1977, pp. 507–515.

Hirsch, R. M., et al. "The Influence of Man on Hydrologic Systems." In *The Geology of North America* 0–1. Boulder, CO Geological Society of America, 1990.

Kaufman, M. M., and Marsh, W. M. "Hydro-ecological Implications of Edge Cities." *Landscape and Urban Planning,* 36, 1997, pp. 277–290.

Keller, E. A. "Pools, Riffles, and Channelization." *Environmental Geology* 2, 2, 1978, pp. 119–127.

Leopold, L. B. *A View of a River.* Cambridge, MA: Harvard University Press, 1994.

Leopold, L. B., Wolman, M. G., and Miller, J. P. *Fluvial Processes in Geomorphology.* San Francisco: Freeman, 1964.

McGuckin, C. P., and Brown, R. D. "A Landscape Ecological Model for Wildlife Enhancement of Stormwater Management Practices in Urban Greenways." *Landscape and Urban Planning,* 33, 1995, pp. 227–246.

Selby, M. J. *Earth's Changing Surface.* Oxford: Clarendon Press, 1993.

Smith, D. S., and Hellmund, C. A. (eds). *Ecology of Greenways.* Minneapolis: University of Minnesota Press, 1993.

Stanford, J. A., and Ward, J. V. "An Ecosystem Perspective of Alluvial Rivers: Connectivity and the Hyporheic Corridor." *Journal of the North American Benthological Society* 12:1, 1993, pp. 48–60.

Swift, B. L. "Status of Riparian Ecosystems in the United States." *Water Resource Bulletin* 20:2, 1984, pp. 223–228.

Williams, G. P., and Wolman, M. G. "Downstream Effects of Dams on Alluvial Rivers". *U. S. Geological Professional Paper* 1286, 1984.

15

SHORELINE PROCESSES, SAND DUNES, AND COASTAL ZONE MANAGEMENT

15.1 INTRODUCTION

Shore erosion, flooding, and property damage are nagging and costly problems in the coastal zone. In the United States and Canada, which together account for more than 200,000 miles of coastline, this problem has grown significantly in the past three decades, not because the oceans and lakes are behaving differently than they did years ago, but because of increased development and use of the coast. This problem has given rise to heavy financial and emotional investment and, for many of the 100 million or more North Americans who live on or near a coast, has led to a bittersweet relationship with the sea. The costs to individuals and society are rising. In response, efforts to manage and protect coastal lands have reached critical levels at community, state/provincial, and national levels. And with the prospects for rising sea levels and increasing coastal population worldwide in this century, the problem is sure to become exceedingly acute.

A call for management   In 1972, the U.S. federal government passed the Coastal Zone Management Act, which provides for the formulation of coastal planning and management programs at the state level. Among the responsibilities of the state programs is the classification of coastlines according to their relative stability, including the potential for erosion. To make such a determination, it is necessary to understand not only the physical makeup of the coast, but also the nature of the forces acting on it. The principal force in the coastal zone is wind waves. They are the source of most shore erosion, sediment transport, and deposition which together shape the beaches, sandbars, cliffs, barrier islands, and related features. These landforms are the settings where the dramatic interplay between land use and the sea is carried out.

Finding a proper fit   Finding the proper fit between land use and the sea demands more than muscling the coastline into an engineered fortification to protect land use facilities. It requires a more multidimensional perspective that gives as much weight to ecosystems, geomorphic systems, scenic and recreational resources, history, and culture as it does to protecting real estate. And it must give serious consideration to the remarkable geographic diversity of the coast, not only its differences in carrying capacity and resilience to land use disturbance, but the historical patterns and types of human settlement. In short, we, as a society, must develop more thoughtful and sustainable approaches to our use of the world's sea coast.

15.2 WAVE ACTION, CURRENTS, AND NEARSHORE CIRCULATION

Most wind waves are generated over deep water where they cannot cause erosion because the motion of the wave does not reach bottom. In shallow water, waves not only touch bottom, but they can also exert considerable force against it. Initially, this force is not very great, but as the wave nears shore, it increases rapidly. Where the shear stress of this force exceeds the resisting strength of the bottom material, displacement of particles occurs. Along coasts comprised of loose sediments, such as sand and silts, large quantities of particles are churned up, especially as the waves break in the shallow water near shore.

Wave base depth   The water depth at which waves first touch bottom and begin to move small particles is called wave base (Fig. 15.1). **Wave base depth** increases with wave size and is roughly proportional to 1.0 to 2.0 times the wave height. Relative to wavelength, wave base falls at a depth between 0.04 and 0.5 wavelength. The shallow-water zone along a coastline is defined by the wave base depth, and since the wave base changes with wave size, the shallow zone actually fluctuates in width with different wave events. In the oceans, however, wave base for large waves is generally considered to average around 30 feet; in the Great Lakes it averages around 10 feet.

Wave size   The ability of waves to erode and transport sediment is a function of the wave size and the size and availability of sediment. In deep water, **wave size** is the prod-

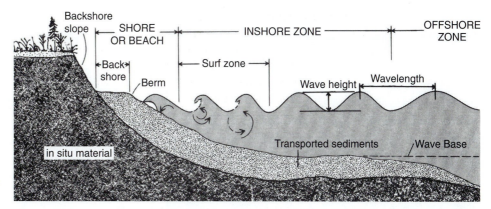

Fig. 15.1 The principal dimensional parameters of waves, the three main shore zones, and their features.

uct of wind velocity, wind duration, and fetch. *Fetch* is the distance of open water over which a wind from a certain direction blows. Fetch varies dramatically depending on the shape and size of the waterbody. It is the main reason why inland lakes cannot generate large waves. The largest waves are produced when wind velocity, duration, and fetch are all large. Forecasts of the size of *deep* water waves can be made using the graph in Fig. 15.2. For example, at a wind velocity of 20 mph and a fetch of 100 miles, wave height will reach 5 feet after about 14 hours duration.

Wave-Size Forecasting Chart

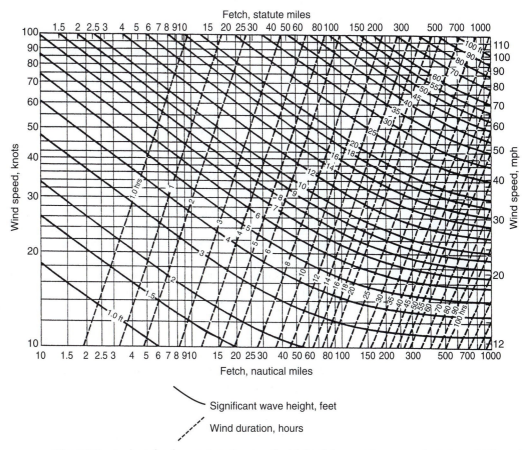

Fig. 15.2 A chart for forecasting the size (height) of deep water waves based on fetch and wind velocity. The duration curves give the minimum time required to reach maximum height for a given wind velocity and fetch.

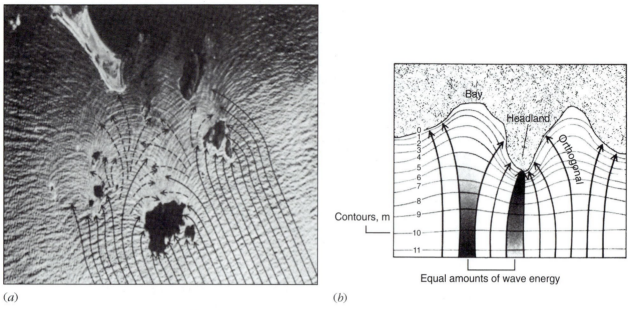

(a)   (b)

Fig. 15.3 Wave refraction patterns (*a*) as revealed by aerial photography, and (*b*) as related to the distribution of energy around a headland and adjacent embayments.

Wave refraction

In shallow water, wave velocity declines with energy loss to friction against the bottom and to turbulence generated by the churning motion of the wave. The rate of deceleration increases toward shore with increasing friction and turbulence, and near shore wavelength shortens rapidly as waves appear to "close ranks" before dissipating themselves on the beach. If a wave approaches shore at an angle, it actually bends or *refracts* as the inner segment slows down much faster than the outer segment. This can be measured by the curved alignment of the wave axis (crest line) in the shallow water zone, and it signals a reorientation of the wave's energy so that its force strikes the shore more head on.

Significance of refraction

Because refraction is controlled by the direction of wave approach and the bottom topography of the shallow-water zone, **refraction patterns** can be forecast based on deep-water wave direction and bathymetric maps. This can also be done by constructing lines called *orthogonals* (perpendicular lines) cross the crest of approaching waves as they appear on an aerial photograph (Fig. 15.3*a*). Assuming wave size (and therefore energy) is uniform along the entire wave crest in deep water, we see that any convergence or divergence of the orthogonals in shallow water represents a change in the relative distribution and the orientation of wave energy. Normally, refraction causes wave energy to become focused on headlands and the seaward sides of islands, and diffused in embayments and on the leeward sides of islands. These are typically the sites of erosion and deposition, respectively (Fig. 15.3*b*).

In order for waves to modify shorelines, sediment must be picked up and moved from one place to another. For this to take place, the force exerted by the motion of the waves must be great enough to dislodge particles. On the outer edge of the shallow-water zone, particle movement is slight, especially with sand-sized sediment, and is limited to a to-and-fro motion with the passage of each wave. Nearer shore, this motion is combined with a lifting action, especially where waves are breaking, that carries particles into the water column and on a turbulent ride before settling back to bottom. Although the individual movements of particles can be in any direction, the net direction of movement after the passage of many waves is usually parallel to the shoreline.

Longshore transport

Two factors account for the parallel, or **longshore transport** of sediment. One is that most waves approach and intercept the coast at an angle; therefore, the direc-

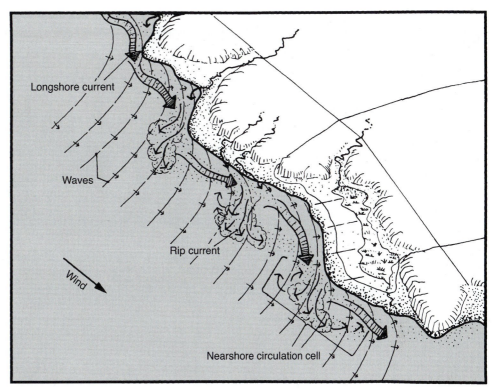

Fig. 15.4 Nearshore circulation showing waves, longshore currents, rip currents, and the transport of sediment from a source (above) to a sink (below).

tion of wave force is oblique to the shoreline. Although waves refract into a more direct approach angle near shore, most retain a distinct angle as they cross the shallow-water zone. Thus, a large component of the energy for sediment transport is set up parallel to the shoreline. The second factor takes the form of a current that flows along the coast. This current, called a *longshore current*, moves parallel to the shoreline in the direction of wave movement at rates as great as 2 to 3 feet per second. Longshore currents are driven by wave energy, increasing in velocity and size (volume) with wave size and duration. When sediment is churned up by waves, the longshore currents transport it downshore before it settles back toward the bottom where it is churned up and transported again.

Nearshore circulation Other types of currents also operate along the shore, and one of the most prominent is the *rip current*. Flowing seaward from the shore across the lines of approaching waves, rip currents are narrow jets of water that intercept the longshore train of sediments carrying part of it into deeper water. Together, rip currents, waves, and longshore currents form **nearshore circulation cells**, moving both water and sediments toward, away from, and along the shore. These cells are in turn nested in larger circulation systems that transport sediments from *source areas*, such as river deltas, to *sinks*, which are deposition areas such as embayments (Fig. 15.4).

15.3 NET SEDIMENT TRANSPORT AND SEDIMENT MASS BALANCE

On most shorelines, waves and currents may change direction with the passage of storms or the seasons, which in turn causes reversals in the longshore transport. Thus, the same sediments, more or less, may be transported past one point on the

Sediment transport measures

shore several times in one year. A measure of this quantity, the total amount of sediment moved past a point on the coast (with no rule against double counting), is called **gross sediment transport**. For obvious reasons, it is not a good indicator of the balance of sediment at a place on the coast, because it tells us neither whether the shore is losing or gaining sediment, nor whether the beach is growing or shrinking. Therefore, another measure is used, called **net sediment transport**.

$$\text{Net } Q_t = Q_p - Q_s$$

where

Net Q_t = net quantity of sediment/year
       Q_p = longshore transport in the primary direction
       Q_s = longshore transport in the secondary direction

Net sediment transport is the *balance* between the sediment moved one way and that moved the other way along the coast. If the longshore system is driven predominantly by waves from one quadrant of the compass, then net sediment transport can be large. By contrast, if the longshore system operates in both directions, then net transport may be small while gross transport may be large. Ultimately, it is the trend that we are concerned with, because it can tell us something about the development of the coastline.

Sediment mass balance

At a local scale of observation, where only a short segment of beach or small waterbody is concerned, we would want to make a more detailed determination of the **sediment balance** (Fig. 15.5). This would include inputs from backshore slope erosion and runoff (R_i), losses (outputs) from wind erosion (W_O), onshore inputs from sand bar migration (O_i) in summer, and offshore outputs from bar migration (O_O) in fall and winter, as well as longshore input (L_i) and output (L_O). The mass of sediment on the beach for any period of time is equal to inputs minus outputs:

$$\text{Sediment mass balance: } L_i - L_o + O_i - O_o + R_i - W_o = O$$

A value greater than zero means that the reservoir of beach sediment has gained mass; less than zero, that it has lost mass. Generally, these two trends are manifested in larger- and smaller-sized beaches respectively. In most places, detailed data are not available

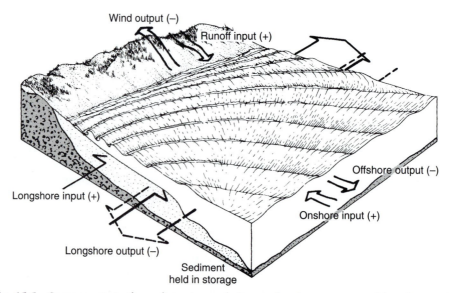

Fig. 15.5 Components in the sediment mass balance of a short segment of beach, a useful model for local planning and management problems.

Data sources for computation of the sediment mass balance, nor is there adequate time or resources in most planning studies to acquire the necessary data. Therefore, in site planning we must usually resort to an interpretation of local records such as old maps, land survey records, and archival aerial photographs, and the testimony of long-time property owners to gain an idea of local trends.

15.4 TRENDS IN SHORELINE CHANGE

Sediment sources Rivers are the primary *sediment source* for longshore transport, providing more than 90 percent of the total sediment supply yearly on the world's coasts. Where river sediments are abundant, virtually all available wave energy may be expended in transporting the sediment load along the coast. However, where river sediments are scarce, such as in Southern California where reservoirs have deprived Pacific beaches of much nourishment from stream sediments, the body of beach sediment is often small and only a fraction of available wave energy is expended in moving it. Moreover, by virtue of its small volume, the sediment mass offers little protection for the *in situ* material under and behind the beach. Under such circumstances, the backshore can be severely eroded and the resulting debris incorporated into the longshore sediment system. As erosion takes place, the shoreline retreats landward. Such coastlines are characterized by features such as sea cliffs, bluffs, or wave-cut banks, and are referred to as **retrogradational**, because over the long run they retreat landward as they give up sediment.

Retreat and erosion   **Retreat** is defined as the landward displacement of the shoreline. It is usually caused by erosion; it may also be caused by a rise in water level, subsidence of coastal land, or any combination of the three. Where retreat is caused by erosion, the amount of *in situ* material actually lost can be computed by multiplying the retreat rate (*R*) times the backshore slope height (*H*) times a given length (*L*) of shoreline (Fig. 15.6):

$$\text{Erosion} = R \cdot H \cdot L$$

Progradation At the other extreme are those coastlines that build seaward, called **progradational** coastlines. Progradation is usually caused by deposition of sediment brought on by decreased wave energy (e.g., fewer storms), increased sediment supply, decreased water depth, and/or construction of sediment trapping structures such as groins.

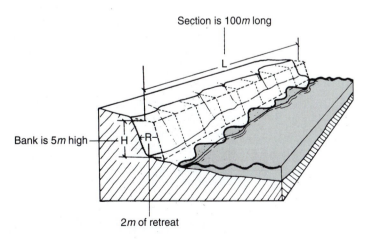

Erosion = 5*m* × 2*m* × 100*m* = 1000 cubic meters

Fig. 15.6 The erosion associated with shoreline retreat is computed from the retreat rate (*R*), the backshore slope height (*H*), and the length of the shoreline (*L*).

Table 15.1 Coastal Change Rates for the United States Based on Historical Records

Coastal Type	Region	Rate (ft/yr)
Mud-flats	Florida	−1.0
	LA–Texas	−7.0
Bedrock	Atlantic	+3.3
	Pacific	−1.7
Sandy beaches	Atlantic	−3.3
	Maine–MA	−2.3
	Mass–NJ	−4.3
	Gulf	−1.3
	Pacific	−1.0
Deltas		−8.3
Barrier islands	Gulf	−2.0
	LA–Texas	−2.6
	FL–LA	−1.7
	Atlantic	−2.6
	Maine–NY	+1.0
	NY–NC	−5.0
	NC–FL	−1.3

Source: Dolan et al., 1983.

Progradational coastlines are characterized by low relief terrain in the form of various depositional features such as broad fillets of sand at the heads of bays, spits and bars near the mouths of bays and behind islands, and beach ridges and sand dunes.

Most depositional features are prone to rapid changes in shape and volume; therefore, short-term trends should generally not be used as the basis for formulating land use plans in the coastal zone. In most instances, huge amounts of sediment are needed to build a small area of land in a sand bar or a bayhead beach, for example, because the bulk of such features are built under water. It is important to remember this if the rate of growth or decline of the feature is to be related to the net sediment transport rate over some time period. The planning and development implications of depositional coastlines are taken up later.

Regional coastal trends It is instructional to examine regional trends in coastal change based on historical data such as old maps and charts. One such study was conducted for the United States and it revealed that as a whole the U.S. coast (including the Great Lakes) is retreating by an average rate of 2.6 feet per year. Further, the rate varies from an average of nearly 6 feet of retreat on the Gulf Coast to less than 1 inch a year on the Pacific Coast. Table 15.1 gives the rate for various regions and coastal (landform) types. Such data are meaningful only in a general way inasmuch as they indicate the direction (+ or −) and the rate of change that, for representative stretches of shoreline, can more or less be expected in the future. Locally, however, they may be meaningless because rates and directions of change vary with the coastal type, local sediment systems, harbor facilities, and the incidence and effect of forceful events such as hurricanes. Generally speaking, predictions of trends in shoreline change, especially longterm rates, should be viewed cautiously, especially in light of the impending sea level and weather changes related to global warming.

15.5 SAND DUNE FORMATION AND NOURISHMENT

Another trend associated with active shorelines, especially prograding ones, is the formation of sand dunes. Sandy beach deposits formed by wave and current action are highly prone to wind erosion. Once entrained by wind, the sand may be heaped up

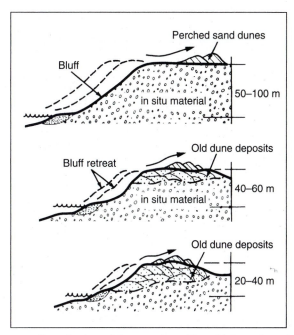

Fig. 15.7 Three types of perched sand dune settings. The old dune deposits in the lower diagram formed originally as low elevation sand dunes. The situation portrayed in the upper diagram is common to the Oregon and Great Lakes coasts, among other places.

into dunes and driven inland up to a mile or more. The fields of sand dunes that result represent a separate subsystem in the larger sediment system of the coastal zone.

Coastal dunes   **Sand dunes** are a common and integral part of every major coastline from the Arctic to the tropics. Their formation and maintenance are directly dependent on two key factors: (1) an ample supply of erodible sand; and (2) a source of wind energy to drive the sand landward. The source of sand for the vast majority of coastal dunes is the beach in front of them. Broad, sandy beaches with little or no plant cover are clearly the best sources. A second source of dune sand is wave-cut banks and cliffs of sandy composition. These two sources give rise to two different classes of coastal sand dunes: (1) *low-elevation dunes* that begin near the backshore and rise gradually landward; and (2) *perched dunes* that begin above water level near the brow of a bluff or sea cliff (Fig. 15.7).

Coastal aerodynamics   The shore is a windier place than locations a little inland from it. The reason has to do with the movement of air at the base of the atmosphere, called the **atmospheric boundary layer**. This layer of air, which measures about 1000 feet deep, drags over the earth's surface and is slowed by the resulting friction. When the boundary layer moves over open water, a relatively smooth surface, it is less impaired by friction than it is over land; therefore, it is generally faster over water. In addition, as the air flows from water onto the land, it also tends to accelerate at the coastline. This is caused by the constriction of the boundary layer as the land rises in elevation at the coast. The air must be squeezed through a smaller space, and in order to maintain continuity of flow, wind velocity increases, providing in turn more energy to erode and transport sand. This phenomenon is especially pronounced along high banks and sea cliffs where it may give rise to the formation of perched sand dunes (Fig. 15.8). It also accounts for the erosion of sand along high dune ridges located several hundred meters inland from the shore.

Dune formation   On most beaches, dune formation begins in the backshore zone with the development of a low ridge of sand paralleling the shore. As the ridge grows, blowouts form at selected spots from which tongues of sand migrate landward. These function

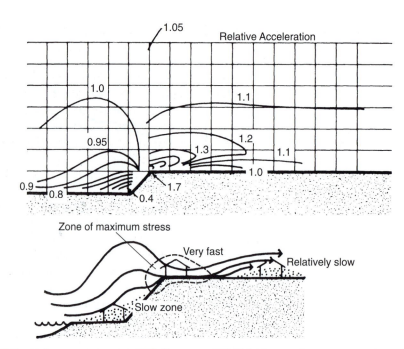

Fig. 15.8   Airflow over a coastal bluff or sea cliff. Velocity accelerates as much as fourfold from the toe to the brow of the slope. Below, corresponding development sites and wind streamlines.

as ramps for transport of sand wafted from the beach. The advancing edge of the dune deposit buries vegetation, soil, and sometimes wetlands as it moves inland. From the blowout to the leading edge of the dune, the form (in planimetric view) is one approximating an elongated U and in most places is called a *hairpin dune* (Fig. 15.9).

As the dune moves inland, it not only grows longer but deeper as sand is added to the main body of the deposit. On some barrier islands and sand spits, dunes may

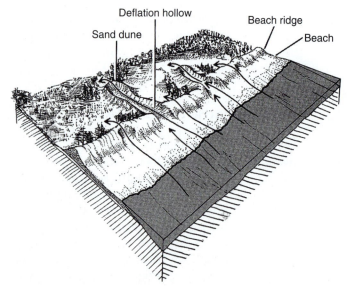

Fig. 15.9   The most common setting and forms associated with coastal sand dunes. The hairpin form of the dune follows the main line of airflow inland from the beach, driving the dune over the landscape.

Dune migration migrate all the way across the landform to the inner shore. On most shores, however, the available wind energy is insufficient at some distance inland to transport enough sand to overcome vegetation. Therefore, the dune ceases to advance, and plants take over the surface. If the process of dune formation is repeated many times along a segment of coastline, the deposits tend to overlap and merge together forming a complex dune ridge called a *barrier dune*. The barrier dune generally marks the highest landward extent of the dunefield (the total mass of dune features landward of the beach). The barrier dune is usually steep on the leeward (inland) side, where wind deposits are held in place by vegetation, and gentle on the windward side where sand is transported upslope from the blowouts.

Dune management dilemma Coastal sand dunes can present difficult land use and management problems. They are extremely attractive to residential, resort, and recreational development by virtue of the open vistas, the juxtaposition of open sand, grass-covered and wooded surfaces, and of course, the access to the shore. They are, however, very fragile environments where vegetation, sand deposits, and slopes may be so delicately balanced that only slight alterations in one can lead to a chain of changes such as renewed wind erosion, blowout formation, and dune reactivation.

Dunes as sediment systems What is often not clearly understood about coastal dunes is that they are themselves small sediment systems with source areas, transportation zones, and sediment sinks. In order for dune systems to maintain themselves, nourishment must be sustained. This requires maintenance of source areas as well as the flow of air from source areas to sinks (deposition zones). When development takes place in dunefields, source areas may by reduced with the planting of groundcover on barren surfaces as a part of landscaping and erosion control programs. In addition, transportation zones may be disrupted with the building of roads and structures in the backshore and lower dune areas. In the area of the barrier dune complex, a favorite location for siting homes and resorts, construction inevitably requires alteration of *metastable slopes* held in place by vegetation. (See Section 3.4 in Chapter 3 for a discussion of metastable slopes.)

Managing dune systems With respect to reductions in sand dune nourishment, the main problem is the weakening of the dune building and movement processes and the encroachment of vegetation over the dunefield. Successful *erosion control* programs that stabilize source areas (such as backshore slopes) may so limit the supply of sand that the interplay between geomorphic processes and vegetation—which gives the dunefield much of its special character—becomes largely one-sided and the dunefield becomes overgrown and inactive. In that case, the dunefield is transformed into sand hills, or what are sometimes called *fossil dunes*. Management programs that recognize wind erosion, sand movement, and a changeable landscape as the most desirable components of dune environments—in the same way that wave action and beach dynamics are viewed as attributes of attractive shorelines—may be advisable in many coastal areas.

15.6 APPLICATIONS TO COASTAL ZONE PLANNING AND MANAGEMENT

Despite calamities from storms, erosion, and flooding, population and development are on the rise in coastal areas worldwide. In the United States more than 50 percent of the population now lives in coastal counties, and development pressure there is 40 to 50 percent higher than in noncoastal counties. Understandably, proposals for development in the coastal zone today are subject to serious scrutiny in the United States and Canada. In order to evaluate a proposal, it is first necessary to know the makeup and dynamics of the coastline involved. This usually involves an inventory of coastal

Coastal inventory

landforms and lithology (composition), and an assessment of recent erosion and deposition trends. It is especially important to map landform indicators of geomorphic trends of coastal change including where the coast is composed of hard and soft materials. Within this framework, more detailed studies can be carried out for the specific segment where the action is proposed. At the heart of such studies, especially when shoreline structures are involved, is an analysis of the sediment budget.

Sediment transport analysis

Two types of methods are available in net sediment transport analysis of the longshore system: shoreline change and wave energy flux. The most accurate method is based on measurements of volumetric changes in a shoreline. This usually involves comparing old charts and aerial photographs with newer ones, or making detailed measurements of sediment accumulation behind a shoreline barrier such as a breakwater. The latter necessitates making a topographic survey at the time of construction and one some years later, and then measuring the net change between the two. Where harbor entrances are maintained by sand bypassing operations such as dredging, the amount of sediment that must be transferred from one side of a barrier to the other each year can be determined by this method (Fig. 15.10). The *second method* is based on computations of wave energy and established correlations between wave energy flux and sediment transport rates. This method requires wave data for the location in question, and since such data are usually scarce, the wave energy computations are difficult to prepare.

Land use planning implications

Upon completion of the coastal inventory and estimates of the sediment budget, two important questions can be answered: (1) What is the nature of the system we are dealing with in terms of the amounts of sediment being moved and the directions of movement? and (2) What is the relationship among the features, processes, and trends of the coast, such as the development of prograding shorelines, and the nourishment of dunefields? Answers to these questions can then be used to test the feasibility of development schemes, evaluate community plans, or guide the formu-

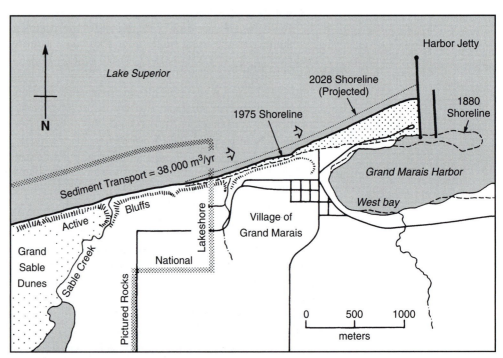

Fig. 15.10 Trends in shoreline change, 1880–1975 with a projection to 2028. Note change as a result of the harbor jetty (built in the 1960s) has been negative on the east and positive on the west. Net sediment transport is eastward.

lation of coastal zone management plans. Among the problems commonly faced by private development is how to deal with "soft" environments such as sand dunes, wetlands, and erodible beaches. Coastal communities, on the other hand, are frequently faced with serious conflict when it comes to the need to protect coastal resources while accommodating economic development schemes involving harbor improvements, marinas, and related facilities.

Site planning procedure   Because the coastal zone is so attractive for residential development, it is necessary to examine briefly some of the problems of site planning near shorelines. In evaluating a site for development, it is important, as recommended previously, to first examine the position of the site in the larger sediment system and determine whether it is in a retrograding, prograding, or stable part of the coastline. Next, it is necessary to determine the composition of the site, its topographic configuration, and whether it belongs to a special coastal habitat such as a delta, estuary, or salt marsh. Low elevation, soft shorelines such as barrier islands along the Gulf and Atlantic coasts, which are periodically cut back and overwashed by hurricane waves, are very risky development sites. The same holds for each of the soft environments mentioned above. On the other hand, sites defended by bedrock are usually safe from rapid erosion, though the bedrock may pose significant problems in building construction, wastewater disposal, and stormwater drainage.

Shoreline oscillation   Shorelines composed of unconsolidated (soft) materials should be treated carefully where development calls for permanent structures because most are prone to seaward/landward oscillations over various time intervals. Minor oscillations occur seasonally with winter/summer changes in the sediment mass balance. Over larger time periods, the magnitude of fluctuations can be expected to increase with the average return period; that is, big changes occur with a lower frequency than the small ones. Studies show that most oscillations are related to climatic fluctuations in the range of 20 to 60 years or more that are marked by periods of increased storminess and shore retreat with intervening periods of quieter weather and beach building.

Indicators of change   But even with records of climate-related fluctuations, it is difficult to determine the proper setback distances for buildings because the magnitude of shore fluctuations is often different for different segments of a continuous shoreline and in many areas records lack sufficient detail to help figure out local trends. In some instances, however, vegetation and slope conditions can provide insight into trends and fluctuations. Exposed roots, tipped trees, and undercut banks are signs of shore erosion and retreat, (Fig. 15.11*a*), whereas stable backshore slopes and numerous young plants on the backshore are signs of progradation. In fact woody plants can be used to measure the rate of progradation.

Geobotanical clues   If tree stands of different *age classes* can be identified on the backshore, they may reveal how far landward the sea or lake has advanced since the stands became established. In Fig. 15.11*b*, for example, the presence of 75- to 100-year-old trees beginning at a distance of 100m (330 ft) from the shore indicates that the area beyond this point has been free of transgression by waves for at least 75 to 100 years. In contrast, seaward of this point the tree ages tell us that the area has been eroded away at least once in the last 75 to 100 years and has subsequently reformed. Nearest shore the vegetation reveals that the frequency of destruction and rebuilding is evidently 25 years and less. This technique is best suited to areas where one stand gives way sharply to another, because the absence of a transition between neighboring stands is a sign that the older stand has been cut back by erosion and the younger stand subsequently established on new deposits.

Despite their appeal as places to build and live, erodible, low-elevation coastal terrain is generally not suitable for development. Time and time again, nature demonstrates its unsuitability to us, particularly in areas exposed to the open sea. This suggests that such terrain should best be left to open space, special easements, nature

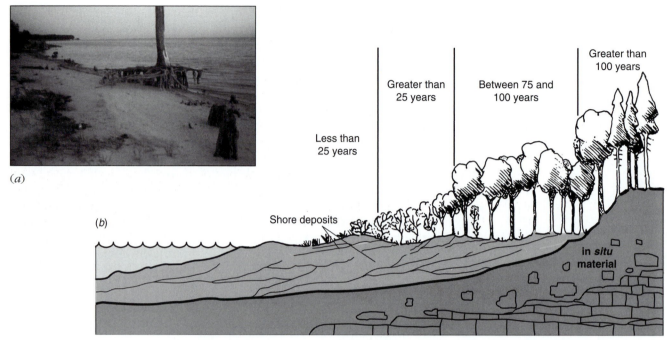

(a)

(b)

Fig. 15.11 (*a*) Evidence of a retreating shoreline in exposed roots and dead trees. (*b*) The extent of past landward fluctuations in the shore are sometimes revealed by the age patterns of tree stands.

Dealing with the unsuitable site

preserves and parks. Where commitments have already been made to development programs, the planner's role is to devise layout schemes in which structures are not only least prone to damage but themselves least interfere with coastal processes and seasonal changes. As a rule, this requires large setbacks not only because of the threat of wave erosion but also because of storm surges and flooding. The ocean-front development along the barrier islands of the Atlantic and Gulf coasts is repeatedly damaged or destroyed by hurricane-driven erosion and flooding because, among other things, setbacks are much too small (Fig. 15.12). Unfortunately, the harsh lessons of the hurricanes have not curbed development or improved setbacks as property owners, motivated in part by support from the National Flood Insurance Program, rebuild in the aftermath of hurricanes such as Hugo (1989), Andrew (1992), Fran (1996), and Isabel (2003). Unable to guide land use toward safer and less costly locations and configurations, planning officials are turning to **risk management** programs that include early warning systems, evacuation plans, and disaster relief plans.

Elevated sites and wind stress

For development sites situated near the crest of a sea cliff or backshore slope, the threat of wave erosion may not be great; however, the effects of wind and slope instability may be significant. The pattern of onshore airflow over a coastal slope produces a zone of low velocity on the lower slope and accelerated velocity on the upper slope relative to wind velocities at comparable elevations over the water. For long, continuous coastal slopes, the velocity increase from the slope foot to the slope crest is about four times for slopes of 4:1 inclination or greater. In addition, as the air crosses the brow of the slope it appears to cling to the ground (or tree canopy) for some distance inland, and then the fast air separates from the surface (Fig. 15.8).

Risks of great vistas

Since the force of wind increases approximately with the cube of velocity, these zones are in reality greater risks for site planning than raw wind data would lead us to believe. Sites situated near the brow of the slope are subject to severe wind stress during storms and, in cold climates, to high rates of heat loss and deep penetration

Fig. 15.12   Damage to the beach environment and residential property from Hurricane Fran which struck a massive blow to the North Carolina coast in September, 1996. The ocean shore lies in the foreground. Safer ground lies across the road behind the houses.

of ground frost. Downwind from the brow, sand and snow accumulate where airflow separates; this zone is often the site of sand dune and snowbank formation. If the brow of the slope is forested, sand and snow accumulation often take place along the brow (Fig. 15.8). Clearly, each zone presents limitations on development, and the nearer the crestslope where the coastal vistas are best, generally the more serious the limitations. Therefore, we can safely say the placement of structures on or close to the brow of the slope is not advisable as a general practice. It is also necessary to add that development is feasible on stable sites only where proper site analysis and appropriate engineering and architectural solutions have been worked out. This would necessitate slope analysis to determine stability against mass movement and erosion, soil assessment for stormwater and wastewater drainage, and wind velocity profiling to establish the aerodynamic conditions. Among many other things, the design of structures and landscaping on sites with exposures to strong winds should conform to the aerodynamic forces for best performance.

15.7 SITE MANAGEMENT CONSIDERATIONS

Besides the various guidelines for planning and development already described, there are a number of additional considerations to be made in managing coastal sites. Most of these relate to the backshore area and how to maintain the stability of banks and bluffs. The backshore zone represents a buffer between *in transit* (shore) material and *in situ* (terrestrial) material (see Fig. 15.1). The primary objective in site management is maintenance of the backshore slope against disturbance by land use and erosion by various processes.

The susceptibility of backshore banks and bluffs to erosion and failure is not simply a matter of exposure to storm waves and strong winds. These forces are serious considerations, but terrestrial processes such as runoff and stabilizing factors such as bank vegetation are also very important. Stormwater is an especially difficult problem because of the accelerated erosive force of runoff passing over a steep slope.

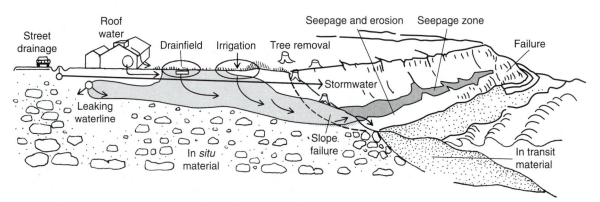

Fig. 15.13 A schematic diagram illustrating common sources of bank stability problems related to land use alterations of drainage and vegetation.

Bank erosion and failure

Stormwater from impervious surfaces such as shoreline roads and residential complexes can cut deep gullies into banks, weakening them and leaving them more susceptible to wave erosion. Another drainage problem is soil saturation from septic system drainfields, lawn irrigation, and leaking water and sewer lines. This water often concentrates on the lower slope where it reduces bank resistance to failure (slumps and slides) and wave erosion (Fig. 15.13). In one incident of coastal slope failure at Palos Verdes Hills in Southern California, artificial contributions of water from 150 residences were estimated at 32,000 gallons per day. Elimination of lawn irrigation and drainfield infiltration coupled with prudent stormwater management would have greatly reduced slumping and the resultant damage to homes and infrastructure.

Vegetation and footpaths

Vegetation is a source of bank stability that must be managed with care. Not only do root systems provide resistance against erosion and slope failure, but plants draw large amounts of water from the soil, thereby reducing saturation and seepage. A common set of problems with residential development is (1) removal of trees to improve vistas, (2) introduction of exotic species that are often less effective as bank stabilizers, and (3) applications of excessive irrigation water to maintain introduced plants. Another destabilizing factor is the development of footpaths and the placement of structures such as stairways, viewing decks, and beach houses on or near banks. Footpaths not only destroy vegetation but human traffic weakens soil, pushes large amounts of material downslope, and opens the slope to erosion by runoff and wind. Structures on slopes also alter and weaken vegetation, and their footings are often points of concentrated erosion related to turbulence in runoff and wave swash.

Seawalls

Finally, we must address the pressing question of structural controls such as seawalls, breakwaters, and groins. Huge expenditures are made in the United States and Canada on engineered structures in an effort to improve navigation, reduce shore erosion, and protect coastal real estate. Most erosion controls not only prove to be ineffective in halting erosion in the long run, but are often a source of serious damage to the coastal environment, especially shore and nearshore habitats and recreational and aesthetic resources. Seawalls, for example, which are commonly used to defend individual properties, often increase the scouring effects of waves resulting in deeper water immediately near shore. Eventually the wall fails, but before it does it may cause increased erosion on adjacent unprotected properties as waves converge on the projecting seawall and refract to the right and left.

Groins, jetties, and sand bypassing

Groins, jetties, and breakwaters are designed to trap sediment and build wider beaches or protect harbor openings by interrupting the longshore flow on sediment (Fig. 15.10). Since these structures function like dams on rivers, they commonly deprive areas downshore of their sediment supply, thus leading to beach erosion and shoreline retreat. This problem is so serious around harbor entrances that sand must

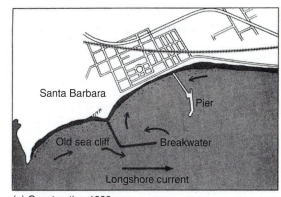

(a) Construction 1928

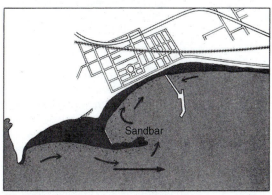

(b) 1948

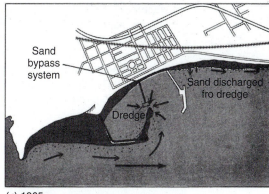

(c) 1965

Fig. 15.14 Sand accumulation caused by the harbor facilities at Santa Barbara, California, and the sand bypass system constructed to transport the sand and clear the harbor entrance.

be mechanically transferred from one side of the entrance to the other, a process called *sand bypassing* (Fig. 15.14). The lesson for planners is to approach the problem of beach maintenance at a broader scale, one that incorporates the larger sediment system, its sources, sinks, and transport zones into shoreline management programs. Where, however, it is necessary to protect navigation facilities, the expense of sand bypassing must inevitably be added to the costs of managing the coastal zone.

Beach nourishment   Increasingly, *artificial nourishment* of the longshore system is being used to help maintain beaches. This process involves feeding a beach with a supplemental supply of sand by pumping or barging it from another source. Miami Beach is often cited as a successful example of a beach nourishment program. Here the shore was fed with 25 million tons of sand that have maintained a beach for more than 2 decades.

Beach nourishment is favored by many communities, but it is not without problems, not due least of which is the big financial cost and the disturbance of the littoral environment. The following case study explores another method of shoreline restoration and stabilization that is showing signs of promise in Puget Sound.

15.8 CASE STUDY

In Search of a Green Alternative to Conventional Engineered Shore Protection: The Rootwall Concept

Elliott Menashe

Shoreline home sites afford stunning views and are coveted properties, but they can also be unstable and hazardous places to live. Finding a balance between land use pressures and shoreline health and stability is an on-going challenge. Common development practices, such as insufficient setbacks, clearing and grading, vegetation removal, and stormwater runoff, contribute concurrently and cumulatively to increased incidence of the erosion, slope destabilization and environmental degradation which in turn plague communities, planners, and landowners themselves. Ironically one of the most profound and damaging coastal modifications is the widespread construction of conventional engineered structures designed to protect shorelines from erosion. Known variously as seawalls, bulkheads and revetments, they are widely employed to stop erosion and hold banks in place. However, shorelines are dynamic systems and it is becoming evident that building permanent structures and attempting to impose a static equilibrium to these dynamic landforms is both doomed to failure and damaging to coastal ecosystems.

The scope of the environmental impacts associated with these "hard structures" is also becoming better understood. Research findings indicate that shoreline armoring can actually increase erosion and contribute to more severe slope failures by effectively lowering the level of beaches and interrupting the movement of sediments along shorelines. Biological impacts may include reduction of terrestrial organic matter, loss of shoreline vegetation, and resultant reductions in biological diversity and abundance of intertidal biota. Equally important are the losses to the shore's recreation value and aesthetic appeal.

Various "soft-shore" alternatives are being explored. The rootwall is one such promising alternative which is being explored in the Puget Sound region of the

A failing seawall on a beach cleared of large woody debris and trees. Notice the accentuated erosion of the bank on the right.

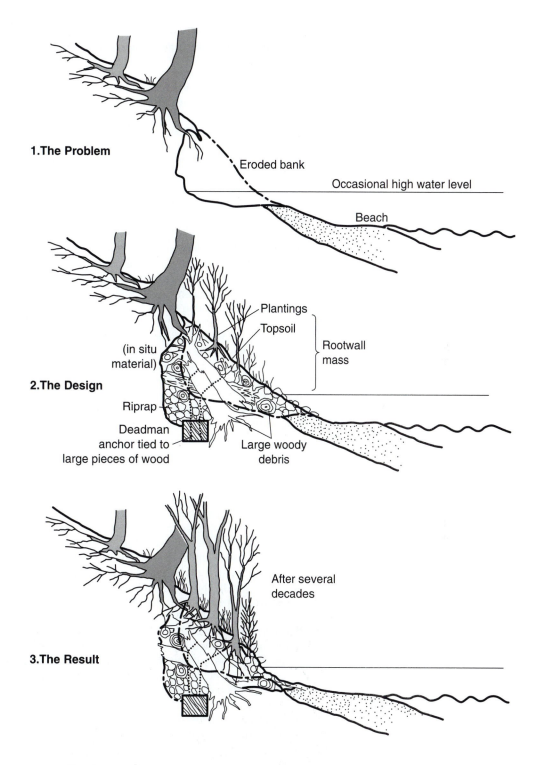

1. The Problem

Eroded bank

Occasional high water level

Beach

2. The Design

Plantings

Topsoil

Rootwall mass

(in situ material)

Riprap

Deadman anchor tied to large pieces of wood

Large woody debris

3. The Result

After several decades

Pacific Northwest. The rootwall employs large-tree and root masses, called large woody debris (LWD), as primary structural components to provide immediate protection from wave attack and slope erosion. The concept is based on the observed role of large trees, both living and dead, in helping stabilize shorelines in heavily forested coastal zones. Trees and their root masses, not only add strength to backshore slopes, but adapt to changing conditions as the bank and shore retreat and change form around them. Moreover, when a tree topples onto the beach its root and trunk mass become natural armor for the shore, dissipating wave energy, capturing sediment, and providing habitat for new plants such as dune grass and sand cherry. Indeed an integral element of the rootwall is the incorporation of desirable native vegetation into the structure's design to encourage the establishment of plant communities which have been disrupted by past human activities such as the logging of old-growth forests and land development. The purpose of the rootwall is not to prevent all erosion or coastal change. Rather, it is designed to reduce the severity and scale of erosional processes related directly or indirectly to human activities. In addition, the rootwall helps re-establish the ecological character of the shoreline.

Along forested shorelines large woody material is an important structural and biological component of the nearshore environment. Accumulations of large pieces of woody debris, notably whole trees, can support a wide range of biotic functions, including maintenance of crucial linkages between terrestrial and marine areas. LWD which lodges behind backshore berms provides the framework from which complex micro-habitat features and stable shore forms are created, reducing beach sediment suspension and removal by waves, prograding of beaches, and maintaining stable rates of sediment yield to coastal systems. Smaller woody debris and cut logs without rootmasses, on the other hand, are too buoyant and mobile to resist tidal action and storm surges. This material rarely remains in place long enough to provide the stabilizing benefits of LWD, and during storms, can become water-borne and function as battering rams.

The implications of deliberately creating LWD structures which mimic the beneficial natural assemblages found on shorelines is obvious. Additional benefits would include improvement of nearshore and distant-view aesthetics, the low cost of woody debris (much of it can be salvaged from site clearing operations inland of the coast), and the development of high quality coastal wildlife and fishery habitat features. The transfer of LWD from site clearing operations (burnpile) to marine shorelines provides the added benefit of re-establishing the link between the forests and the marine environment which has been interrupted for nearly 100 years. Drift logs in inland waterways, a constant hazard to navigation, would provide an additional source of LWD for rootwall structures, converting a liability into an asset. Since available LWD is now much smaller than that commonly found on beaches before the 1850s, the structural elements of the rootwall would have to be anchored in beach substrates.

Versions of rootwalls are currently being proposed for selected Puget Sound shorelines as an alternative to conventional bulkheads. They appear to be best suited to low-to-moderate energy beaches and locations sporadically subject to wave attack. Undoubtedly it will take a number of years of field experiments to come up with a successful rootwall design. However, based on the success of engineered logjams in river systems as a means of erosion control and habitat restoration, we are encouraged by the concept. It is important to add that rootwalls are not a stand alone device. They can be used in conjunction with other "soft-shore" protection measures, such as beach nourishment. In addition, plantings of native trees and shrubs within and behind the LWD structure should be a crucial element of this biostructurally engineered system. Finally, design and construction of a rootwall structure would be site specific and require close collaboration among several professionals, including landscape architects, coastal geomorphologists, geotechnical engineers, marine construction contractors, and marine vegetation restorationists.

Elliott Menashe is owner and principal of Greenbelt Consulting, Clinton, Washington. He specializes in vegetation management and restoration of coastal areas. ∎

15.9 SELECTED REFERENCES FOR FURTHER READING

Allen, James R. "Beach Erosion as a Function of Variations in the Sediment Budget, Sandy Hook New Jersey, U.S.A." *Earth Surface Processes and Landforms* 6, 1981, pp. 139–150.

Bascom, W. N. *Waves and Beaches.* Garden City, NY: Doubleday, 1964.

Coastal Engineering Research Center. *Shore Protection Manual.* Washington, DC: U.S. Government Printing Office, U.S. Army Coastal Engineering Research Center, 1973.

Davies, J. L. *Geographical Variation in Coastal Development.* London: Longman, 1977.

Dolan, R. et al. "Erosion of the U.S. Shorelines" in *CRC Handbook of Coastal Processes and Erosion.* (Komar, P. D., ed). Boca Raton, FL: CRC Press, 1983.

Gutman, A. L. et al. *Nantucket Shoreline Survey.* Cambridge, MA: MIT Sea Grant Program, 1979.

Healy, R. C., and Zinn, J. A. "Environmental and Development Conflicts in the Coastal Zone." *Journal of the American Planning Association* 51:3, 1985, pp. 299–311.

Inman, D. L., and Brush, B. M. "The Coastal Challenge." *Science* 181, 1973, pp. 20–32.

Jarrett, J. T. "Sediment Budget Analysis Wrightsville Beach to Kure Beach, North Carolina," Coastal Engineering Research Center Reprint 78–3, U.S. Army Corps of Engineers, 1978, pp. 986–1005.

Kuhn, G. G., and Shepard, F. P. "Beach Processes and Sea Cliff Erosion in San Diego County, California" in *CRC Handbook of Coastal Processes and Erosion,* (Komar P. D., ed). Boca Raton, FL: CRC Press, 1983.

Marsh, W. M. "Nourishment of Perched Sand Dunes and the Issue of Erosion Control in the Great Lakes." *Journal of Environmental Geology and Water Science,* 16:2, 1990, pp. 155–164.

Marsh, W. M., and Mewett, A. M. *A Framework Plan for Coastal Zone Management,* East Central Vancouver Island. Regional District of Comox-Strathcona, Courtenay, British Columbia, 2002.

Zenkovich, V. *Processes of Coastal Development.* trans. D. Fry. Edinburgh: Oliver and Boyd, 1967.

16

SUN ANGLES, SOLAR HEATING, AND ENVIRONMENT

16.1 INTRODUCTION

The prudent landscape planner should be as sensitive to solar radiation as a part of environmental analysis as he/she is to water supply, stormwater, slope, or soil stability. Solar radiation is a key consideration in all types of residential design, first because it is a primary source of light and energy. Houses and landscaping designed to take advantage of solar heating opportunities in winter and solar screening in summer are not only less expensive facilities to operate but are often more comfortable and pleasing as living environments. Beyond the home and yard, solar radiation is an essential consideration in many aspects of environmental planning and management, including microclimate, soil moisture, and plant ecology.

Analysis of solar radiation begins with examination of sun angle, and its variations with topography and the seasons. This is essential to an understanding of the distribution of solar energy in the landscape. However, the translation of solar variables into meaningful information for planning decisions can be a difficult exercise and, unfortunately, one that few practitioners are able to accomplish effectively. Approached properly, the problem often requires computation of radiation balance and heat budgets, taking into consideration radiation gains and losses, surface reflection, ground materials, and energy flows in the form of ground heat, sensible heat, and latent heat. Our discussion here begins with rudimentary sun angle concepts and then goes on to solar heating of the landscape and its implications for local environments, including microclimates and landscape ecology.

16.2 SUN ANGLE AND INCIDENT RADIATION

Sun angle

The angle formed between sunlight approaching the earth's surface and the surface itself is called the **sun angle**. To envision this angle, think of straight rays of light striking a flat surface such as an airfield. A more direct angle, one that is closer to 90 degrees, causes a greater concentration of solar radiation on the surface. Conversely, smaller angles have weaker solar intensities.

Incident radiation

Understandably, sun angle is an important factor in heating the earth's surface. For example, visually compare the areas bombarded by the beam in diagrams (*a*) and (*b*) of Figure 16.1. The beam in (*a*), which is the same strength as the beam in (*b*), spreads over more surface area because it strikes the surface at an angle. Since (*a*) spreads over more area, its density, or **incidence**, is lower. We can show this with a simple computation that involves dividing the quantity of energy in the beam (S_i) by the area (A) that it strikes:

$$SI = \frac{S_i}{A}$$

where

SI = solar radiation incident on the surface, cal/cm^2 • min
S_i = quantity of energy in the beam, cal/min
A = surface area intercepted by the beam, cm^2 or m^2

Because the earth is curved, most solar radiation enters the atmosphere and hits the surface at an angle. As one goes farther poleward, the angle becomes smaller and the beam of radiation becomes more diffuse and thus less intense. Figure 16.1*c* shows this by contrasting a beam at the equator (B) with one near the Arctic Circle (A). Beam A not only spreads over more surface area, but it also passes through a greater distance of atmosphere, thereby giving the atmosphere a greater opportunity to reflect and scatter radiation before it reaches the ground.

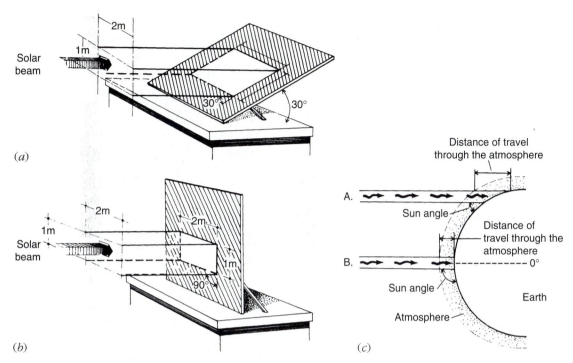

Fig. 16.1 Variations in radiation intensity on an inclined surface (*a*), and a vertical surface (*b*); diagram (*c*) showing the relationship between sun angle and the curvature of the earth including the relative thickness of the atmosphere to incoming solar radiation at high and low latitudes.

16.3 VARIATIONS IN SUN ANGLE WITH SEASONS AND TOPOGRAPHY

Seasonal change   To understand sun angles more completely, we must take some additional factors into account: one is seasonal change in the earth's tilt with respect to the sun. Because of the inclination of the earth's axis (23.5 degrees off vertical), the earth appears to tip toward and away from the sun as it orbits around the sun. This produces seasonal changes in sun angle for all locations on the earth.

Four seasonal sun angles are important at any location on the planet, and they tend to correspond to the seasons in the midlatitudes (Fig. 16.2). For the Northern Hemisphere, the highest and lowest angles occur each year on June 20 to 22 and December 20 to 22, respectively. These dates are called *summer* and *winter solstices*. In fall and spring the sun angles are intermediate, and there are two dates on which they are exactly intermediate, March 20 to 22 and September 20 to 22, called the *equinoxes*.

From summer solstice to winter solstice, the sun angles for any location in the middle latitudes (defined, for convenience, as the zone between 23.5 degrees and 66.5 degrees latitude) vary by 47 degrees. Sun angle readings are normally given as the high noon position of the sun represented by the angle formed between one's outstretched arm pointed at the sun and the horizon of the landscape.[1] Figure 16.3 illustrates the principal sun angles in the year for 50 degrees north latitude.

Computing the sun angle for any latitude and date involves three basic steps. First, *Computing sun angle*   the *declination of the sun* must be known. This is the latitude on the earth where the

---

[1] This would actually be an approximation because solar radiation is refracted (bent) somewhat as it passes through the atmosphere.

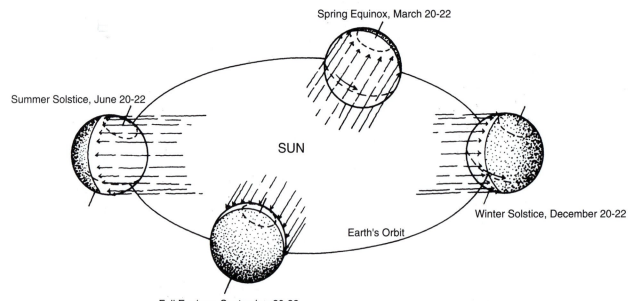

Fig. 16.2   Seasonal changes in sun angle and the orbital path of the earth about the sun. The 23.5 degree angle of tilt is constant throughout the Earth's revolution.

sun angle is vertical (90 degrees) on a given date. For this information we consult the graph in Fig. 16.4*a*. Next, the zenith angle must be determined, which can be done by counting the number of degrees that separate the latitude of the location in question from the declination. The scale in Figure 16.4*b* may be helpful in making this count. *Zenith angle* is the angle formed between a vertical line (perpendicular to the ground) and the position of the sun in the sky. In Figure 16.3, it is the angle (ZA) between the sun and the broken line at the time of the summer solstice. The last step is to subtract the zenith angle from 90 degrees. This gives us the **sun angle**. The following example shows the steps to be followed in computing a sun angle:

- ■ Location   = 50 degrees north latitude (given)
- ■ Date   = June 15 (given)
- ■ Declination of sun = 23 degrees (Fig. 16.4*a*)

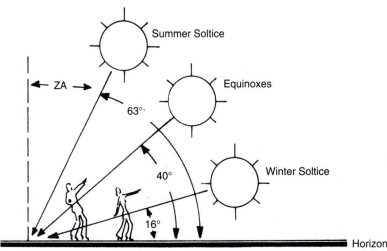

Fig. 16.3   The annual changes in sun angle for 50 degrees north latitude.

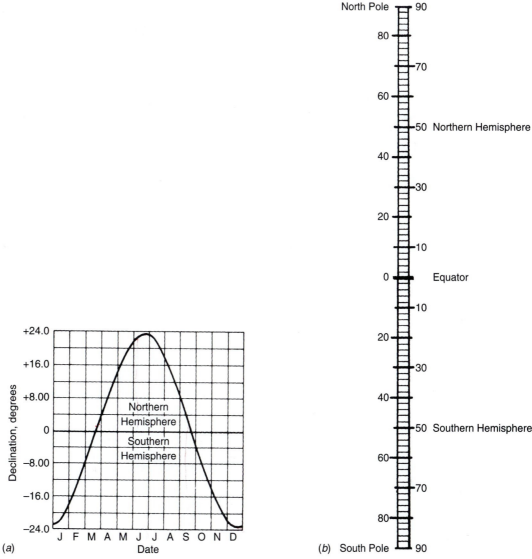

Fig. 16.4 (*a*) Sun declination chart. Read the chart by finding the date in question on the bottom line; then follow the nearest vertical line upward to the curved graph line. At that point, take the nearest horizontal line to the side of the chart and read the appropriate number. Be sure to read Northern or Southern Hemisphere. (*b*) Zenith angle chart. Find the latitude in question; then find the declination of the sun and count the number of degrees between the two. This will give you the zenith angle.

- ■ Zenith angle, *ZA*   = 27 degrees (Fig. 16.4*b*)
- ■ Sun angle, *SA*   = 90 degrees minus *ZA*, that is, 90°–27°
- ■ *SA*   = 63 degrees

Landscape effects on SA   Once the sun angle of a location is known, we can move to the local scale and examine the influence of the landscape—that is, how the sun angle varies with hills, valleys, buildings, and the like. Hillslopes and roofs that face the sun are brighter and warmer than those that face away from the sun. In addition, the angle changes from dusk to dawn so that slopes with an eastward component to their orientations are favored by the morning sun and those with westward components are favored by the afternoon sun.

Ground SA
To compute the influence of an inclined surface on local sun angle (let us call it **ground sun angle**), we must first determine (1) the sun angle on flat ground for that latitude; (2) the direction in which the slope faces; and (3) the angle of the slope (that is, its inclination in degrees). If the slope faces the noon sun, the sun angle on the slope is equal to the flat ground angle plus the angle of the slope. If the product is greater than 90 degrees, then subtract it from 180 degrees to get the appropriate angle. For slopes that face away from the sun, the sun angle is equal to the flat surface sun angle minus the angle of the slope. If the product is negative, the slope is in shadow.

$$\text{Ground sun angle} = SA \pm \alpha$$

where

Ground sun angle (SA_g) = sun angle on slope face
SA = sun angle on flat ground
α = angle of slope in degrees

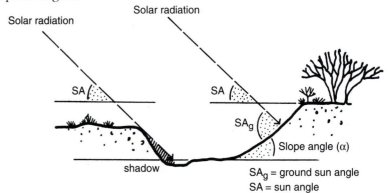

16.4 RADIATION BALANCE AND SOLAR HEATING

To determine the amount of solar heating on a surface, it is first necessary to understand that at ground level solar radiation can be disposed of in only two ways: reflection and absorption. The reflective capacity of a surface, called **albedo**, is expressed as the percentage of incoming solar (shortwave) radiation rejected by the surface:

Albedo

$$A = \frac{S_o}{S_i} \times 100$$

where

A = albedo
S_o = incoming shortwave (solar)
S_i = outgoing shortwave (solar)

All earth materials reflect a portion of the solar radiation that strikes them, but the values vary widely as Table 16.1 reveals.

Solar gain
The solar energy absorbed by a surface, which we will call **solar gain**, is equal to incoming shortwave (S_i) less the amount reflected (S_o):

$$\text{Solar gain} = S_i - S_o$$

This quantity represents energy added to the absorbing material in the form of heat, and it in turn will produce a rise in the material's temperature. The actual amount of temperature rise for a given amount of energy added will vary according to the material's composition. This means that equivalent amounts of heat in two different mate-

Table 16.1 Albedos for Various Surfaces

Material	Albedo, Percent
Soil	
Dune sand, dry	35–75
Dune sand, wet	20–30
Dark (e.g., topsoil)	5–15
Gray, moist	10–20
Clay, dry	20–35
Sandy, dry	25–35
Vegetation	
Broadleaf forest	10–20
Coniferous forest	5–15
Green meadow	10–20
Tundra	15–20
Chaparral	15–20
Brown grassland	25–30
Tundra	15–30
Crops (e.g., corn, wheat)	15–25
Synthetic	
Dry concrete	17–27
Blacktop (asphalt)	5–10
Water	
Fresh snow	75–95
Old snow	40–70
Sea ice	30–40
Liquid water	30–40
30° lat. summer	6
30° lat. winter	9
60° lat. summer	7
60° lat. winter	21

Source: From William D. Sellers, *Physical Climatology* (Chicago: University of Chicago, 1974). Used by permission.

Solar radiation to heat rials, say, water and soil, will not yield the same temperature. This is explained mainly by differences in a property called *volumetric heat capacity* (or *specific heat*), which for water is high compared to that for sand. (See Table 18.1 in Chapter 18.)

Taking into consideration the fact that solar radiation usually strikes surfaces in the landscape at an angle, it is necessary to combine the concept of incident radiation flux (flow over the receiving area) with albedo to determine the solar heating for a surface. This can be done computationally using just three variables: the ground sun angle (based on latitude, date, and surface inclination or slope), the intensity of solar radiation, and the albedo of the surface:

$$SH = S_i (1 - A) \sin SA_g$$

where

SH = solar heating in cal/cm^2 • min
S_i = incoming solar radiation in cal/cm^2 • min
A = albedo (1 – A gives the percentage absorbed)
SA_g = ground sun angle in degrees

Examples of solar heating It is instructive to see how important slope and albedo are in the solar heating of a varied landscape. For example, given the surfaces (at 45 degrees north latitude) represented by the profile in Figure 16.5, the rates of solar heating at noon on the equinox would be as follows:

1. Building Roof

- slope = 45°
- orientation = south
- albedo = 10%
- SA_g = 90°
- S_i = .78 cal/cm² • min

Solution:

$$SH = .78 (1 - .10) \sin 90°$$
$$= .78 (.9) 1.0$$
$$= .70 \text{ cal/cm}^2 \bullet \text{min}$$

2. Concrete Wall

- slope = 30°
- orientation = north
- albedo = 27%
- SA_g = 15°
- S_i = .78 cal/cm² • min

Solution:

$$SH = .78 (1 - .27) \sin 15°$$
$$= .78 (.73) .26$$
$$= .15 \text{ cal/cm}^2 \bullet \text{min}$$

3. Plowed Field

- slope = 5°
- orientation = south
- albedo = 22%
- SA_g = 50°
- S_i = .78 cal/cm² • min

Solution:

$$SH = .78 (1 - .22) \sin 50°$$
$$= .78 (.78) .77$$
$$= .47 \text{ cal/cm}^2 \bullet \text{min}$$

4. Sandstone Slope

- slope = 25°
- orientation = south
- albedo = 40%
- SA_g= 70°
- S_i = .78 cal/cm² • min

Solution:

$$SH = .78 (1 - .40) \sin 70°$$
$$= .78 (.60) .94$$
$$= .44 \text{ cal/cm}^2 \bullet \text{min}$$

The concrete wall gains the least energy, about one-third that of the field and slope and about one-fifth that of the roof. We should note in passing that this pattern of solar heating in the landscape could be changed dramatically with some simple

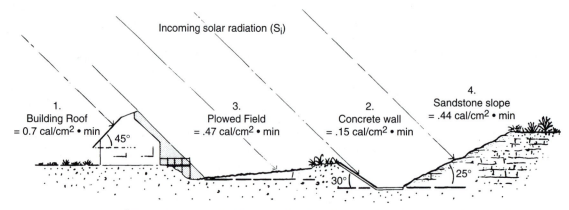

Fig. 16.5 Variation in solar heating related to slope and surface materials at four locations.

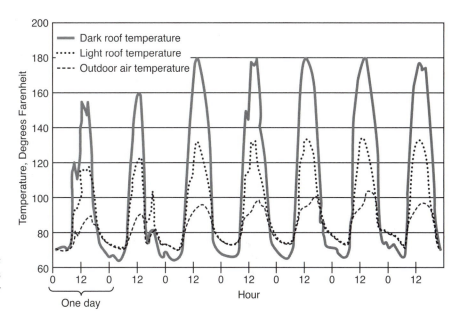

Fig. 16.6 Variations in the temperature of light-colored and dark-colored roofs based on field measurements. The lower line represents air temperature.

Thermal microclimate

and about one-fifth that of the roof. We should note in passing that this pattern of solar heating in the landscape could be changed dramatically with some simple design alterations. For example, light colored materials would cool the roof significantly (Figure 16.6), and the shade of a plant cover would help bring down ground temperatures on the plowed field and sandstone slope.

The influence of these variations on ground-level climate, called **microclimate**, depends on many additional factors, including (1) how much of the solar energy absorbed by a surface is returned to the air over it as heat (either as sensible heat or by longwave (infrared) radiation, which in turn may heat the air); (2) local wind conditions (which account for the flushing rate of heated air from surfaces); and (3) the size of the area covered by each heating surface (which determines the relative balance of thermal influences among the different surfaces in an area). Under calm air conditions, the layer of air over surfaces such as these will begin to develop a pattern of temperatures roughly corresponding to the pattern of solar energy absorbed.

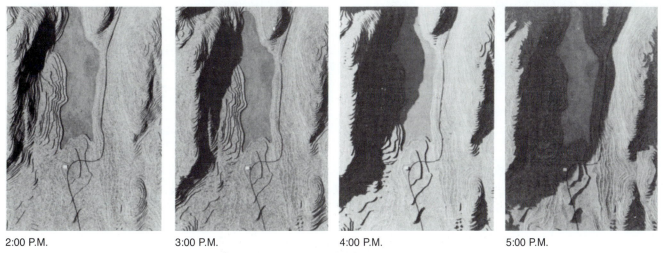

2:00 P.M.   3:00 P.M.   4:00 P.M.   5:00 P.M.

Fig. 16.7 Afternoon patterns of sunlight and shadow at Jordan Pond Valley, Acadia National Park, Maine, on August 1, based on a simulation model. The shadow slope will begin cooling by midafternoon following by cool air drainage in the late afternoon and evening.

may carry over well past the period of peak solar radiation and into the evening, given that regional weather systems do not obliterate it.

16.5 IMPLICATIONS FOR LAND USE, VEGETATION, AND SOIL

Land use impacts

To gain an idea of the impact of land use change on the gain of solar energy by the landscape, we can compare the differences in slope (both angle and orientation) and surface materials before and after development. This would involve first mapping the predevelopment slopes of various angles, orientations, and compositions, measuring their areas, and then computing their total solar gain over some time period. These figures would be summed for the entire project area and compared to the parallel figure based on the same computation for the postdevelopment landscape. Although many additional factors, such as means and extremes in air temperature, wind, precipitation, and humidity, would have to be taken into account to determine the climatic significance of a change in solar gain, such a comparison does provide one measure of the relative impact of different land uses and development schemes on the environment at ground level.

Vegetation patterns

Variations in incident radiation owing to differences in the orientation and inclination of slopes can have a profound influence on vegetation patterns and soil conditions. In semiarid mountainous areas, such as parts of Colorado, New Mexico, California, and British Columbia, the more direct sun angles on south-facing slopes result in greater surface heating and, in turn, higher rates of soil moisture evapora-

(a)   (b)

Fig. 16.8 (*a*) Differences in vegetation on north- and south-facing slopes. The north-facing slopes sustain forest cover, whereas south-facing slopes are limited mainly to grasses. (*b*) Shadow zone along north-facing cliffs on Lake Superior helps creates a cool microclimate conducive to certain arctic and subarctic plants.

tion and plant transpiration than on north-facing slopes. The resulting difference in soil moisture is often great enough to cause marked differences in vegetation on north- and south-facing slopes. The photograph in Fig. 16.8*a* shows one such example from southern Colorado, where moisture stress limits trees to north-facing slopes.

Slope erosion and landforms

In even drier areas, where only herbs and shrubs can survive, the plant cover on south-facing slopes is often measurably lighter than that on north-facing slopes. Because more ground is exposed, erosion by runoff may also be higher on south-facing slopes, resulting in higher densities of gullies and lower slope angles. In addition, there may be a difference in the abundance of certain species, with the more drought-tolerant species making up a higher percentage of the plant cover on south-facing slopes.

Plant species

Differences in plant species related to the influence of slope on incident radiation can also be found in humid regions, though the examples are rarely as obvious as those in dry areas. In the midlatitudes, combinations of heat and light may ensure the survival of certain plants on extreme slopes. For instance, north-facing cliffs along the south shore of Lake Superior harbor species of ferns and mosses that are separated by hundreds of miles from the main bodies of their populations in arctic and subarctic regions (Fig. 16.8*b*). Apparently, the low light intensities and cool temperatures along these cliffs have favored the survival of these plants since the end of the last continental glaciation about 10,000 years ago.

16.6 IMPLICATIONS FOR BUILDINGS AND LIVING ENVIRONMENTS

The placement and size of buildings and trees in cities can seriously affect the reception of solar radiation. With rising concern over solar energy and building cooling, this issue has gained significance in urban planning and design. "Shadow corridors" and "solar windows" (or gaps) are two of the most common solar features of cities

Solar windows and corridors

(Fig. 16.9). **Solar windows** are narrow spaces between tall buildings through which the solar beam passes to ground level. Depending on the orientation and spacing of the buildings, the shaft of light may illuminate a patch of ground for only a short time each day, making it difficult to maintain street plants and virtually impossible to utilize solar radiation as a source of energy.

Shadow corridors are elongated zones, bordered by a continuous ridge of tall buildings that block the sun. In the most extreme situations, direct (beam) solar radiation is never received in such environments: the only light comes from diffused sky radiation and radiation reflected from nearby buildings.

The length of a shadow cast by a building or tree is a function of the height of the object and the sun angle; computations can be made using the formula:

$$S_l = \frac{b}{\tan SA}$$

where

S_l = shadow length
b = height of the object
SA = sun angle

tangents (tan) for angles 5° – 85°			
5° = .087	25° = .466	45° = 1.0	65° = 2.14
10° = .268	30° = .577	50° = 1.19	70° = 2.75
15° = .268	35° = .700	55° = 1.43	75° = 3.73
20° = .364	40° = .839	60° = 1.73	80° = 5.67
			85° = 11.43

This formula is traditionally used in site planning in areas of excessive heat and intensive solar radiation because it is necessary to provide for shade in pedestrian areas, parking lots, on building faces, plazas, and the like. (also see Section 17.6 and Fig.

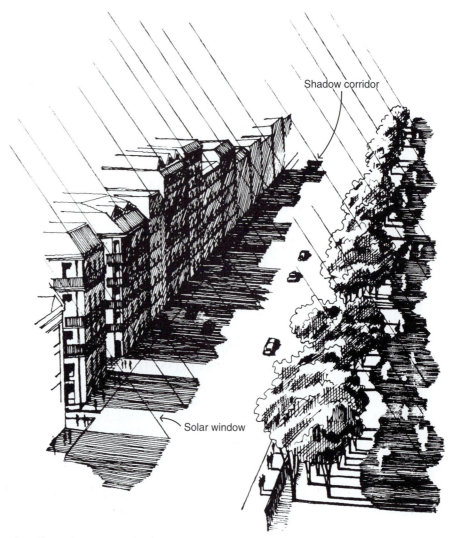

Fig. 16.9 The pattern of solar radiation in the urban environment as altered by tall buildings.

17.12). The need for shade is generally greatest in the hours between 11 A.M. and 4 P.M. when high solar intensities are coupled with high air ground temperatures (Fig. 16.10).

Tree shade   The efficiency of trees in providing shade is related to several factors including tree placement, foliage density, and canopy size and structure. The greatest shade efficiency is provided by forests with multiple storied structures such as the one shown in Figure 17.10 where there is an 80 percent reduction in solar radiation. In residential areas where only a single story of yard and street trees is possible and canopy size and density are the chief concerns, species such as red oak and sugar maple (in the North) and live oak (in the South) are particularly effective. Properly placed, a healthy, mature shade tree can reduce beam (direct) solar radiation by 50 percent or more.

Winter microclimate   While shade can be a distinct advantage for local pedestrians and residents of cities prone to hot summers and frequent heat waves, in many northern cities, such as Minneapolis, Detroit, Calgary, and Montreal, shadow corridors in winter encourage the buildup of ice and snow, making foot travel hazardous (Fig. 16.11). Moreover, the solar gain is very poor in these zones for living units with northerly exposures, resulting in cooler room temperatures and higher heating costs. This is

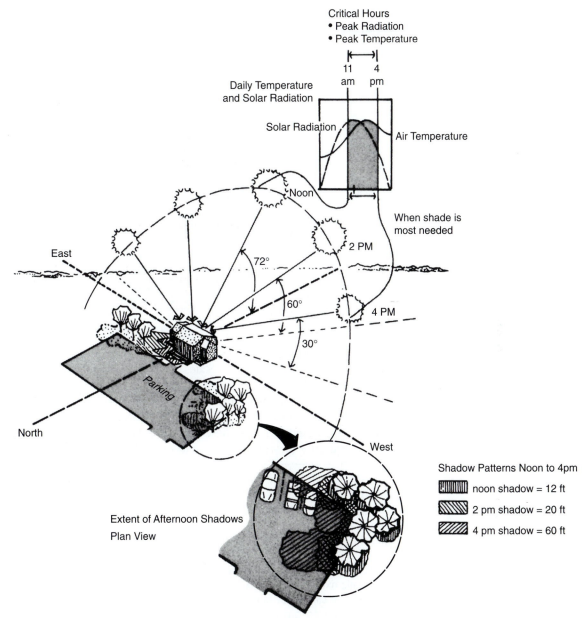

Fig. 16.10 Shadow patterns between noon and 4·p.m. associated with a building and trees near a parking lot. Shade is most critical between 11 a.m. and midafternoon, when air and ground temperatures are highest.

especially significant in light of findings in Great Britain and the United States concerning illness and death among the elderly brought on by hypothermia.

Hypothermia   **Accidental hypothermia**, a disorder characterized by low body temperature (near 90°F), slowed heartbeat, lowered blood pressure, and slurred speech, can be brought on in persons over age 70 by room temperatures as modest as 65°F, inadequate clothing, and prolonged periods of physical inactivity. Solar exposure may make a difference of several degrees in room temperatures, especially during cold spells, and in turn can tip the balance between hypothermia and a normal state of health in the elderly (Fig. 16.11). The United States National Institutes of Health estimate that more than 2 million elderly people in the United States are vulnerable to accidental hypother-

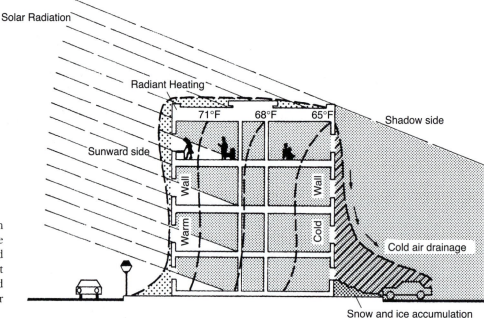

Fig. 16.11 Schematic diagram illustrating the effects of sun angle on winter living conditions in and around a northern apartment building. The north wall is cold on both the interior and exterior of the building.

elderly (Fig. 16.11). The United States National Institutes of Health estimate that more than 2 million elderly people in the United States are vulnerable to accidental hypothermia. Undoubtedly, many of these persons inhabit buildings whose orientation, design, and neighborhood exclude or greatly restrict access to direct solar radiation in living spaces. On the other hand, these same conditions may be an advantage during the summer because they are not prone to excessive heating.

16.7 CASE STUDY

Solar Considerations in Midlatitude Residential Landscape Design

Carl D. Johnson and Brian Larson

One of the primary objectives in residential and urban design is mitigation of the climatic extremes in spaces occupied by humans. In architecture the focus is on the internal climate of buildings, which is achieved through air conditioning, light control, and so on. In landscape architecture the primary concern is with outdoor spaces, and climatic modification is attained through the use of vegetation, siting of buildings, the use of different ground materials, and topographic features, either as they exist or as they could be constructed.

In the continental midlatitudes, discomfort from the cold poses a major restriction to the use and enjoyment of outdoor space. Therefore, in the design of modern residential complexes, it is desirable to achieve some modification of microclimate to encourage greater use of patio and yard space. Understandably, the level of modification that can be expected through landscape planning is relatively modest, particularly when set into a Minnesota or Quebec winter. On the other hand, small modifications of marginally cold or cool weather, such as that of spring and fall, are indeed possible, and days that would otherwise be uncomfortably cool can in fact be made to be quite pleasant through sensitive planning and design.

Newport West is a low-density townhouse development located on the northeast edge of Ann Arbor, Michigan. The development site is characterized by hilly terrain with a major swale running through it. The east-facing and south-facing slope of the swale, an old farmfield fringed by trees, was selected for the building site because it offered the greatest opportunity to optimize microclimate conditions and conserve building energy.

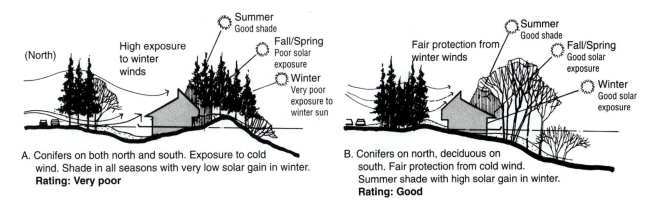

A. Conifers on both north and south. Exposure to cold wind. Shade in all seasons with very low solar gain in winter.
Rating: Very poor

B. Conifers on north, deciduous on south. Fair protection from cold wind. Summer shade with high solar gain in winter.
Rating: Good

1. The Alternatives; A, B, C

C. Mixed tree cover on north and south. Exposure to cold wind. Low solar gain in winter. Roof exposed to summer sun.
Rating: Poor

pockets were designed to provide comfortable outdoor spaces in fall and spring, and to afford habitats for exotic plants such as azaleas and rhododendrons. The townhouses were constructed with large south-facing windows to receive solar radiation and augment interior heating in fall and winter. On the southwest sides of the units, facades were protected from excessive heating by the afternoon summer sun with full crown deciduous trees.

After construction and landscaping were completed and the units were occupied, a set of temperature readings were taken in early March to determine the effec-

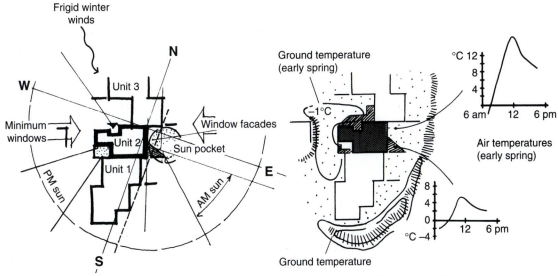

2. The Layout: Microclimate Vectors and Design

3. The Results: Air and Ground Temperatures

After construction and landscaping were completed and the units were occupied, a set of temperature readings were taken in early March to determine the effectiveness of the design of modifying microclimate. Ground temperatures varied substantially depending on solar exposure, and were 7°C higher at the 5 cm depth on inclines near south-facing walls compared to surfaces near north- and northwest-facing walls. In the patio spaces, air temperatures varied with shade and beam radiation receipt; in the southeast-facing sun pocket the daily high temperature (at surface level) on one bright day was nearly 10°C higher than in permanently shaded areas nearby. On cloudy days and windy days the difference was negligible, however.

An examination of the climatic records shows that in the U.S. Midwest a total of 10 to 20 days in fall and spring can be classed calm and sunny with uncomfortably cool ambient air temperatures. This brief study suggests that these sorts of conditions can be improved in near-building spaces through climate-oriented building design and siting. It also suggests that conditions for exotic plants are more favorable in sun pockets, although it should also be recognized that summer heat may be excessive, and protective shading with deciduous trees may be necessary. The study also implies that residential units that offer both warm and cool outdoor exposures are preferable to those with single exposures in the warm/cold climates because they increase the opportunities for the seasonal use of outdoor space. This consideration is growing increasingly important as people are faced with smaller residences situated in community or neighborhood clusters with limited yard space.

Carl D. Johnson was a founding partner of Johnson, Johnson and Roy, Inc., Ann Arbor, Michigan, and Brian Larson is a senior landscape architect in Austin, Texas. ∎

16.8 SELECTED REFERENCES FOR FURTHER READING

American Institute of Architects Research Corporation. *Solar Dwelling Design Concepts.* Washington, DC: U.S. Department of Housing and Urban Development, 1976.

Buffo, John, et al. "Direct Solar Radiation on Various Slopes from 0 to 60 Degrees North Latitude." *U.S.D.A. Forest Service Research Paper PNW-142,* 1972.

City of Davis (California). *A Strategy for Energy Conservation.* Davis, CA: Energy Conservation Ordinance Project, 1974.

Griggs, E. I. , et al. "Guide for Estimating Differences in Building Heating and Cooling Energy Due to Changes in Solar Reflectance of a Low-Sloped Root." ORNL Report 6527, Oak Ridge National Laboratory, 1989.

Heisler, G. M. "Effects of Individual Trees on Solar Radiation Climate of Small Buildings". *Urban Ecology,* 9, 1989, pp. 337-359.

Huang, J. et al. "The Potential of Vegetation in Reducing Summer Cooling Loads in Residential Buildings." *Journal of Climate and Applied Meteorology* 26, 1987, pp. 1103–1116.

Land Design/Research, Inc. *Energy Conserving Site Design Case Study, Burke Center, Virginia.* Washington, DC: U.S. Department of Energy, 1979.

Marsh, William M., and Dozier, Jeff. "The Radiation Balance." In *Landscape: An Introduction to Physical Geography.* Reading, MA: Addison-Wesley, 1981, pp. 21–35.

National Institute on Aging. "A Winter Hazard for the Old: Accidental Hypothermia." Washington, D.C.: U.S. National Institutes of Health, Department of Health, Education and Welfare, 1981, Pub. no. (NIH) 78-1464.

Sizemore and Associates. *Methodology for Energy Management Plans for Small Communities.* Washington, DC: U.S. Department of Energy, 1978.

Sterling, Raymond, et al. *Earth Sheltered Community Design.* New York: Van Nostrand Reinhold, 1981.

Tuller, S. E. "Microclimatic Variations in a Downtown Urban Environment."

17

MICROCLIMATE, AIR POLLUTION, AND THE URBAN ENVIRONMENT

17.1 INTRODUCTION

Urban development can cause significant changes in atmospheric conditions near the ground. In extreme situations, such as in the heavily built-up areas of larger cities, these changes extend hundreds of meters into the atmosphere and are of such magnitudes that they produce a distinct climatic variant, the urban climate. Generally speaking, the urban climate is warmer, less well lighted, less windy, foggier, more polluted, and often rainier than the regional climate in which it is situated.

The dimensions of urban climate

Within the urban landscape microclimatic variations can also be considerable. Air quality may be exceptionally poor along transportation corridors and in industrial sectors. Residential sectors may be warmer than average in summer and in the inner city some areas may get so hot during heat waves that people perish from heat stress. On the other hand, areas between tall buildings may receive little or no beam radiation, obtaining much of their energy in winter from the heat loss of adjacent buildings.

These variations can be important considerations in urban planning and design. Documentation of the desirable climatic effects of vegetated areas, for example, helps provide a rationale for the inclusion of parks and greenbelts in urban master plans. Today, transportation planning in American and Canadian cities invariably includes air quality guidelines and goals, and proposals for industrial development must include forecasts on gaseous and particulate emissions and plume patterns under different atmospheric conditions. But that is not the case with urban thermal conditions, and it is not the case in the developing world where most urban growth will take place in this century.

Within a few decades, several urban centers will exceed 50 million, and one, Shanghai, will approach 100 million or more. Growth will be accompanied by an infusion of automobiles, rising air pollution, massive areas of hard surface, and unprecedented levels of consumption, all of which will combine to drive urban climate to greater and greater extremes. In some sectors summer heat levels will be especially extreme, particularly when combined with the forcing effects of global warming. This chapter summarizes what we know about urban climate and offers some recommendations for climate consideration in urban planning and design.

17.2 THE URBAN HEAT ISLAND

Urbanization transforms the landscape into a complex environment characterized by forms, materials, and activities that are vastly different from those in the rural landscape. Not surprisingly, the flow of energy to and from the urban landscape is also different. As a whole, the receipt of solar radiation is substantially lower, while the generation of sensible heat (heat of dry air) at ground level is greater in cities compared to the neighboring countryside. Furthermore, the rate of heat loss from the

Urban heat balance

urban atmosphere through outflows of infrared radiation and flushing of heated surface air is lower. On balance, the increase in sensible heat output coupled with lower rates of heat loss is more than enough to offset the reduced input of solar radiation, resulting in somewhat higher air temperatures in urban areas throughout most of the year and much higher temperatures on selected days. The spatial pattern of these temperatures is often concentric around the city center, producing a **heat island** in the landscape. The geographic extent and intensity of the urban heat island varies with city size (based on population) and with regional weather conditions. In general, large cities under calm, sunny weather produce the strongest heat islands (Fig. 17.1).

Urban boundary layer

The overall structure of the urban atmosphere can be envisioned as a large dome centered over the urban mass. This body of air is called the **urban boundary layer**. Because of a heavy particulate content, it is highly efficient in back-scattering solar

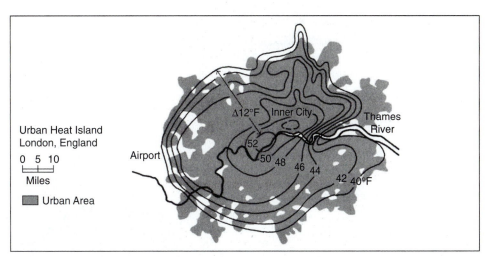

Fig. 17.1 The configuration of the urban heat island over London, England, on a winter day. The difference in temperature from the city center to the perimeter is 12°F.

radiation, often effecting a reduction as high as 50 percent in the lower 5000 feet above a city. A growing amount of research findings also reveal that increased cloudiness and precipitation are associated with the urban atmosphere and that these trends carry downwind to neighboring areas in the urban region. Thus, it appears that the domelike structure of the boundary layer is pronounced only during relatively calm atmospheric conditions, whereas during a steady airflow across the region, the dome is tipped downwind and develops a plume that diffuses over a regional airshed.

Surface heating At ground level the causes underlying the formation of the urban heat island, and indeed the urban climate in general, are many and complex. *First*, the materials of the urban landscape possess different reflective and thermal characteristics than those of the rural landscape. For many cities, the albedos of street and roof materials are lower than the landscapes they displaced, resulting in higher levels of solar absorption, and for cities in forested regions, the urban surface is less shaded than the tree-covered one it supplanted. The *volumetric heat capacities* (or specific heat) of street and building materials are lower than those of materials such as moist soil in the rural landscape (see Table 18.1 in Chapter 18). This means that urban surfaces generally reach a higher temperature with the absorption of a given quantity of radiation and, in turn, heat the overlying air faster. *Second*, the **Bowen Ratio**, which is a measure of the heat released from a surface in the sensible form relative to the latent form, is much higher in cities owing to the limited areas of open water, vegetation, and exposed soil. With a paucity of vapor sources, *latent heat flux* from the surface is relatively low; conversely, *sensible heat flux* is relatively high, giving rise to higher air temperatures. Added to this is the heat released from artificial sources (automobiles, buildings, etc.). In the midlatitudes, these sources typically contribute more energy to the interior of a large city such as New York City in mid-winter than the solar source does.

Heat loss Heat inputs represent only one side of the system. The other side is, of course, heat outputs, or losses, from the urban atmosphere. The principal consideration in this regard is wind speed because wind is responsible for flushing away the envelope of heated surface air. Because of the aerodynamic roughness added by tall buildings, cities tend to have much lower wind speeds at ground level; therefore, heated air tends not to be flushed away as readily as it is in rural landscapes. Furthermore, the urban atmosphere tends to retain more heat because of a higher carbon dioxide content. On balance, then, the urban landscape yields and retains more heat, thereby accounting for the overall heat island effect.

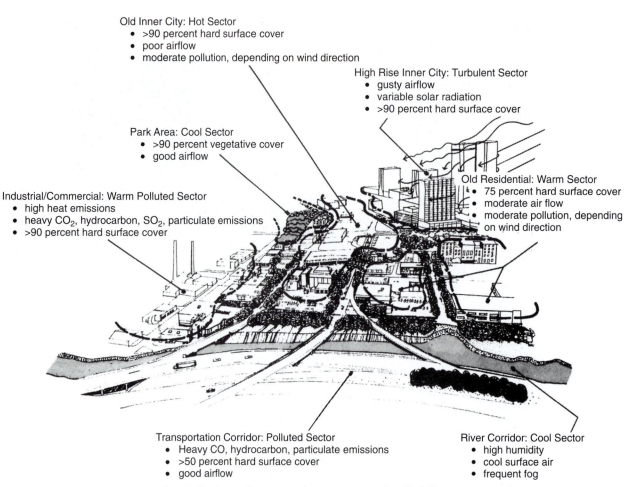

Old Inner City: Hot Sector
- >90 percent hard surface cover
- poor airflow
- moderate pollution, depending on wind direction

High Rise Inner City: Turbulent Sector
- gusty airflow
- variable solar radiation
- >90 percent hard surface cover

Park Area: Cool Sector
- >90 percent vegetative cover
- good airflow

Old Residential: Warm Sector
- 75 percent hard surface cover
- moderate air flow
- moderate pollution, depending on wind direction

Industrial/Commercial: Warm Polluted Sector
- high heat emissions
- heavy CO_2, hydrocarbon, SO_2, particulate emissions
- >90 percent hard surface cover

Transportation Corridor: Polluted Sector
- Heavy CO, hydrocarbon, particulate emissions
- >50 percent hard surface cover
- good airflow

River Corridor: Cool Sector
- high humidity
- cool surface air
- frequent fog

Fig. 17.2 Microclimate conditions associated with different sectors of a city. Conditions vary with surface cover, solar radiation, airflow, and air pollution among other things.

17.3 MICROCLIMATIC VARIATIONS WITHIN THE URBAN REGION

Although the climate of an entire city is an important issue for air pollution control boards and regional transportation planners, the most important issues to the urban planner, landscape architect, and architect are climatic variations within the urban area. Perhaps the easiest variation to visualize is that associated with solar radiation around tall buildings, but other parameters including temperature, wind, fog, and pollution also show considerable variation within the urban landscape (Fig. 17.2). With the exception of air pollution, the nature and significance of these variations are generally not well documented. But it is widely agreed that the extremes in these atmospheric conditions within the urban landscape affect the health, safety, and comfort of a significant number of people in most cities. The following paragraphs provide brief summaries of five key parameters of the urban climate.

Solar shadows and exposure

Solar Radiation   Beam radiation is intercepted by buildings, and, depending on sun angle and building height, a shadow of some size is created. Where buildings are closely spaced, a **shadow corridor** may form (see Fig. 16.9 in Chapter 16). For a single building site, incoming solar radiation varies with season, wall orientation, and time of day, and in cold climates, there may be distinct differences in heating and comfort from the sunny to the shadow sides of buildings such as the one shown

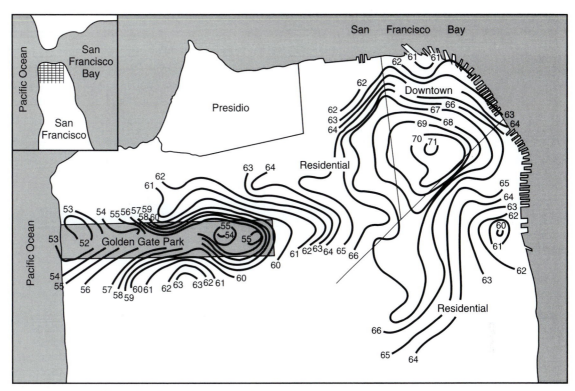

Fig. 17.3 The influence of a large park on the temperature of San Francisco contrasts sharply with the heat island over the heavily built-up downtown area.

in Figure 16.11. Similarly, building sides with high solar exposure may be subject to extreme heating in summer, especially during heat waves when high ambient air temperatures combine with solar heating. In the 1995 heat wave that killed 700 people in Chicago, more than half the victims lived on the top floor of their buildings, where, of course, solar heating was most extreme. On the other hand, pockets sheltered from beam radiation by tall buildings may perpetually be in shade and lighted only by diffuse solar radiation. These pockets can be cool and damp, especially where airflow is poor. Coupled with the refuse that often collects in back lanes and alleyways, the resultant environmental conditions can be very unsavory.

Thermal differentiation **Temperature** Most cities are geographically diverse in surface materials, physical forms, and activities, and we would expect settings as different as people parks and industrial parks to develop markedly different temperature regimes. Studies show, however, that this is so only where thermal variations are not masked by strong regional weather systems or extreme local influences on climate. The latter is exemplified by a small park of vegetation in the midst of an inner city; whatever modification in temperature is achieved by the park is masked by the thermal umbrella of the surrounding mass of buildings.

Cool pockets Thermal modification of the urban heat island by a large park or greenbelt can be significant, however. One set of readings made in San Francisco shows how extreme the cooling related to a large park can be (Fig. 17.3). The temperature difference between Golden Gate Park and the downtown near Union Square was more than 15° F. Other investigations have shown older residential areas with mature trees to be cooler than new residential areas and other urban surfaces. In Washington, D.C., it is not uncommon for the corridor of parks and water features along the Potomac River Valley to be cooler during summer days and evenings than the heavily built-up areas on either side of it.

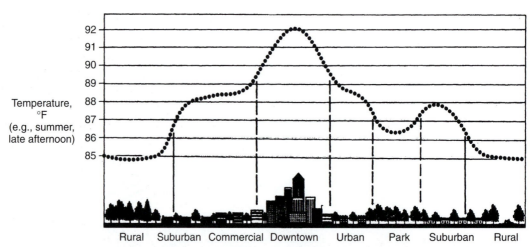

Fig. 17.4 A schematic diagram depicting the nature of the urban heat island and its boundary on the urban fringe.

Heat island border

On the perimeter of the city, the urban heat island may decline sharply where the urban landscape quickly gives way to the rural landscape. This is illustrated by the London, England, example shown in Figure 17.1. Pictured in a temperature profile, this sort of border is characterized by a *cliff* in the graph line. From a planimetric perspective, the border configuration appears to be very irregular in detail with cool inliers represented by parks and river corridors and warm outliers represented by large shopping centers and industrial parks (Fig. 17.4).

Wind and Convective Mixing The general influence of a city on airflow is to reduce wind speed at levels near the ground. This can be illustrated by comparing the profiles of wind speed over urban and rural surfaces. The elevated topography of the urban environment displaces the **wind velocity profile** upward, leaving a thicker layer of slow-moving air near the ground (Fig. 17.5). At a more detailed level of observation, however, large variations in wind speeds can be found within rela-

Wind velocity profile

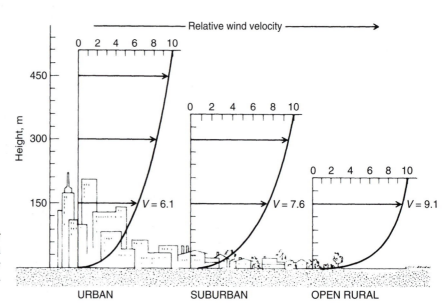

Fig. 17.5 Profile of wind velocity over urban and rural landscapes. Although ground-level wind velocities are markedly lower in cities, turbulence tends to be higher because of tall buildings.

Fig. 17.6 Airflow over and around buildings. Highest velocities are reached on the windward brow and across the roof of the tallest building. A strong flow of air is also deflected down the building face (A), but a calm zone develops in the space between the building (B). Accelerated flow (C) is associated with the canyon between large buildings.

Airflow around buildings

tively small areas. Much of this variation is related to the size, spacing, and arrangement of buildings. Three example are noteworthy. First, in the case of an individual building, the structure represents an obstacle to airflow and, in order to satisfy the continuity of flow principle, wind must speed up as it crosses the building. In a two-dimensional model, the highest speeds are reached on the windward brow of the building and across the roof. Air is also deflected from the brow down the face of the building (labeled A in Fig. 17.6); on the leeward side, speeds decline and streamlines of wind spread out and some descend toward the ground.

Where two tall buildings of similar heights are spaced close to each other, the streamlines of fast wind do not descend to the ground but are kept aloft by the roof of the second building. This gives rise to a small pocket of calm air between the buildings where mixing with the larger atmosphere of the city is limited (labeled B in Fig. 17.6). Depending on local conditions, the air in such pockets, as we noted earlier, may be measurably different from the surrounding atmosphere.

The third example involves the alignment of buildings and streets. Streets bordered by a continuous mass of tall buildings have the topographic character of canyons and, if aligned in the direction of strong winds, tend to channel and constrict airflow. This produces higher wind velocities at street level, especially during gusts, and increased turbulence along the canyon walls (labeled C in Fig. 17.6).

Contributing factors

Fog and Precipitation The incidence of fog in cities may be twice that of surrounding country landscape. Most of this is usually attributed to the abundance of condensation nuclei (particulates) from urban air pollution. Local concentrations of fog are more common in selected areas, especially under calm atmospheric conditions coupled with strong night-time cooling at the surface. Several contributing factors can be identified besides air pollution, and one is related to the availability of water vapor near the ground. In low-lying coastal areas and in river valleys, the concentration of vapor may be appreciably higher than elsewhere in the city. In addition, cold air drainage into low-lying areas promotes fog development. Conversely, heated buildings and hard surfaces may locally reduce fog development because they tend to limit the normal rate of fall in night-time air temperatures.

Precipitation is also greater in some urban areas. In the Detroit metropolitan region, for example, annual precipitation in some areas is as much as 8 inches greater than the mean annual values in the surrounding nonurban areas. Other urban centers

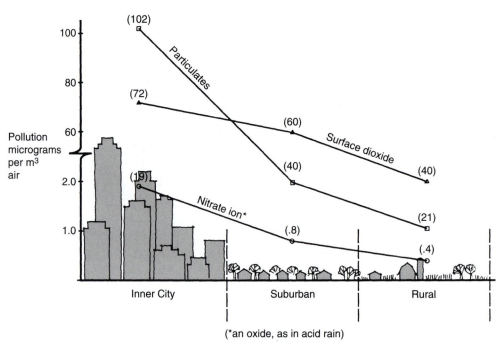

Fig. 17.7 Air quality changes from the inner city to suburbia and the rural landscape beyond.

such as St. Louis and Chicago may induce greater precipitation downwind from the city, whereas other large cities have little apparent influence on precipitation rates even though the incidence of fog and clouds may be greater. The chief cause of increased precipitation is undoubtedly the higher incidence of particulates in the urban atmosphere. Instability (rising air) caused by the urban heat island may also be a factor.

Air Pollution   Although a body of heavily polluted air may blanket an entire urban region under certain atmospheric conditions, pollution levels are on the average *Spatial distribution* higher in the inner city than in surrounding suburban areas (Fig. 17.7). In addition, pollution levels on many days vary sharply from one quadrant or sector of an urban area to another. Two factors account for this: (1) the site-specific nature of many pollution sources such as power plants, highway corridors, and industrial plants; and (2) short-term changes in the mixing and flushing capacity of the urban boundary layer. During windy and unstable weather, pollutants are mixed into the larger mass of air over the city and flushed away, thereby limiting heavy concentrations, if any, to relatively small zones downwind of discharge points.

During calm and stable atmospheric conditions, however, pollutants tend to build up over source areas, and if these conditions are prolonged, the concentrations coalesce to form a composite mass over the urban region. This is especially pronounced with *particulates*, considered the most harmful pollutant to human health, *Urban dust dome* which often form a visible **dust dome** over cities (Fig. 17.8). The largest particles are discharged from sources such as power plants, foundries, and construction projects, and tend to settle close to their sources. Near factory and power plant exhaust stacks, the fallout pattern can be traced in the distribution of dirt blanketing the neighboring landscape. Intermediate-sized particles usually remain suspended for several days and spread over a much greater volume of air than large particles. They and smaller particles contribute to the formation of huge dust domes that may envelop an entire urban region. The class of smallest particles, which are less than 1.0 micrometer (one-millionth of a meter) in diameter, may remain suspended in the atmosphere for sev-

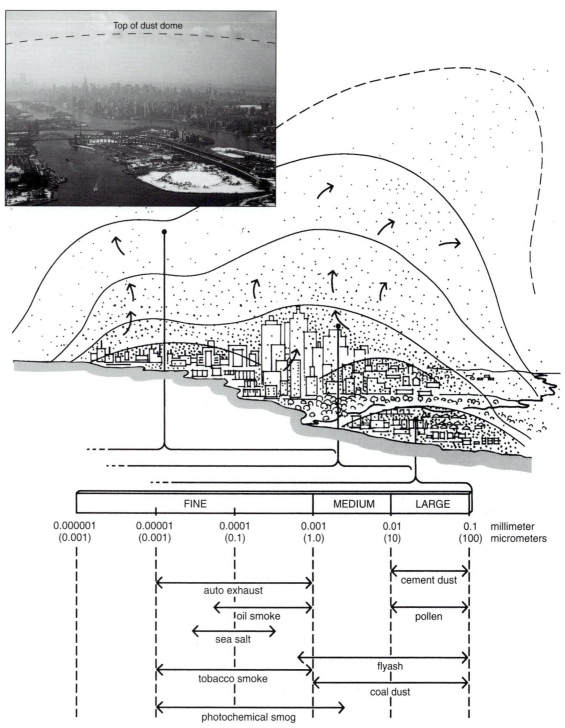

Fig. 17.8 A schematic portrayal of a dust dome over a large city during relatively calm atmospheric conditions. Larger particulates tend to be concentrated closer to their sources. The inset photograph shows the hazy conditions associated with the dust dome over New York City on a winter day.

eral weeks or months, because the slightest motion of the air moves these particles upward and laterally. Most gaseous pollutants also have long residence times in the atmosphere. If they are lighter than air, such as methane and CFCs, they may float upward in the atmosphere, where they remain until they are altered chemically and/or recycled through the earth's surface environments.

Pollution mass balance   The mass balance of pollutants for a given volume of atmosphere can be estimated based on the total rate of pollutant emission and the rate of removal by airflow. Removal includes both lateral and vertical components; therefore, it is easy to imagine how heavy the buildup of pollution can become during a prolonged thermal inversion when airflow in all directions is negligible. Moreover, stagnation of polluted air increases the prospects for oxidation and photochemical processes involving sulfur dioxide, nitrogen oxides, and hydrocarbons, leading to the formation of **smog** containing sulfuric acid, nitric acid, and noxious gases such as ozone. Under severe episodes of air pollution, the only realistic management option (other than regulating people's outdoor activities) is short-term reduction in emissions. Officials in cities such as Los Angeles have actually restricted automobile traffic and industrial activity to avert a health disaster. Such decisions depend not only on the gross level of air pollution, but also on the levels of critical pollutants, especially hydrocarbons, oxides of nitrogen, sulfur dioxide, and airborne particles (Table 17.1).

Regional effects   The plumes of polluted air generated from metropolitan areas are known to extend tens, hundreds, and in extreme cases, thousands of miles beyond their source areas. The effects on regional climate are not well documented, but it is known that they are more pronounced in certain regions and seasons and are characterized by increased cloudiness, precipitation, and turbulent weather. Another regional effect has

Table 17.1   Major Air Pollutants and Their Sources

Pollutant	Source	Effects
Carbon monoxide and dioxide	Gasoline-powered vehicles Industry using oil and gas Building heating using oil and gas	CO_2 promotes heat retention and atmospheric warming CO enters human bloodstream rapidly, causing nervous system dysfunction and death at high concentrations
Sulfur oxides (sulfur dioxide and sulfur trioxide)	Industry using coal and oil Heating using coal and oil Power plants using coal, oil, and gas	Irritate human respiratory tract and complicates cardiovascular disease Damage plants, especially crops Promote weathering of building skin materials
Nitrogen oxides (nitric oxide and nitrogen dioxide)	Gasoline-powered vehicles Building heating using oil and gas Industry and power plants	Irritate human eyes, nose, and upper respiratory tract Damage plants Triggers development of photochemical smog
Hydrocarbons (compounds of hydrogen and carbon)	Petroleum-powered vehicles Petroleum refineries General burning	Toxic to humans at high concentrations Promote photochemical smog
Oxidants (secondary pollutants in smog including ozone)	Vehicle exhausts Industry Photochemical smog	Irritate eyes Damage plants Promote cancer
Particulates (liquid or solid particles generally smaller than 100 micrometers)	Vehicle exhausts Industry Building heating General burning Spore- and pollen-bearing vegetation	Some are toxic to humans Some pollens and spores cause allergic reactions in humans Promote precipitation formation

recently been added—**acid rain** in southeastern Canada and northeastern United States. Much of this is caused by the formation of sulfuric acid from the combination of atmospheric moisture and sulfur trioxide in polluted air. Because of the regional flow of weather systems across the Midwest, the acidic moisture is carried from industrial areas and precipitated in Ontario, Quebec, and New England where the biota of thousands of lakes and ponds have been damaged by the increase in water acidity.

17.4 AIR POLLUTION MANAGEMENT IN THE URBAN REGION

Air quality regulation

Under the **U.S. Clean Air Act** (originally enacted in 1970 and last amended in 1992), the U.S. Environmental Protection Agency (EPA) was directed to set air quality standards to protect human health with an adequate margin of safety. The resultant standards, called the **National Ambient Air Quality Standards (NAAQS)**, defined two levels of air quality standards: (1) primary standards designed to protect the health of people; and (2) secondary standards designed to protect against environmental degradation. The Clean Air Act amendments also set air quality standards for certain land use areas. Three classes of areas were defined for the purpose of preventing air quality deterioration beyond certain base performance levels.

- Class I. Areas such as national parks and wilderness areas have the highest standards (that is, where least deterioration is allowed).
- Class II. These are broadly defined as intermediate areas and have intermediate standards.
- Class III. Industrial urban areas have the lowest standards.

In practice, all areas are defined as Class II unless they are moved into Class I or Class III by the EPA.

In the 1970s and 1980s, pollution control in the United States focused on the least expensive measures and largest sources of pollution. Two major areas of activity, large industrial facilities and motor vehicles, were pressed with most of the responsibility for cleanup. Emission standards were implemented for automobiles that included exhaust system devices called catalytic converters on all new vehicles. In addition, lead was banned from use as a gasoline additive in new automobiles. Fuel economy standards, called **CAFE standards** (corporate average fuel economy standards) were also mandated. As a result of these actions, pollution rates per automobile dropped significantly and lead levels in air and water declined over most of the United States. The Great Lakes, for example, saw lead concentrations drop dramatically between 1970 and 1980. On the other hand, the number of automobiles in use increased significantly between 1970 and 1988 and the total miles traveled by Americans annually doubled (from about 1000 billion to over 2000 billion), both of which offset much of the gains made by reduced emissions and better mileage (Fig. 17.9).

Emission standards

In the late 1980s, further amendments were added to the Clean Air Act. These amendments address three major areas:

Current targets

- Reductions in sulfur dioxide emissions from power plants with a target of 10 million tons reduction per year over the 1990s.
- Reductions in hazardous air pollutants (mainly synthetic compounds such as benzene, and PCBs).
- Improved air quality in selected areas, mostly urban areas, where ambient air quality standards are consistently being violated.

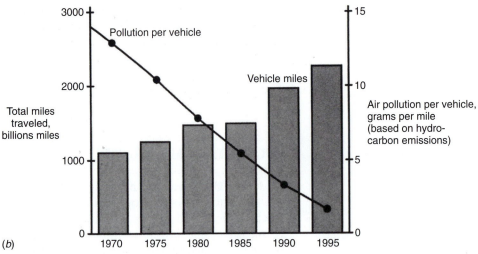

Fig. 17.9 Trends in automobile miles traveled and pollution emissions per automobile in the United States, 1970–1995. The improved emissions performance is being offset by increased travel.

The Los Angeles experience

In Southern California, where regional air pollution is extremely critical, various new management concepts are being considered. The Los Angeles Air Management District is implementing a program based on a market incentives concept. Under this program, polluters are allocated a pollution allowance, that is, a fixed quantity of allowable emissions. The total allowance for the district is designed to meet air quality standards set by the EPA. The novelty of the program is that it allows polluters—excluding the more than 8 million vehicles on the road daily—to buy and sell allocations. This rewards those who are able to achieve lower emission rates by allowing them to sell allocations for profit to less efficient operations. The overall balance, given both more efficient and less efficient operations, is designed to lower pollution levels over the district. Automobile pollution, on the other hand, is being addressed by strict implementation of state emissions standards.

The Los Angeles situation is an extreme example of the urban air pollution problem faced by virtually every metropolitan center in the United States. The principal source of the problem is increased automobile traffic brought on by population growth, increased use of cars, and longer travel distances. Expressways in and around large cities typically carry an astonishing 200,000 to 300,000 vehicles daily.

Automobiles, roads, and sprawl

Many cities have sought various means of reducing the traffic load, including the use of ride sharing and mass transit systems. At the same time, however, the states and the federal government continue to build expressways, which encourages more automobile travel over greater distances as well as continued urban sprawl.

In fact, highway-induced sprawl is now giving rise to a new urban form, the **edge city**, at key interchanges around and between urban regions. Served exclusively by automobile and truck traffic, edge cities foster even greater travel, especially among commuters. Daily one-way travel distances between home and workplace of 50 miles or 75 miles are not uncommon today. The 1990 U.S. census revealed that the trend toward long commuting distances continues to rise; today nearly 25 percent of the workers in the United States actually drive to another county to work compared to 15 percent in 1960. On balance, the United States, and Canada as well, have developed an urban transportation system that is very expensive for the environment, the citizen, and local units of government. It relies overwhelmingly on cars and trucks to move people and materials and encourages long daily travel distances.

This in turn not only promotes massive fuel consumption and air pollution, but necessitates a huge and expensive highway infrastructure, sprawling land use patterns, and highly inefficient communities.

17.5 CLIMATE IN URBAN PLANNING

Scientific understanding

Planners and designers widely recognize climate as an important element in urban planning, but few have been able to incorporate variables such as wind, heat, and solar radiation effectively into the information base for decision making. Several factors are responsible, and one of the most important is the level of scientific understanding of microclimate in the built environment. Although architects and engineers understand many of the influences of climate on a building, for example, wind stress, solar exposure, and corrosion of skin materials, comparatively little is known about the influence of buildings, especially building masses, on microclimate. In contrast to urban hydrology, for example, technical planning is able to provide fewer models and less accurate forecasts to guide the urban planner in setting the heights of buildings, the balance between hard surfaces and vegetative surfaces, the widths of streets, and the like.

Finding regulations

A second factor is the general lack of *planning regulations* pertaining to climate. Although it is widely recognized that urban climate affects the health and well-being of people, resulting among other things in greatly increased medical costs, few ordinances have been enacted establishing climate performance standards for environments outside buildings. As noted, the exceptions are in the area of air quality. National regulations on industrial and automotive emissions seek to improve living conditions in cities. In local transportation projects involving federal funds, a transportation master plan is required that takes air quality into consideration. In locales that are subject to severe episodes of air pollution, such as Los Angeles County, local agencies are responsible for regulating the outdoor activity of schoolchildren, and, under emergency conditions, for reducing automotive and industrial activity. Beyond examples related to air quality, however, planning agencies pay little attention to climatic parameters such as temperature, airflow, fog, and radiation. Prospects for change in this state of affairs are not good where a direct relationship to human safety or to capital costs is not apparent, that is, where direct savings to individuals, companies, agencies, or institutions are not clearly evident. A case in point was the emergence of solar energy-oriented communities in the 1980s where ordinances on "solar rights" were legislated because access to the sun could be given some economic value.

Climate in EIS

Climate is invariably addressed in environmental impact statements, but it usually consists of descriptions of existing conditions with some "forecasts" about potential changes given a proposed action. Only cases involving air quality changes are a source of serious concern and may be the basis for recommending against or altering a proposed action. As for changes in physical components of climate, no guidelines or performance standards have been established to aid planners in formulating plans and reviews. As a result, in most cases no one is quite sure how much importance should be ascribed to a suspected change in some aspect of physical climate. Thus the issue is often relegated to the bin of unused information.

17.6 CLIMATIC CRITERIA FOR URBAN PLANNING AND DESIGN

Striking balance

Throughout the world cities are growing and urban leaders are being pressed to accomodate new facilities, technologies, and people. This demands that the urban planner and designer be on constant watch for opportunities to improve pedestrian movement, commuter traffic, land use, air quality, wastewater disposal, and so on.

The overriding challenge is to achieve a proper balance between the economic functions that are necessary to the city's existence and an environment that allows for the health and well-being of the populous. Planning to improve climate and the quality of the atmosphere is a good example of this challenge, judging from the on-going debate in the United States over whether to relax pollution standards in order to improve the economy.

Five climatic factors influence the comfort and health of people in most urban environments; air temperature, humidity, solar radiation, wind, and air pollution. While little can be done in designing cities to combat regional atmospheric conditions, such as those governed by the movement of air masses, measures can be taken to minimize thermal extremes and high levels of air pollution associated with microclimates within the city. Basically, only four types of climatic controls or changes are possible through urban planning and design, given the goal of improving living conditions in cities prone to excessive heat and air pollution:

Planning objectives

1. *Reduce summer solar radiation* by shading critical surfaces, for example, pedestrian walks, waiting areas, and busy streets.

2. *Reduce the abundance of concrete and asphalt*, and increase the amount of vegetation and open water. This will create higher volumetric heat capacities and greater rates of latent heat flux, thereby lowering air temperatures.

3. *Increase airflow at ground level* to flush heated and polluted air away from the city.

4. *Reduce pollution* by decreasing emission rates, improving flushing rates, and locating discharge points to minimize impact on heavily populated sectors.

Scale and orientation

In applying climatic factors to urban design, it is important first to consider the scale of the problem. For problems of citywide scope, the location, structure, and layout of streets, building masses, and industrial parks must be weighed against airflow patterns, sources of pollution, such as existing traffic corridors and industrial parks, and the ratio of open to developed space. In inland cities the maintenance of ground-level airflow in summer is very important; therefore, inner city street corridors should be wide, aligned with prevailing winds, and kept free of major obstacles.

Where vacant land is available in and around built-up areas, it should be converted to vegetation, and, to as great an extent as possible, vegetation should be expanded into inner city areas along streets, in pocket parks, and on rooftops. Where pollution is a problem, seasonal patterns of airflow and weather should be considered in locating industry, power plants, and the like. In the midlatitudes, winter weather often produces plume patterns that direct polluted air toward the ground. Where this phenomenon is known to occur, polluting activities should be situated so as to minimize their impact on residential areas.

At the scale of individual blocks of buildings, attention must be given to orientation with respect to airflow and solar radiation and to building sizes and forms. The relative heights of buildings in masses must be taken into account. For example, to minimize the nuisance and danger of wind gusts to pedestrians, studies have shown that the taller of two adjacent buildings should not exceed the shorter by more than twofold. Generally speaking, the taller the buildings, the more attention should be given to the effects of spacing, orientation, form, skin materials and other factors on microclimate.

Vertical zonation opportunities

In considering the vertical dimension of urban climate, it is important to ask what level (elevation) is most appropriate for different human activities. Clearly, ground level in the inner city has several distinct disadvantages including heavy air pollution and competition with automotive traffic. Similarly, high elevations pose the hazard of high-speed winds that can damage structures and impair human safety. At the middle level, however, in the four- to ten-story range, air is generally cleaner than that at ground level but substantially less windy than that at higher elevations, providing

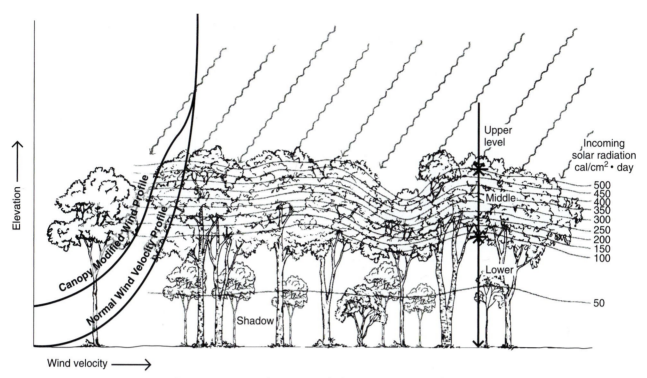

Fig. 17.10 Vertical zonation of climate the canopy of a large forest and the related levels of solar radiation. The middle level holds promise for human habitation in cities.

a somewhat healthier and more comfortable climate. Therefore, where heat and pollution at ground level are a problem it seems that rooftop spaces and balconies in this zone offer promise for expanded human use if, as New Yorkers and others are finding, appropriate landscaping can be introduced to ensure shade and safety. This concept is similar to the model of bioclimatic zonation in large, tropical forests where the middle level of the canopy is the optimum climatic zone for many creatures, the top being too windy and hot and the floor too humid and shaded (Fig. 17.10).

Rooftop storage of stormwater may also be desirable in modifying the urban climate. Upon evaporation, large quantities of sensible heat are taken up and released with the water vapor, thereby cooling the roof surface and the air over it. In Texas, for example, as much as 80 inches of water can be evaporated from open water surfaces in the average year—twice the average annual precipitation for most of the state. Therefore, it is easily possible to dissipate the total annual quantity of stormwater if the water can be held on rooftops.

Streetscapes and heat stress   At the street scale, consideration must be given to the potential for thermal stress on people in waiting and walking spaces. This is especially critical in cities that record official temperatures above 90°F on many days per month in summer, because in thermal microclimates such readings translate into temperatures above 100°F (Fig. 17.11). In addition to high temperatures, intensive solar radiation, poor airflow, high humidity, and physical exertion also contribute to **heat stress** or heat syndrome. Thus, along pedestrian corridors with high solar exposures, poor air circulation, and long walking distances, the potential is great for heat syndrome among walkers. To avoid this, shaded rest stops with good ventilation should be provided at appropriate locations. The distribution and location of stops should be based on origin and destination patterns for different walkers, elderly and disabled especially.

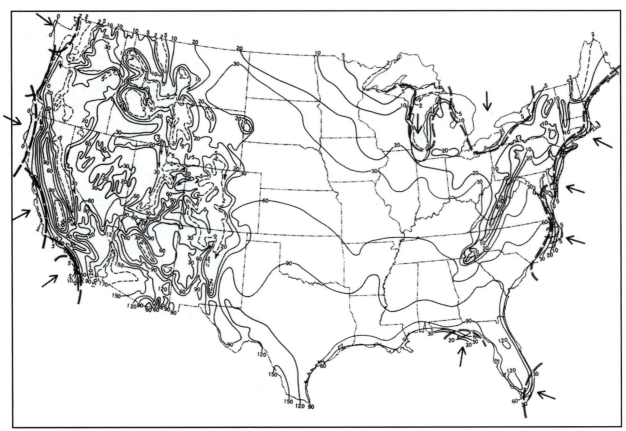

Fig. 17.11 The distribution of the average number of days per year with air temperature reaching 90°F or more. The arrows identify relatively cool coastal locations.

Finally, it is necessary to evaluate the performance of completed urban design projects based on microclimatic factors related to health, safety (e.g., wind), energy costs, and aesthetics. With respect to heat stress, performance standards can be based on general criteria such as those presented in Figure 17.12, and would require field

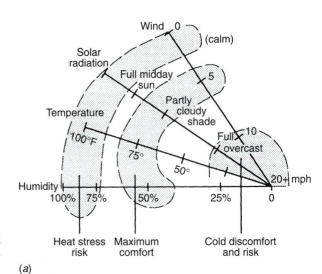

Fig. 17.12 (*a*) A generalized climatic comfort chart that can be used to evaluate urban environments for their suitability for human occupation.

(*a*)

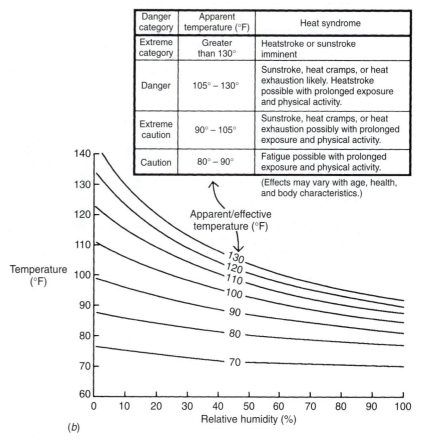

Danger category	Apparent temperature (°F)	Heat syndrome
Extreme category	Greater than 130°	Heatstroke or sunstroke imminent
Danger	105° – 130°	Sunstroke, heat cramps, or heat exhaustion likely. Heatstroke possible with prolonged exposure and physical activity.
Extreme caution	90° – 105°	Sunstroke, heat cramps, or heat exhaustion possibly with prolonged exposure and physical activity.
Caution	80° – 90°	Fatigue possible with prolonged exposure and physical activity.

(Effects may vary with age, health, and body characteristics.)

Fig. 17.12 (*b*) A heat stress chart based on air temperature and relative humidity.

measurements of humidity, wind, temperature, and solar radiation in the various types of spaces occupied by people at critical times of the day. Following design evaluation, modifications should be made to improve performance.

17.7 CASE STUDY

■ Modifying Urban Climate and Reducing Energy Use Through Landscape Design

W.M. Marsh

Almost everywhere cities are growing, and as they expand, the size and intensity of the urban heat island increases. While this is taking place, regional climate is heating up with global warming. Peak temperatures in Los Angeles have increased by 5 degrees F or more since 1940, primarily in response to growth of the urban region. Researchers estimate that the added cost of air conditioning for the 5 degree increase is at least $150,000 per hour. Similar trends in other large U.S. cities are producing increases in energy costs for air conditioning of $50,000 to $100,000 per hour. A rough estimate of the total air conditioning costs related to the urban heat island in the United States is $1 million per hour or $1 billion per year.

To reduce the urban heat island significantly would require major changes in the way we design, operate, and live in cities. To begin with we can reduce air conditioning demands by making just two modifications in the urban landscape: (1) more light colored, that is, higher albedo, building materials, and (2) more shade

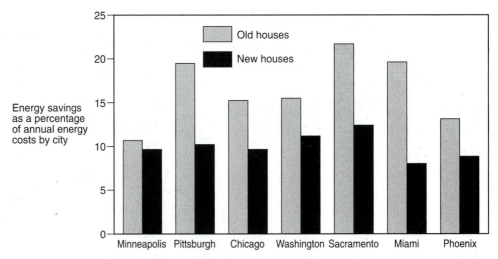

(*a*) Energy savings for air conditioning resulting from an
increase in albedo from 30 to 40 percent (from Taha, 1988).

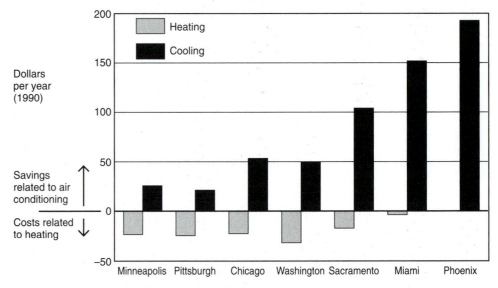

(*b*) Energy savings and added cost window costs do take
into foliage changes and wind buffering (from Huang 1990).

trees, that is, less direct solar radiation on houses and streets. Studies in California show that a mature tree cover can reduce the daytime temperatures in residential neighborhoods by up to 9 degrees F, and tree shade on residential buildings can lower air conditioning costs by 40 percent or more, depending on housing type. A change from dark to light roof and walls can reduce home air conditioning costs for some houses in some cities by as much as 22 percent. Extended over an entire city, the use of lighter materials could reduce heat island temperatures by as much as 5 degrees F. Because less fossil fuel would be burned to generate power for cooling, the indirect benefits of these energy-reducing measures would also include lower carbon dioxide inputs to the atmosphere and healthier living conditions with cleaner, cooler urban air.

Based on computer simulations, researchers found a difference in the effects of increased albedo between older, poorly insulated houses and newer houses with better insulation. In older houses, an increase in albedo from 30 to 40 percent produced an energy savings for air conditioning of 11 to 22 percent. In a well-insulated house, the same albedo change produced an energy savings of 8 to 13 percent (Graph A). Although few data are available for summer/winter comparisons, it is apparent that the real advantage in changing urban landscapes to higher albedo materials is found in southern cities. And, for northern cities some consideration must also be given to the heating advantage of lower roof albedos in winter.

To some extent, the same holds true for tree shade. Ideally, we should design for more shade in summer and less shade in winter as is illustrated in Case Study 16.7. Therefore, the choice of tree types and their placement along streets and around residential buildings are very important. The general advantages and disadvantages of summer and winter tree shade in terms of energy use in northern and southern cities are shown in Graph B. Again, the data are based on computer simulations. As landscape architects know all too well, community officials are often reluctant to add sizable street trees to infrastructure projects. With these findings, however, it is now possible to carry the debate beyond the usual case for the role of trees in creating livable neighborhoods and into the realm of dollar savings, to say nothing of the implications for improving urban climate. Indeed, the cost of trees and their maintenance can be measured against energy savings over the life-cycle period of the trees.

At the scale of residential lots, research reveals that the least expensive way to reduce air conditioning costs is to shade the air conditioning unit itself because these machines are less efficient at high temperatures. The shade from trees, tall shrubs, or a vine-covered trellis can reduce temperatures around the air conditioner by 6 or 7 degrees F, which can improve the efficiency of the air conditioner by as much as 10 percent. In addition to shade, siting of the air conditioner unit to take advantage of the cooler, north-facing side of the building is an obvious first step. Finally, it is critical in locations with heavy air conditioning needs to provide shade on windows and walls with high solar exposure. Here it is possible with an efficient shade system to reduce residential air conditioning costs by 40 percent or more. Trees are the best source of wall and window shade not only because they sharply reduce solar radiation but because they also cool the air around them through transpiration. Trees should be planted so that, near maturity, the limbs reach within 5 feet of the east or west walls and 3 feet of the south wall. This will help create a cool envelope between the trees and the house, thereby reducing building skin temperatures and lowering heat penetration. ∎

17.8 SELECTED REFERENCES FOR FURTHER READING

American Society of Landscape Architects Foundation. *Landscape Planning for Energy Conservation*. Reston, VA: Environmental Design Graphics, 1977.

Berry, Brian J. L., and Horton, F. E. *Urban Environmental Management: Planning for Pollution Control*. Englewood Cliffs, NJ: Prentice–Hall, 1974.

Chandler, T. J. *The Climate of London*. London: Hutchinson and Co., 1965.

Duckworth, F. S., and Sandberg, J. S. "The effects of Cities Upon Horizontal and Vertical Temperature Gradients", *Bulletin American Meteorological Society* 35, 1954, pp. 198-207.

Ellis, F. P. "Mortality from Heat Illness and Heat-aggravated Illness in the United States." *Environmental Research* 5:1, 1972, pp. 1–58.

Federer, C. A. "Trees Modify the Urban Microclimate." *Journal of Arboculture* 2, 1976, pp. 121–127.

Huang, Y. J. et al. The Wind-Shielding and Shading Effects of Trees on Residential Heat and Cooling Requirements. *ASHRAE Transaction*. Atlanta, GA: Society of ASHRAE, 1990.

Landsberg, H. E. "The Climate of Towns." In *Man's Role in Changing the Face of the Earth*. Chicago: University of Chicago Press, 1956, pp. 584–606.

Marsh, William M., and Dozier, Jeff. "The Influence of Urbanization on the Energy Balance." In *Landscape: Introduction to Physical Geography.* Reading, MA: Addison-Wesley, 1981.

Oke, T. R. *Boundary Layer Climates.* New York: Halsted Press, 1978.

Oke, T. R. "Towards a Prescription for the Greater Use of Climatic Principles in Settlement Planning." *Energy and Buildings* 7, 1984, pp. 1–10.

Olgyay, Victor, *Design with Climate.* 4th ed. Princeton, NJ: Princeton University Press, 1973.

Quayle, R., and Doehring, F. "Heat Stress." *Weatherwise*, 34, June 1981, pp. 120–124.

Taha, N. G. *Site-Specific Heat Island Simulations: Model Development and Application to Microclimate Conditions.* LBL Report No. 24009. Berkeley, CA: Lawrence Berkeley Laboratory, Univ. of California, 1988.

Thurow, C. *Improving Street Climate Through Urban Design.* Chicago: American Planning Association, Planning Advisory Service Report 376, 1983.

18

GROUND FROST, PERMAFROST, LAND USE, AND ENVIRONMENT

18.1 INTRODUCTION

Practically everywhere in the landscape we can see the direct or indirect influences of ground heat. The germination of many seeds depends on ground temperature. Evaporation of soil moisture is influenced by soil heat and urban heat islands intensify with the heating of ground under dry, unshaded streetscapes. Permafrost, which occupies 25 to 30 percent of the land area of this planet, is a form of ground frost that can place severe stress on most modern land uses. In North America, the largest areas of permafrost lie in Canada and Alaska. The Trans-Alaska Pipeline and several similar projects in Canada have brought national attention to permafrost environments. Chief among the concerns is the impact of development on the tundra, one of the least disturbed of the major ecosystems on earth. This concern becomes more acute each decade with rising political and economic pressure to open up permafrost lands to oil drilling, iron ore mining, and other extractive activities (Fig. 18.1).

Though less serious than permafrost, seasonal ground frost can be an important consideration in planning and engineering facilities in the midlatitudes. In particular, water pipes and sewer lines must be laid below the frost line, and building foundations and roadbeds must be designed to minimize disruption and damage from frost. Despite modern road designs, frost is a major source of pavement damage, a fact vividly reflected in the contrasting condition of paved roads and the associated costs of highway maintenance in northern and southern states.

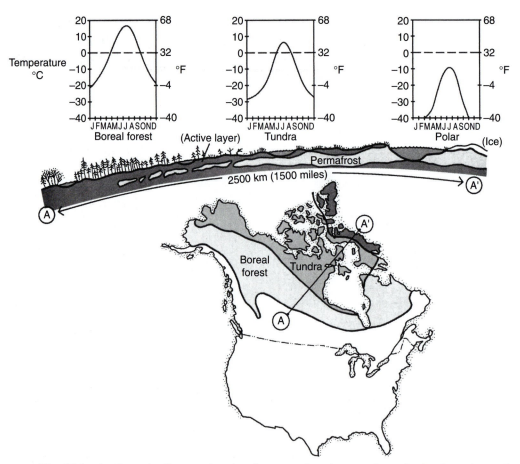

Fig. 18.1 A schematic diagram showing the permafrost layer under polar land, tundra, and boreal forest and the atmospheric temperatures regimes associated with each.

18.2 DAILY AND SEASONAL VARIATIONS IN SOIL HEAT

Soil heat flow

The thermal condition of the soil is a product of the atmospheric climate over it. As weather and climate change, so changes ground heat, particularly in response to solar heating and air temperature fluctuations. Differences in temperature between the soil and the base of the atmosphere set up vertical energy exchanges. When the ground surface is relatively warm and the underlying soil cool, then heat flows into the soil. When the soil is warmer than the atmosphere, heat flows out of the soil into the overlying air. Hardly ever are the soil and atmosphere at the same temperature, because the atmosphere is subject to such rapid, large temperature changes, whereas the soil is slow to change temperature, especially at depth.

The diurnal cycle

We can examine soil heat flow in various time frames, beginning with a *diurnal period* or 24-hour day/night cycle. On a summer day, for example, the ground surface may heat to a temperature of more than 100°F, while just a foot or two below, the soil temperature may be only 70°F. The heat flow under these conditions is downward, of course, but because soil is not a good conductor, the total penetration is only 2 to 3 feet into ground before the sun sets and the supply of surface heat is lost. The heated surface continues to give up heat to both underlying soil and the overlying air, and at some point later in the night, the surface cools to a temperature even lower than that of the underlying soil, perhaps 50°F. The thermal gradient is now reversed from the daytime gradient and the heat flow reverses as well. The heat gained by the soil during the day flows upward at night. This variation in soil heat and daily directions of flow goes on to a lesser or greater extent on most days of the year at all terrestrial locations experiencing light and dark phases. Depending on the intensity of surface heating and the ability of the soil to conduct heat, there is a maximum depth, called **diurnal damping depth**, that marks the limit of daily temperature change for every soil (Fig. 18.2).

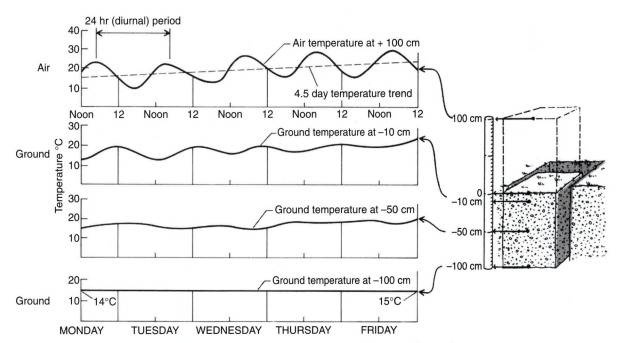

Fig. 18.2 Ground temperatures at three depths and their relation to air temperature over a 4.5-day period. The diurnal damping depth appears to lie close to 50 centimeters, for beyond that depth the daily variation in surface temperature is not apparent.

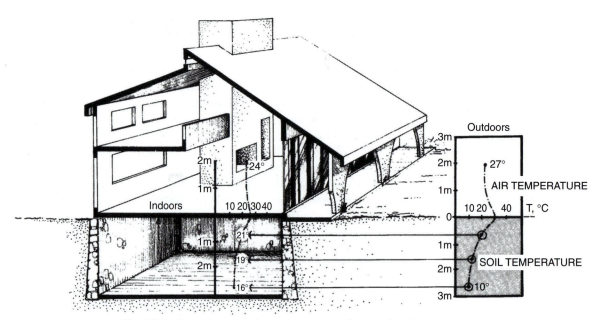

Fig. 18.3 Summer ground temperatures at depths of 1, 2, and 3 meters, compared to those at the same depths in the basement of a house.

Seasonal heat flow

The ground temperature also varies with the seasons. If we examine the average surface temperatures for summer, fall, winter, and spring, it is apparent that from winter to summer the soil should be heating up, while from summer to winter it should be cooling down. The depth of the seasonal change is much greater than that of the diurnal change, so the *seasonal damping depth* is much greater, on the order of 10 feet in the midlatitudes. However, owing to the time it takes heat to reach this depth, the soil does not reach its maximum temperature until a month or more after the surface has reached its maximum. Thus ground heat and surface heat are always out of phase with each other. The heat seasons in the soil lag behind the heat seasons on the surface and in the atmosphere. Therefore, the soil is always cooler in summer and warmer in winter than the surface. These facts have some important implications for building design and energy conservation. In regions with warm summers and cold winters, such as southern Canada and the U.S. Midwest, subterranean structures have a distinct thermal advantage over above-ground structures. Basements, for example, are cooler in summer, and in winter, a basement is less expensive to heat than a comparable structure above ground (Fig. 18.3).

18.3 CONTROLS ON SOIL HEAT AND GROUND FROST

Thermal parameters

The rate at which heat flows into and out of the soil depends on two main factors: (1) the *thermal gradient* (temperature differential) between the soil at some depth and the surface; and (2) the soil's *thermal conductivity* which is determined by its composition (Table 18.1). In the first column of Table 18.1, the thermal conductivities are given for nine different earth materials. Notice that sand and clay conduct heat better than organic material, and that conductivity increases with soil moisture content. Organic matter is a very poor heat conductor, about 10 times slower than sand and clay, whether wet or dry. As a result, organic matter often serves as an effective thermal insulator, which helps explain why permafrost is particularly prominent and lasting in areas of muck and peat soils.

Table 18.1 Thermal Properties of Some Common Earth Materials

Substance	Thermal Conductivity[a]	Volumetric Heat Capacity[b]
Air		
Still (at 10°C)	0.025	0.0012
Turbulent	3,500–35,000	0.0012
Water		
Still (at 4°C)	0.60	4.18
Stirred	350.00 (approx.)	4.18
Ice (at −10°C)	2.24	1.93
Snow (fresh)	0.08	0.21
Sand (quartz)		
Dry	0.25	0.9
15 percent moisture	2.0	1.7
40 percent moisture	2.4	2.7
Clay (nonorganic)		
Dry	0.25	1.1
15 percent moisture	1.3	1.6
40 percent moisture	1.8	3.0
Organic soil		
Dry	0.02	0.2
15 percent moisture	0.04	0.5
40 percent moisture	0.21	2.1
Asphalt	0.8–1.1	1.5
Concrete	0.9–1.3	1.6

[a] Heat flux through a column 1 m² in W/m when the temperature gradient is 1°K per meter.
[b] Millions of joules needed to raise 1 m³ of a substance 1°K.

Thermal diffusivity

The rate at which a given temperature, such as the freeze line (0°C), actually moves into the soil is somewhat different than is suggested by the conductivity value. This rate, called *thermal diffusivity*, is a product of the volumetric heat capacity (given in the second column of Table 18.1) and the conductivity of the soil. Diffusivity is highest at moisture contents between 8 and 20 percent. Thus, in saturated soils, frost penetration is usually not as great as it is in damp soils, owing to the higher heat capacity of soggy soil.

Ground cover

The landscape also plays a major part in ground frost penetration, in particular, vegetation, snow cover, and land use. Among other things, snow cover and vegetation dampen surface wind velocities, which tends to reduce soil heat loss. Land use, on the other hand, has a variable effect. For instance, a building reduces heat flow from the soil, whereas a barren highway usually helps to increase it (Fig. 18.4a). The combined effects of land use, vegetation, and snow cover can be dramatic. Where forest cover has been cleared for agriculture or urban development and snow has been removed by plowing and/or wind, soil heat loss and frost penetration are maximized. The graphs in Fig. 18.4b illustrate the influence of snow cover for a grass-covered site in Minnesota. The key factor is the wind chill effect of fast moving cold air directly on or very close to the soil surface.

Snow cover

For most parts of North America, the depth of frost penetration is not well documented and is usually estimated from climatic records. The standard map of frost depth in the coterminous United States is based on representative winter temperatures and does not take into account factors such as snow cover or vegetation (Fig. 18.5). As a result, field measurements of frost penetration will often show appreciable deviation from this map for any winter. A snow-covered swamp in Maine may receive no ground frost, whereas a nearby airfield may receive 6 feet or more.

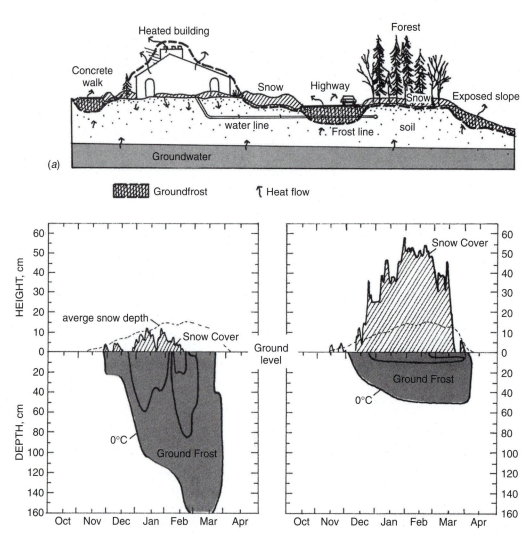

Fig. 18.4 (*a*) Schematic illustration of the influence of vegetation, land use, and snow cover on ground frost distribution and depth. (*b*) Measured frost penetration related to snow cover. In the first graph snow cover was light and frost penetration great; the second illustrates the opposite condition.

Estimating frost depth   When snow cover *is* taken into account, the following formula and graph can be used to estimate the depth of frost penetration in the northern United States and southern Canada. The formula combines snow depth and heating degree days to give degrees temperature per inch (or centimeter), denoted as T. Taking the result of a computation using this formula, we find that the maximum depth of the zero-degree isotherm for the winter is read from the graph, in the manner illustrated for 20°F per inch (4.4°C/cm). For this example, the depth of the zero-degree isotherm would be about 105 cm (41 inches).

$$T = \frac{\Sigma HDD_{\text{OCT-MAR}}}{\Sigma (S \bullet n)_{\text{NOV-MAR}}}$$

where

T   = degrees temperature per inch (or cm) (To find the frost depth, this figure is read into the vertical scale of the graph.)

where (cont.)

HDD = heating degree days, summed October through March
S = average monthly snow depth, inches
n = number of days with snow cover of 0.5 inch or more ($S•n$ is computed month by month and summed November through March.)

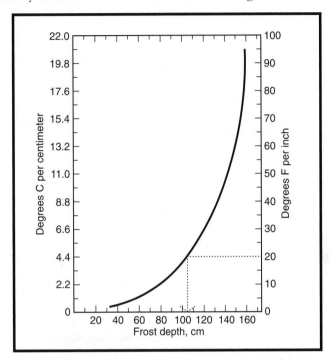

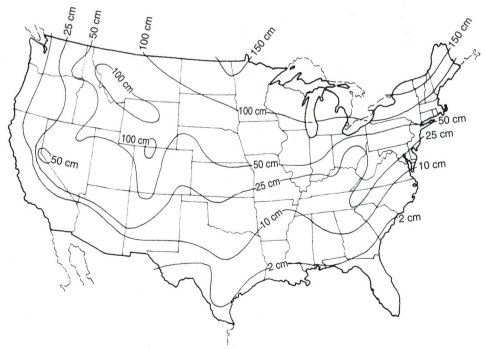

Fig. 18.5 Expected ground frost penetration by the end of winter. Departures from these values may be considerable, depending on the year and local snow, soil, and land use conditions.

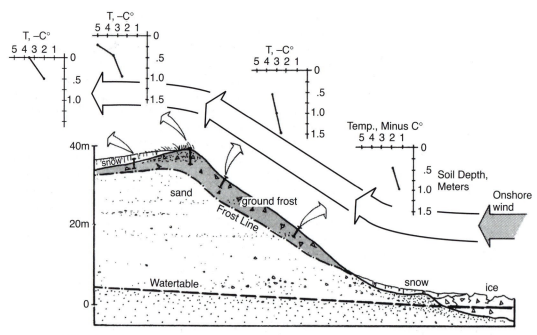

Fig. 18.6   Frost penetration on a sandy slope exposed to northerly wind blowing off Lake Superior. Wind velocity increases upslope, and heat loss increases, with highest wind velocity producing greatest frost penetration in the upper slope (also see Fig.15.8).

Slope exposure   In addition, the orientation and exposure of the ground can be critical; on barren slopes a southward exposure may make an appreciable difference in radiation receipts (see Fig. 16.7*a* in Chapter 16), whereas north-facing slopes may not only receive less radiation but also lose ground heat more rapidly because of exposure to northerly winds (Fig. 18.6). A related consideration on slopes is the upslope increase in wind velocity which induces accelerated rates of soil heat loss and frost penetration at higher elevations. The crests of high windy slopes may develop the deepest frost layers, especially if they are north facing.

Table 18.2   Ground Frost Susceptibility

	Low	*Medium*	*High*
Soil type	Organic	Wet mineral (clay, loams, sand)	Well-drained mineral (loams, sand)
Soil moisture	Saturated	Moist (near field capacity)	Damp (less than field capacity)
Vegetation	Heavy forest	Grass, low shrubs	Barren
Wind exposure	Low exposure to cold, fast wind (usually S, SW, SE facing slopes)	Intermediate exposure (such as E, NE, W facing slopes)	High exposure to cold, fast wind (usually N, NW facing slopes)
Snow cover	>50 cm (Nov.–Mar.)	10–50 cm (Nov.–Mar.)	<10 cm (intermittent cover throughout winter)
Solar exposure	South-facing slope >20%	Flat ground or locally irregular terrain	North-facing, shaded
Land use/cover	Parks, wetlands, stream corridors, woodlots	Pasture, planted fields, golf courses	Airfields, highways, parking lots, barren fields

On balance, then, we must take many factors into account when we attempt to forecast the pattern of ground temperature fluctuations and frost penetration, especially in areas of varied terrain. Unfortunately, mathematical models that integrate many variables are difficult to manipulate, require data from the field in order to be set up, and do not lend themselves to mapping programs. In environmental inventories for impact studies, master planning, or constraint studies we must instead often turn to simpler and less expensive methods in identifying areas susceptible to heavy frost penetration. One of these is the map overlay method in which individual maps showing the influences of topography, vegetation, snow cover, exposure, land use, and soils are superimposed on one another and the resulting combinations are designated high, medium, or low susceptibility. Each of the maps is usually broken down into categories that are numerically coded prior to overlaying; for example, soils might be classed as moist organic (1), moist mineral (2), and well-drained mineral (3), three being most susceptible to seasonal frost penetration. The results of this method do not indicate how deep frost penetration should be; rather, they indicate only the relative penetration, and may be used to isolate areas where more detailed analysis can be carried out (Table 18.2).

Mapping frost potential

18.4 PERMAFROST

From the southern border of the United States, the depth of ground frost penetration increases northward to a point in Canada where the inflow of summer heat is inadequate to melt the deeper winter frost completely away. The layer of frozen ground that remains is **permafrost**. Under persistently cold ground conditions, permafrost can penetrate hundreds of feet downward until its advance is offset by heat flow from the earth's crust. In varied terrain, permafrost first appears in isolated pockets on north-facing slopes where solar heating is weakest or under thick insulating layers of organic soil. In North America, such pockets of permafrost are reported as far south as 50° north latitude in the Canadian Shield and even farther south at high elevations in the Rocky Mountains, but their occurrence is very sporadic and usually hidden under several meters of soil material. These pockets mark the southern fringe of a broad permafrost region called the *discontinuous zone* that stretches across North America from Laborador in eastern Canada to the Bering Strait in western Alaska.

Discontinuous permafrost

Northward the patches of permafrost grow much broader and thicker and are overlain by a layer of soil called the **active layer**, which freezes and thaws seasonally. Near the Arctic Circle, the discontinuous zone gives way to the *continuous zone* where permafrost extends uninterrupted over vast areas of land, and in the extreme, reaches depths of a 1000 feet or more. The thickness of the active layer also changes northward with shorter and cooler summers. In the discontinuous zone it is generally 6 to 12 feet thick, but poleward declines to a very thin layer or disappears altogether in the upper reaches of the continuous zone (Fig. 18.7). Although permafrost is decidedly a terrestrial phenomenon, it is also known to extend under the shallow waters of the Arctic Ocean, an apparent relict condition of the Ice Age when sea levels were significantly lower.

Continuous permafrost

Nowhere is the seasonal flux of soil heat more apparent than in permafrost regions. In summer, the active layer develops with the penetration of heat from the surface (Fig. 18.8a). As the ground thaws the soil becomes saturated with meltwater, shallow ponds develop called *thaw-lakes*, and local runoff systems are activated. But with the onset of cold weather in fall, the heat flow reverses in the upper active layer, surface water freezes, local runoff ceases, and frost begins to penetrate the soil from the surface. Since the lower active layer is still thawed at this time, heat flows both

Summer melting

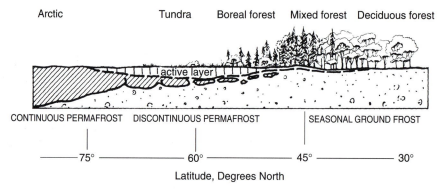

Fig. 18.7 Ground frost zones between 30° and 80° N latitude, and the vegetation associated with each.

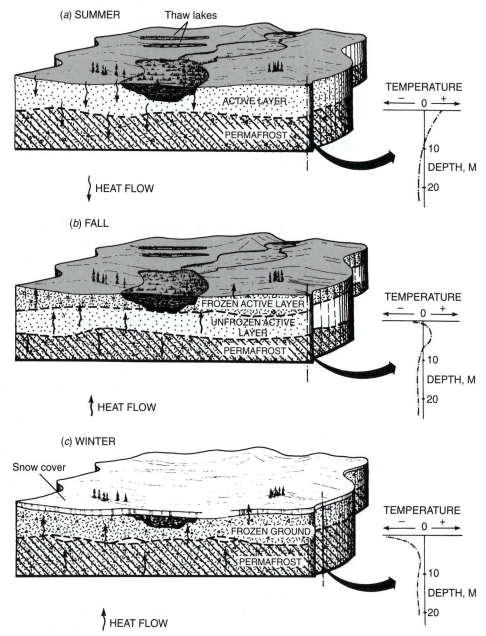

Fig. 18.8 The seasonal models of ground heat flow associated with permafrost illustrating the dynamics of the active layer.

Freeze up upward and downward from this relatively warm zone sandwiched within the frozen ground (Fig. 18.8*b*). The temperature profile assumes a spoon shape at this time; but in the ensuing months of winter, when the active layer freezes out completely and surface temperatures fall far below 0°C, the thermal gradient is fully reversed from that of summer, and the heat flow is upward. Paradoxically, the permafrost layer is the primary source of heat for the landscape during winter (Fig. 18.8*c*).

18.5 LAND USE AND FROZEN GROUND

Modern land uses have proven to be problematic in most permafrost regions. In the past half-century many military installations, railroads, highways, and communities have been built in Alaska, northern Canada, and Russia and have provided ample evidence to illustrate the nature of the problem. The energy flow (both heat and solar radiation) at ground level is altered first with the clearing of vegetation and surface grading. When a foreign material such as concrete or asphalt is placed on the ground,

Ground subsidence the thermal regime of the active layer is altered further, making it grow colder or warmer. In soils where ice comprises a large part of the soil bulk, thawing can reduce the permafrost (because of a volume reduction of about 9 percent as water changes from ice to liquid) and cause the ground surface to subside. This process leaves sink-hole-like depressions in the surface known as **thermokarst**. As the depressions form they collect water, thus causing further alteration of vegetation and soils that in turn may advance thawing and thermokarst development. When heated buildings, utility lines, or oil lines are set on the ground without adequate insulation to check the flow of heat into the active layer, ground subsidence can be more dramatic and very damaging to both structures and environment. In Russia, it is speculated that pipeline oil leaks from permafrost-related damaged and poor line maintenance are responsible for spilling millions of barrels of oil into the tundra, boreal forest, and Siberian streams every year (Fig. 18.9).

Drainage problems Other problems experienced by land use in permafrost regions include inadequate drainage in summer and unstable surface materials subject to settling and shifting. Along the Trans-Alaska Pipeline, for example, service roads were built on elevated gravel beds to minimize their thermal influence on the underlying permafrost, but the roadbeds often interrupted summer drainage, raising water levels in ponds and wetlands that may flood roadways. Drainage problems coupled with ground subsidence, water supply, and waste disposal problems have placed severe limitations on urban development in permafrost regions. In North America, the northern limit of urban development roughly coincides with the 0°C mean annual temperature line, which falls a few hundred miles south of the permafrost zone (Fig. 18.10).

Facility damage Frozen ground is also a problem outside permafrost areas. In northern Europe, southern Canada, and the northern United States, as well as in many other areas of the world, ground frost causes highway buckling, damage to building foundations, and freezing and breakage of water pipes. In the United States and Canada, building codes generally recommend that foundations be set below frost depth (usually given as 4 to 5 feet in the northern tier of states and in southern Canada) to minimize frost heaving and damage. But despite design codes and special engineering, watermains frequently freeze in extremely cold years in Minnesota, Wisconsin, Michigan, and Ontario, costing cities millions of dollars in repairs. In highway construction, frost heaving caused by the growth of ice lenses in the roadbed, called *ice segregation*, is a serious problem. To prevent it, gravel-based roadbeds are required because gravel does not transmit capillary water upward from the underlying soil fast enough to allow ice lenses to form under the cold concrete or asphalt.

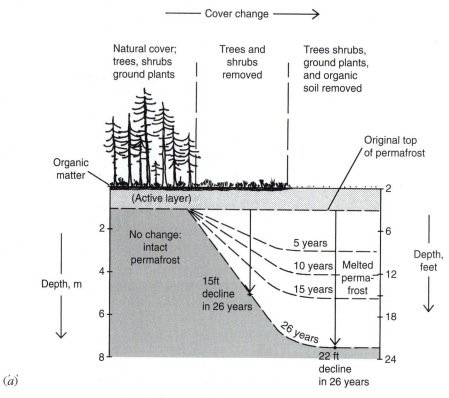

(b)

Fig. 18.9 Above (a) measured change in permafrost depth in response to land clearing. Below (b) subsiding caused by melting of the permafrost from the heat generated by a furnace in this house.

18.6 PLANNING AND DESIGN APPLICATIONS

With the exception of engineering design standards, such as those previously mentioned, little formal attention has been given to ground frost in community planning outside permafrost regions. Within permafrost regions the picture is quite the opposite, though permafrost is by no means universally recognized and addressed in planning methodology and practice, even in the most hostile settings. Although modern

The problem

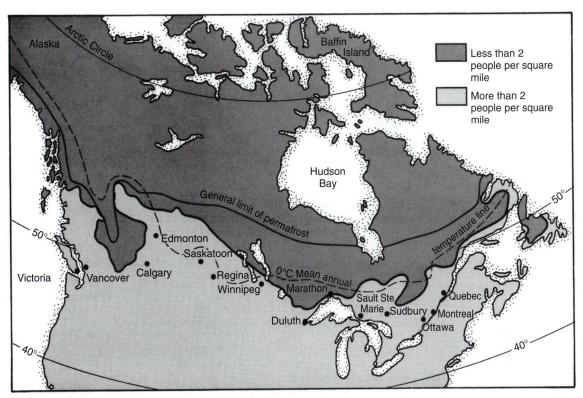

Fig. 18.10 The southern limit of permafrost, the 0°C mean annual temperature line, and the northern fringe of urban development in Canada.

engineering makes it possible to construct and successfully operate many types of infrastructure systems in permafrost environments, the design and management methods and techniques needed to mitigate the environmental impacts of these systems are not yet fully developed. At least part of the problem here is inadequate knowledge of the full range of impacts because of the sizable time lags that occur between an action and the resultant environmental impacts. In other words, some impacts from decades-old projects such as the Trans-Alaska Pipeline are still unfolding.

In the Fairbanks, Alaska, planning region, the Natural Resources and Conservation Service (NRCS) reviews development proposals with an eye to potential permafrost problems. The first level of evaluation involves checking the location of the proposed development against the distribution of soils known to have permafrost problems. Drawing on the results of permafrost research in the Fairbanks area, the NRCS has been able to classify the soils of the Goldstream, Saulich, Ester, and Lemeta series as those with greatest susceptibility to permafrost (Fig. 18.11). For projects that would involve these soils, the review may be taken to a second level of evaluation, which includes an examination of the types of activities and facilities actually proposed. In some cases the project may be compatible, because it is judged to be neither prone to damage from the environment nor itself of a significant threat to the environment. In other cases, modifications may be recommended, such as changes in building sites or the use of special engineering technology for footings and utility lines (Fig. 18.12). In still other cases, the project may be viewed as incompatible with the environment and not recommended for approval.

The NRCS recommendation is then passed on to the staff of the planning agency in charge, where it is combined with recommendations from other technical fields and interest groups to form a general recommendation on the proposal. This state-

Plan review and evaluation

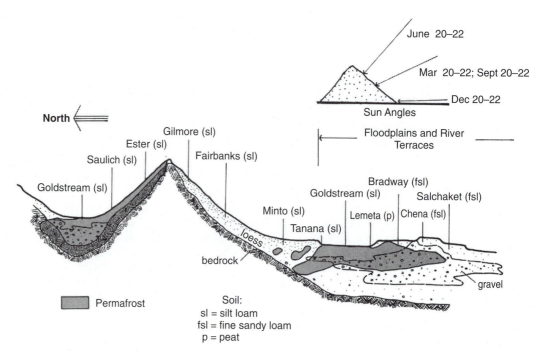

Fig. 18.11 Schematic diagram showing the locations of soils and their susceptibility to permafrost in the Fairbanks, Alaska, region.

ment is then studied and discussed by a decision-making body such as a planning commission, county board, or city council; a vote is taken; and the proposal is approved, denied, approved with specified reservations that ask the applicant to agree to certain changes, or sent back to the applicant or the reviewer for further work and documentation.

Fig. 18.12 Design modifications in residential development in response to permafrost, Norman Well, Northwest Territory, Canada. Homes, water lines, and sewage lines are built above ground to minimize thermal disturbance of the permafrost.

18.7 CASE STUDY

■ **Permafrost and the Trans-Alaska Pipeline**

Peter J. Williams

Remarkable as it now seems, the earliest proposals in the Alaska Oil Pipeline Project were to construct the pipeline in a conventional manner, burying it in the ground for virtually the entire distance of 800 miles from Prudhoe Bay to Valdez. In fact, the most basic problem was that the warm pipeline might thaw the underlying permafrost. Wherever the permafrost contained quantities of ice in addition to that within the soil pores, settlement or subsidence would inevitably follow thawing. At one planning stage serious consideration was given to using a chilled pipeline: cooling the oil so that the pipe could be buried without thawing the permafrost. This proposal was rejected because of the effect of low temperatures on the oil, which would not flow satisfactorily. The interaction of a chilled pipeline with the soil, which was to present problems for subsequently proposed gas pipelines, was apparently not foreseen.

Probably three-quarters of the proposed route overlay permafrost, and at least half of this was estimated to contain ice, which on melting would cause settlement. The effect of the warm pipeline was such that up to 30 feet of soil could be expected to thaw in the first year. The amount would vary, of course, depending on the pre-existing ground temperatures and the type of soil.

Near Prudhoe Bay the permafrost is some 2000 feet thick. Southward it becomes generally thinner, although the thickness is very variable. Thawing around a buried, warm pipeline would progress downward, although at a decreasing rate, for many years. The degree of consequent settlement, of subsidence, would depend on the amount of "excess" ice in the thawed layer, but quite often there would be several meters displacement. Obviously, such effects left unchecked would cause great disruption of the pipeline.

As soil surveys proceeded much more ground ice was discovered than was initially predicted. This, coupled with findings as to the amount of thaw that would occur, led gradually to the decision to build more and more of the pipeline above ground elevated on pile supports, rather than buried in the ground as first envisaged. The air passing beneath an elevated line would dissipate most of the heat from the pipe and greatly reduce the thawing of the permafrost. Furthermore, the pile supports, or "vertical support members" (known as VSMs), could be designed to permit lateral movements of the pipe as it expanded or contracted with temperature changes.

Air view of service road and Alaska Pipeline under construction, 1975.

Vertical support members with thermal devices to resist heat flow and keep footings frozen in the ground.

Raising the pipe above ground on the VSMs did not ensure that there would be no thawing of the ground. The disturbance inflicted on the ground during the course of construction was sufficient to initiate temperature changes in the soil that could result in thawing to a significant depth, in all except the coldest, most northern areas. Climatic change too, might in places initiate a continuing thawing.

Ultimately, the solution of the problem lay in the so-called thermal VSM. The thermal VSMs are equipped with devices known as heat pipes. These are sealed 2-inch diameter tubes within the VSMs. They extend below the surface and contain anhydrous ammonia refrigerant. In the winter months this evaporates from the *lower* end of the tube and condenses at the top where there are metallic heat exchanger fins. The evaporation process occurs because, during the winter months, the ground is *warmer* than the air outside. The evaporation process itself cools the lower end of the pipe and the surrounding ground, which is the point of the device: by cooling the permafrost in winter its temperature is sufficiently lowered to prevent thaw during the summer. As the mean ground temperature falls, the heat pipes thus prevent the warming that would otherwise occur following disturbance of the ground surface. About 380 miles of pipe were built above ground, and about 80 percent of this length had thermal VSMs.

Peter J. Williams is Professor of Geography, Director of the Geotechnical Science Laboratories at Carleton University, Ottawa, and a specialist in problems of development in cold environments. (From Pipelines and Permafrost: Physical Geography and Development in the Circumpolar North, *Longman, 1979. Used by permission of the author.)* ∎

18.8 SELECTED REFERENCES FOR FURTHER READING

Allen, L. J. The *Trans-Alaska Pipeline*. Alyeska Pipeline Service, 1977, 2 vols.

Baker, Donald G. "Snow Cover and Winter Soil Temperatures at St. Paul, Minnesota." *Water Resources Research Center Bulletin* 37, University of Minnesota, 1971.

Brown, R. J. E. "Influence of Climate and Terrain on Ground Temperatures in the Continuous Permafrost Zone of Manitoba and Keewatin District, Canada." *Third Conference of Permafrost Proceedings,* 14, Edmonton, vol. 1, 1978, pp. 16–21.

Brownson, J. M. J. *In Cold Margins: Sustainable Development in Northern Bioregions.* Missoula, MT: Northern Rim Press, 1995.

Ferrians, O. J., et al. "Permafrost and Related Engineering Problems in Alaska." *U.S. Geological Survey Professional Paper* 678, 1969.

French, H. M. *The Periglacial Environment.* New York: Longman, 1976.

Péwé, Troy L. "Effect of Permafrost on Cultivated Fields, Fairbanks Area, Alaska." In *Mineral Resources of Alaska, Geological Survey Bulletin* 989, 1951–1953, pp. 315–351.

Smith, M. W. "Microclimatic Influences on Ground Temperatures and Permafrost Distribution in the Mackenzie Delta, Northwest Territories." *Canadian Journal of Earth Science* 122:8, 1975, pp. 1421–1438.

U.S. Natural Resources Conservation Service. *Soil Survey: Fairbanks Area, Alaska.* Washington, DC: U.S. Government Printing Office, 1963.

Walker, D. A., et al. "Cumulative Impacts of Oil Fields on Northern Alaskan Landscapes." *Science* 238, 1987, pp. 757–760.

Washburn, A. L. *Periglacial Processes and Environments.* New York: St. Martin's Press, 1973.

Williams, Peter J. *Pipelines and Permafrost: Physical Geography and Development in the Circumpolar North.* New York: Longman, 1979.

19

VEGETATION, LAND USE, AND ENVIRONMENTAL ASSESSMENT

19.1 INTRODUCTION

Perhaps no component of the landscape is more directly related to land use and environmental change as vegetation. Besides being the most visible part of most landscapes, it is also a sensitive "gauge" of conditions and trends in parts of the landscape that are otherwise not apparent without the aid of detailed observation and measurement. The loss of vigor in tree species near highways, for example, may be an indication of impaired drainage or heavy air pollution, thereby drawing attention to environmental impact problems that might otherwise be overlooked. In agricultural regions, changes in shrub and tree species in swales and floodplains may be a response to heavy sedimentation, pointing up the need for erosion control, and in urban regions the invasion of floodplains by alien plant species in response to decreased peak streamflows is pointing up the need to change our approach to stormwater and flood management.

Vegetation also plays a functional role in the landscape since it is an important control on runoff, soil erosion, slope stability, microclimate, and noise. In site design, plants are used not only for environmental control, but also to improve aesthetics, frame spaces, influence pedestrian behavior, and control boundaries. While other landscaping methods and materials can be used for the same purposes, few are as versatile and inexpensive as vegetation. Not surprisingly, much of the work of the landscape architect involves designing planting plans.

Although it is not widely recognized, there are also some negative aspects to vegetation. The most serious are probably the noxious plants that inhabit disturbed areas in and among urban, suburban, and agricultural lands. These are mainly weed plants, such as poison ivy and ragweed, that are poisonous (usually in the form of allergic reactions) to many people. In addition, many introduced plants displace native species and degrade ecosystems. Two salient examples are kudzu, an Asian vine, that has smothered large areas of landscape in the American South, and purple loose-strife, a tall European herb, that is displacing rushes, reeds, and other native plants from wetlands throughout eastern United States, southeastern Canada, and the Midwest.

19.2 DESCRIPTION AND CLASSIFICATION OF VEGETATION

The need

Most planning projects involving alteration of the environment call for a description of vegetation. This may be combined with land use inventories to produce land cover maps or may be treated as an independent task that includes wetland mapping, species inventories, and related activities. In either case, the objective is to document the distribution and makeup of the vegetative cover, and this requires the use of an appropriate plant or vegetation classification scheme. Detailed descriptions of vegetation are virtually a universal requirement of environmental impact statements, and the conscientious investigator typically provides inventories and descriptions that draw on several classification schemes.

Classification systems

Three types of classification systems are used for plants and vegetation. The *floristic* (or Linnaean) system is the most widely used. It classifies individual plants according to species, genera, families, and so on, using the universally recognized system of botanical names. Plants are normally referenced by their species and genus names given in Latin or latinized forms such as *Pinus montocola* (western white pine). *Form and structure* (or physiognomic) schemes classify vegetation or large assemblages of plants according to overall form with special attention to dominant plants (largest and/or most abundant). The traditional classification for vegetation that begins with forest, savanna, grassland, and desert is a form and structure scheme.

Ecological schemes classify plants according to their habitat or some critical parameter of the environment such as soil moisture or seasonal air temperatures. Various ecological-type systems have been devised, but the only one commonly used is that associated with ecosystems. Large-scale or macro-ecosystems are named for their habitat types, or more broadly, for their physiographic associations, for example, wetlands, floodplains, sand dunes, and stream channels. Owing to our interest in habitat conservation planning as a part of species protection programs, the need to address plants, vegetation, and ecosystems in terms of habitat relations is stronger today than ever.

Selecting a system Unless dictated by environmental regulations, the type of description and classification used in a project should be governed by the nature of the problem and the character of the landscape. Increasingly, we are faced with settings where several waves of previous land uses have used, reused, and rearranged the landscape, leaving a hodgepodge of vegetation. Typically, this includes remnant patches of managed forest such as woodlots, belts of planted vegetation such as hedgerows and windbreaks, and artifacts of primeval ecosystems such as isolated wetland pockets (see Fig. 19.5). These features need to be examined not only as biological phenomena made up of species, populations, communities, and so on, but as part of the site's resource base having annual productivity in organic matter, utility in landscape design, and value in managing runoff, soil erosion, and other environmental systems. Therefore, some mix of floristic, form and structure, and ecological schemes is called for in most planning problems.

A five-level scheme The scheme given in Table 19.1 draws on all three classification systems. It is organized into five levels, each addressing a different classification element. Level I is based on overall structure, level II on dominant plant types, level III on plant size and density, level IV on site and habitat, and level V on significant species. Level V is included to provide for rare, endangered, protected, and highly valued species—a standard requirement today for virtually any planning project—as well as plants of value in landscape management or design for a proposed or existing land use.

19.3 TRENDS IN VEGETATION CHANGE

Early clearing The first and most dramatic event in the history of vegetation change was the nineteenth-century clearing of the North American interior. Nowhere in the world has such a large area been cleared and settled as rapidly as southern Canada and the eastern half of the United States. Between 1800 and 1900, more than 500,000 square miles of virgin forest were cut in the United States alone. In less arable regions such as the Appalachians and the upper Great Lakes much of this land has reverted back to forest, but usually with a different balance of species. A large share of the deforested land was also converted to farmland and over large expanses of the Midwest, East, and South, a landscape best described as agricultural parkland emerged (Fig. 19.1).

Remnant habitats Within this landscape, only in nonarable sites such as swamps and deep stream valleys has the original vegetation escaped destruction. But even these sites have been dramatically reduced and altered. In the United States (less Alaska), more than 50,000 square miles of wetland have been destroyed since settlement, and since 1950, agriculture has accounted for 70 to 90 percent of the losses. In addition, the remaining patches of vegetation are often quite different floristically from the original plant cover because they may be too small to support viable populations and because they have been subject to disturbances such as wood harvesting, sedimentation, and flooding that have eliminated certain tree species and ground plants and encouraged alien species.

Decline of farmland A second trend has been the decline of the agricultural landscape, first in the nineteenth century with the failure of farms in marginal areas and second in the

Table 19.1 Multilevel Vegetation Classification

Level I (vegetative structure)		Level II (dominant plant types)	Level III (size and density)	Level IV (site and habitat or associated use)	Level V (special plant species)
Forest (trees with average height greater than 15 ft with at least 60% canopy cover)		E.g., oak, hickory, willow, cottonwood, elm, basswood, maple, beach, ash	Tree size (diameter at breast height) Density (number of average stems per acre)	E.g., upland (i.e., well-drained terrain), floodplain, slope face, woodlot, greenbelt, parkland residential land	Rare and endangered species; often ground plants associated with certain forest types
Woodland (trees with average height greater than 15 ft with 20–60% canopy cover)		E.g., pine, spruce, balsam fir, hemlock, douglas fir, cedar	Size range (difference between largest and smallest stems)	E.g., upland (i.e., well-drained terrain), floodplain, slope face, woodlot, greenbelt, parkland, residential land	Rare and endangered species; often ground plants associated with certain forest types
Orchard or plantation (same as woodland or forest but with regular spacing)		E.g., apple, peach, cherry, spruce, pine	Tree size, density	E.g., active farmland abandoned farmland	Species with potential in landscaping for proposed development
Brush (trees and shrubs generally less than 15 ft high with high density of stems, but variable canopy cover)		E.g., sumac, willow, lilac, hawthorn, tag alder, pin cherry, scrub oak, juniper	Density	E.g., vacant farmland, landfill, disturbed terrain (e.g., former construction site)	Species of significance to landscaping for proposed development
Fencerows (trees and shrubs of mixed forms along borders such as road, fields, yards, playgrounds)		Any trees or shrubs	Tree size; density	E.g., active farmland, road right-of-way, yards, playgrounds	Species of value as animal habitat and utility in screening
Wetland (generally low, dense plant covers in wet areas)		E.g., cattail, tag alder, cedar, cranberry, reeds	Percent cover	E.g., floodplain, bog, tidal marsh, reservoir backwater, river delta	Species and plant communities of special importance ecologically and hydrologically; rare and endangered species
Grassland (herbs, with grasses dominant)		E.g., blue stem grass, bunch grass, dune grass	Percent cover	E.g., prairie, tundra, pasture, vacant farmland	Species and communities of special ecological significance; rare and endangered specie.
Field (tilled or recently tilled farmland)		E.g., corn, soybeans, wheat; also weeds	Field size	E.g., sloping or flat, ditched and drained, muckland, irrigated	Special and unique crops; exceptional levels of productivity in standard crops

Source: Adapted from W. M. Marsh, *Environmental Analysis for Land Use and Site Planning.* Copyright © 1978, McGraw–Hill, New York. Used with the permission of McGraw–Hill Book Company.

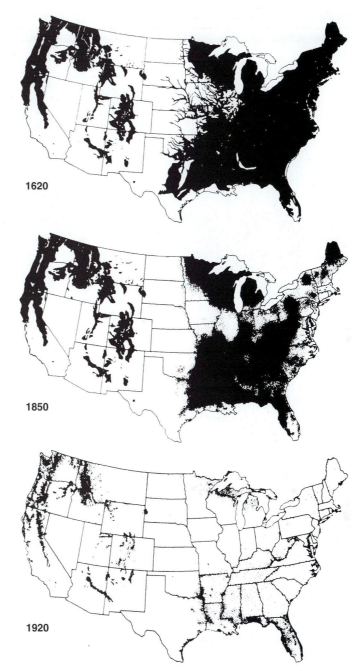

1620

1850

1920

Fig. 19.1 By 1920 vast areas of woodland were transformed into agricultural parkland. Reduction in virgin forest in the United States from the 1600s to the early 1900s. Since 1950 or so farmland has declined and forest has become re-established over large areas beyond urban centers.

The impacts of farming twentieth century with the shift of rural population to the cities. With the abandonment of small farms, much cultivated land has reverted back to natural, or unmanaged, vegetation. The recovery process, however, is often hampered by the degraded condition of the land. With few exceptions (e.g., the farmers of German heritage in Pennsylvania), American farmers in the nineteenth and early twentieth centuries were very hard on the land. Most known soil conservation and crop management practices were ignored, and after a generation or two, fields carved from woodlands

(*a*) 1933   (*b*) 1959

Fig. 19.2   The loss of farmland to suburban development in the Valley Stream area, Long Island, New York, between (*a*) 1933 and (*b*) 1959.

were often abandoned. The plants that invaded these lands were typically second rate, initially little more than weedy species from the field margins, some native, some inadvertently introduced by the farmers. (One estimate places species of foreign origin at nearly 20 percent of the total number of plant species in the northeastern United States and eastern Canada.) On the other hand, these plants helped stabilize the ground and rebuild the soil. Runoff and soil erosion declined and animal habitat was reestablished, though it favored different species than the forest of 50 years before.

Urban sprawl

The next trend was massive urban sprawl after World War II. Initially, much of this growth was absorbed by abandoned farmland, but as the development rate accelerated and land values increased, active farmland was also absorbed (Fig. 19.2). Nearly everywhere that conventional urban sprawl has taken place it has resulted in wholesale destruction of most existing vegetation including the fencerows, woodlots, and orchards of active farmland as well as the second-growth woodland of abandoned farmland. In addition, large tracts of habitat such as riparian woodland corridors have been fragmented and reduced in area. Only in the past few decades has suburban development begun to take a less all-consuming approach toward the landscape. This has been stimulated by three changes: (1) the enactment and enforcement of environmental regulations, particularly wetland and protected species ordinances; (2) public pressure, especially from nongovernmental environmental organizations such as streamkeepers and sportsmen's associations; and (3) the realization by experienced developers that features like woodlots, wetlands, and old hedgerows actually have economic value as real estate, most notably in residential development.

Suburban habitat

As landscapes go, most of suburbia is new and the vegetative cover is still developing, meaning that it is undergoing comparatively rapid change as it adjusts to this new habitat. In most areas, street trees are one or more decades from maturity, hedgerows are still being planted, and property owners are still in the process of making adjustments in yard plants by replacing exotic species with poor survival records with hardier species. One measure of the level of maturity of suburban vegetation is the diversity and abundance of wildlife such as songbirds, squirrels, opos-

Fig. 19.3 Vast areas of abandoned residential neighborhoods in cities such as Detroit are being taken over by weed trees, shrubs, herbs, and fugitive yard plants.

sums, and raccoons. Generally, animal habitat improves with the density, diversity, and productivity of the plant cover. The species diversity of the suburban plant cover, however, usually remains far below that of the rural cover it replaced because of the severe restrictions property owners place on the plant species allowed to grow in yards, and the narrow range of habitat types owing to absence of topographic diversity in most built-up areas.

Urban decay   A fourth trend in vegetation is the deterioration of the managed urban and suburban landscape with urban decay. As inner cities, industrial areas, and old residential neighborhoods decline, the built landscape in many cities is falling into ruin, being abandoned, and taken over by weed species and resilient survivors of old yards and parkways. Vast areas of abandoned neighborhoods in Detroit, for example, are being overgrown by weed trees such as box elder, old shrubbery such as junipers, and noxious plants such as ragweed as vacated houses and streets literally decay (Fig. 19.3). At the other extreme are efforts by cities—including some of the same cities with decaying neighborhoods—to revive their business districts by enriching streetscapes with trees, pocket parks, and greenways along streams and waterfronts (Fig. 19.4).

Managed forests   Beyond the urban scene are the forests managed for lumber and other forest products. This vegetation, too, has been substantially changed not only in the harvesting of original forests but also in the formation of new forests types. Forests managed for a sustained yield of timber based on selective cutting practices, for example, not only favor certain species but eliminate old growth forests from the mix of forest habitats. In addition many native forest species, including shrubs and herbs, have been displaced by introduced species, particularly around settlements and farming areas. And research has demonstrated that nutrients (particularly nitrogen and phosphorus) from atmospheric pollution and agricultural runoff has selectively favored certain shrub and tree species, often introduced ones, resulting in widespread floristic changes even in protected forests. Lastly, sizable areas of planted forests have replaced "natural," managed forests and/or farmland in many areas.

Fig. 19.4   The introduction of vegetation to otherwise barren commercial districts has become a common practice in design programs aimed at revitalizing inner cities.

These forests are essentially monocultures that lack the diversity of species, structures, and habitats of the forests they replaced (Fig. 19.5).

19.4 THE CONCEPT OF SENSITIVE ENVIRONMENTS

Minority environments

The concept of sensitive environments in planning has grown in part from a reaction to the wholesale mistreatment of what we might call *minority environments*, such as wetlands, stream valleys, and sand dunes, whose value cannot be measured accurately by standard economic criteria, that is, how much the land and its ecosystems are worth in the real estate market. It has also grown from improved understanding of the role of such environments in the maintenance and quality of the larger landscape. It is common knowledge in today's society, for example, that wetlands are not only important biologically but play a role in groundwater recharge, and that small

Fig. 19.5 About a century ago the area occupied by uniform stands of planted conifers in this aerial photograph was covered by a diverse hardwood forest.

stream valleys are important riparian habitats as well as sources of water for lakes and wetlands. Accordingly, it is not uncommon to find communities incorporating special provisions for sensitive environments into their master plans based on economic rationale (because it costs more to build in and manage such environments), social rationale (because these environments are valued by people for their scenic and general aesthetic value), and scientific rationale (because they are often necessary to the maintenance of larger environmental systems). Wetlands, stream corridors, and shorelands are examples of landscapes that are widely recognized as sensitive environments.

Wetlands   Prior to the 1950s, wetlands were generally viewed as third-rate landscapes of limited economic value. They were indiscriminately altered and destroyed on a wholesale basis to provide cropland, improve agricultural production, improve navigation, control pests, and provide land for urban development. Wetlands are now widely recognized for their roles as hydrologic, ecologic, recreational, forestry, and agricultural resources. The definition and mapping of wetlands, which is taken up in detail in the last chapter, is generally based on three sets of criteria: vegetation, soils, and hydrology. Although hydrologic processes are usually the controlling force in the origin and formation of wetlands, the definition of wetlands based on hydrologic criteria has proven difficult for a variety of reasons. However, vegetation, especially indicator species, has provided a more useful set of criteria.

Stream corridors   Stream corridors are among North America's most maligned environments. For much of the twentieth century, streams served as hydrologic dustbins for farmlands and settlements, taking on heavy loads of wastewater from sewers and field drains. By the second half of the century, they were often viewed as liabilities to development and subjected to channelization, piping, and damming. This problem and society's effort to recover these prized environments are discussed in Chapter 14. Suffice it to report here that today thousands of grassroots organizations, with the help of governmental agencies are working to restore local streams and the corridors of woodlands and marshes that stretch along them, give them character, and house rich assortments of plant and animal species.

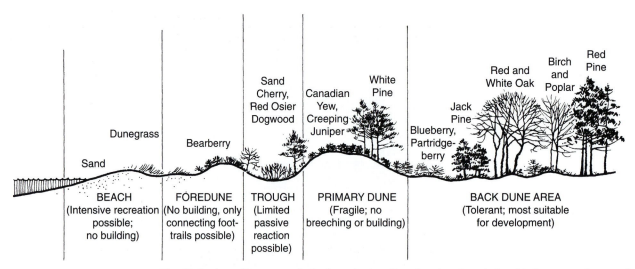

Fig. 19.6   A profile across a belt of sandy shoreline showing the relationship between plants and landforms including an assessment of the relative resistance of microenvironments to disturbance.

Shores and dunes   Coastal areas have long been attractive places to visit, but in the past 50 years development for residential and commercial purposes has increased substantially there. This has been facilitated by the growth of highway and road systems, increased availability of land with the decline of agriculture, and the rising popularity of water-oriented living. Today the coastal zone is gaining population faster than any major geographic setting in the world. Development has led to widespread alteration of coastal environments, especially the "softer" ones such as barrier beaches, beach ridges, sand dunes, coastal marshes, and backshore slopes where prized plant and animal communities are often found (Fig. 19.6). Because of the delicate balance that typically exists between these communities and the hydrologic and geomorphic environment, the imposition of roads, houses, and navigational facilities often leads to loss of entire communities and, in turn, the decline of an environment valued by many for ecological and aesthetic reasons. Furthermore, aggressive development has placed facilities in vulnerable situations that call for protective structures that damage ecosystems, reduce access and degrade scenic quality.

19.5 VEGETATION AS A TOOL IN LANDSCAPE PLANNING AND DESIGN

The place of vegetation as a planning tool has improved significantly in the past several decades. Among other things, the costs of land clearing and landscaping alone have motivated developers to incorporate more existing (predevelopment) vegetation into site plans. In residential areas, a mature shade tree may have an estimated value of $5,000 to $25,000, and the composite assemblage of plantings may improve the real estate value of an average residential lot by $10,000 to $25,000 or more.

Site design   **Landscape Design**   Beyond its direct economic value, vegetation is also recognized for its role in landscape design. Indeed, it can be fairly argued that vegetation is the principal tool of landscape architecture. In site planning, for example, it is regularly used to screen certain land use activities and features, abate noise, modify microclimate, improve habitat and stabilize slopes. As a visual barrier, hedgerows and border trees can help separate conflicting land uses such as residential, commercial,

Table 19.2 Locomotive Noise Reduction With and Without a 200-Foot-Deep Forest

Frequency, Hertz	Noise at 250 ft without Forest	Noise at 250 ft with Forest
31.5 Hz	39 dB	39 dB
63	57	56
125	63	61
250	68	65
500	73	69
1000	74	68
2000	72	64
4000	68	56
8000	61	41
	79 dBA	73 dBA

dB = decibel; unit of measurement of sound magnitude based on pressure produced in air from a sound source

dBA = decibel scale adjusted for the sensitivity of the human ear; a correction factor applied to dB units that takes into account the pattern of sound frequencies perceived by the human ear

Frequency = the pitch of sound measured in cycles per sound; higher pitches have higher frequencies (more cycles per second)

Hertz (Hz) = cycles per second

industrial, and institutional. In this capacity, the density and permanency of the foliage is critical because it controls the transmission of light. In parks, estates, campuses, and similar areas, vegetation is used to frame outdoor spaces, orient pedestrian circulation, and provide the landscape with its essential aesthetic character.

Noise Vegetation can also be used to help control noise. Under barrier-free conditions, the level (magnitude) of sound from a point source decreases at a rate of six decibels with each doubling of travel distance. (From a linear source such as a high-*Decibel modification* way, the decay rate is nearer three decibels per doubling distance.) Placed in the path of sound, vegetation absorbs and diverts sound energy, and is somewhat more effective for sound in the high frequency bands (those above 1000–2000 hertz). In forests the litter layer (decaying leaves and woody materials) appears to be most effective in sound absorption. Table 19.2 gives an example of the reduction in locomotive noise associated with a 250-foot-wide belt of forest. For the most effective use of vegetation in noise reduction, it is best to combine plantings with a topographic barrier such as a berm or an embankment.

Microclimate The influence of vegetation on ground-level climate can be very *Ground-level climate* pronounced. A plant cover effectively displaces the lower boundary of the atmosphere upward from the ground onto the foliage. A microclimate of some depth is thus formed between the foliage (for example, under the tree canopy) and the ground where solar radiation, wind, and surface temperatures are lower than those over a nonvegetated surface. Heat exchange between the landscape and atmosphere is also influenced by vegetation. The *Bowen Ratio* (sensible heat to latent heat flux) is lower because of transpiration, producing somewhat lower air temperatures over vegetated than nonvegetated surfaces (see Fig. 17.10).

As a barrier to airflow, vegetation tends to force wind upward, thereby increasing the depth of the zone of relatively calm air over the ground, forming a *boundary sublayer*. The thickness (or roughness length) of the sublayer increases with the height and density of the vegetation. Under a mature fir forest it is usually around 2.5 meters (8 feet) deep, whereas in grass it is a hundred times less at 2.5 centimeters (1 inch) deep (Fig. 19.7). A related effect can be found on the downwind side of a vegetative barrier

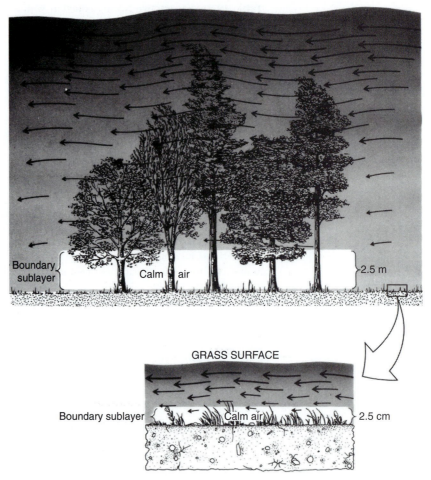

Fig. 19.7   The difference in the thickness, or roughness length, of the zone of relative calm air under coniferous trees and grass. This layer is important to the formation of ground-level microclimates.

Shelter belt   (shelter belt) where a calm zone forms under the descending streamlines of wind. The breadth of this sheltered zone also varies with the height and density of the barrier; however, significant wind reduction can generally be expected over a distance of 10 to 15 times the height of the barrier (Fig. 19.8). In snowfall areas, this zone is subject to the formation of snow drifts, the length of which can be estimated using this formula:

$$L = \frac{36 + 5h}{K}$$

where

L = snow drift length in feet
h = barrier height in feet
K = barrier density factor (50 percent density is equal to 1.0, and 70 percent is equal to 1.28.)

Air Pollution   The influence of vegetation in reducing contaminants in polluted air is not well documented for urban areas, but the existing evidence suggests that it is relatively small. Plants are known to absorb certain gaseous pollutants, for example, carbon dioxide, ozone, and sulfur dioxide, but it is apparently limited to the air imme-

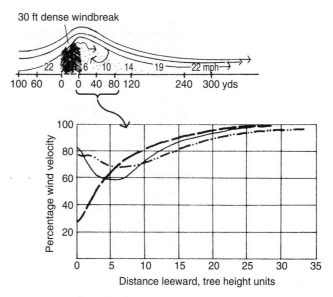

Fig. 19.8 Above, pattern of wind velocity across a barrier. Below, curves representing the percentage change in wind velocity downwind from a tree barrier. Greatest depression of wind velocity can be expected within 10 tree lengths of the windbreak.

Particulate sinks diately around the leaf and thus has only minuscule effects on these pollutants in the larger urban atmosphere. Heavy herb covers and dense stands of shrub and tree-sized vegetation with full covers of foliage act as *sinks* (catchments) for airborne particulates, but their net effectiveness is questionable because a sizable percentage of particulates initially caught appears to reenter the atmosphere within hours or days. Overall, vegetation appears to be most effective in trapping large particles in air moving laterally within several meters of the ground. In areas of heavy air pollution, a more pressing question may be that of the impact of pollutants on the health and survival of vegetation. Ozone and sulfur dioxide are the pollutants causing greatest damage to woody plants; other pollutants such as fluorides, dust, and chlorine are also known to cause damage, but it is usually localized around the point of emission.

Erosion Control We have long recognized the value of plants in erosion control and soil stabilization. Even a light herb cover can reduce wind erosion to negligible levels; indeed, a thin cover of lichens is credited with bringing erosion control to *Wind and runoff control* much of the North American Dust Bowl of the 1930s. In the universal soil loss equation, which we examined in Chapter 12, plant cover is one of the principal factors with both canopy cover and ground cover used in estimating soil loss rates. In combating gullying, the most insidious form of erosion, plant-based **biotechnical** approaches are very effective. *Contour wattling* on side slopes and *brush dams* in channels, for example, are particularly noteworthy as a means of slowing runoff and capturing sediment while establishing soil-stabilizing seedling masses.

Social Value People have long recognized the desirability of vegetation in neighbor-*Social preference* hoods, towns, and cities. This is related not only to the perceived role of plants in climate control, noise abatement, and the like, but also to sociocultural norms that place value on living plants and the habitats they create. Residential preference surveys bear this out when people identify parks and green spaces as important reasons for choosing one neighborhood or community over another. Real estate data also support this because wooded and landscaped lots consistently bring higher prices than those without vegetation or with unkempt vegetation. In senior citizen communities people appear to enjoy better health where plants are abundant, especially plants requiring their attention. Indeed, the case can be made that vegetation is the environmental glue that transforms settlements, neighborhoods, and institutions into communities.

19.6   APPROACHES TO VEGETATION ANALYSIS

In most landscapes the distribution of plants can be highly variable, even at the local scale. The reasons for the variation are often complexly tied to existing environmental (mainly physiographic) conditions as well as to past events such as fires, floods, and land use change, and to the geographic availability of species that inhabit the area. Which of this myriad of variables exerts the greatest control on the composition and distribution of the plant cover is usually a difficult question to answer.

Spatial correlations

In searching for explanations for the distributions of plants, three basic types of studies or approaches are used. *One approach* is to map the distribution of plant types and selected environmental features and then examine the two distributions to see what correlations can be ascertained. For example, a comparison of topography and tree species may show that certain species consistently appear in stream valleys (see Fig. 14.8). What such a relationship means is not revealed by the correlation, but it may provide clues about what questions should be raised for analysis. The floors of stream valleys are usually wetter, subject to more flooding, and comprised of more diverse soils than upland settings. Plants must spend much of the year under conditions of saturated soil and/or standing water. Moisture tolerance may thus prove to be a good candidate for detailed analysis of plants that are found in stream valleys, especially if the other settings that support different vegetation in the study area are appreciably drier (Fig. 19.9).

Defining controls

A *second approach* begins with an examination of the environment in an effort to identify those features and processes that may influence plant distributions. The purpose here is to identify factors known to exert control on certain plants, and by virtue of their distribution, hypothesize which plants should and should not be found there. The analytical part of this approach involves testing the proposed or expected distributions by making measurements in the field and then investigating the detailed relations between the affected plant(s) and the environmental (or habitat) variable (Fig. 19.10). An example of this approach is provided by Pitcher's thistle, a protected plant that grows on sand dunes along the Great Lakes. To complete its life cycle, the plant requires a habitat of shifting sand intermediate in the dunefield between areas of wind erosion and heavy sand deposition. Therefore, a landform is the key to finding Pitcher's thistle: you must first find active coastal sand dunes and next find the sites of intermediate geomorphic activity within the sand dune system.

Plants as indicators

The *third approach* uses key plants as indicators of environmental conditions and events. Based on existing knowledge of the habits and tolerance levels of selected plants, the controlling forces in the environment can often be identified according to which plants are found in an area. At the simplest level, this entails determining the presence or absence of certain types of vegetation over an area. In the example previously cited, the presence or absence of Pitcher's thistle at different locations within a dunefield can be used as an indicator of geomorphic activity, specifically sand erosion and deposition patterns. At a more general level consider a region where forests are the predominant natural vegetation but there are sites where forest cover is absent. This condition could indicate that (1) levels of *stress* (particularly solar radiation, water, heat, or nutrients) are too great for trees; (2) the *resources* of the site, for example, soil cover and water supply, are limited; and/or (3) a recent *disturbance* such as a tornado or some land use activity destroyed the forest at the

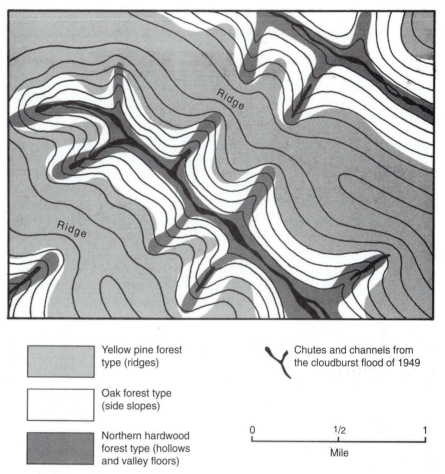

Fig. 19.9 The distribution of forest types in the central Appalachians related to soil moisture conditions associated with landforms and a major runoff event from a massive cloudburst.

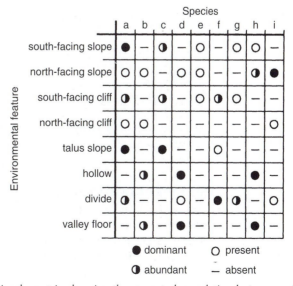

Fig. 19.10 A simple matrix showing the expected correlation between selected plant species and selected processes and features of the environment.

Table 19.3 Vegetation Indicators of Site Conditions

Climatic Region	Absence of Plant Cover	Sparse Herb and Shrub Cover	Thick Herb and Shrub Cover	Brush and Small Trees	Blade and Reed Plants	Highly Localized Tree Cover
Humid (Eastern North America, Pacific Northwest, South)	• Bedrock at or very near surface • Active dunes • Recent human use, cultivation, etc. • Recent fire • Recent loss of water cover	• Bedrock near surface • Recent or sterile soils • Dunes, fill • Recently disturbed (fallow, fire, flood) • Active slopes/erosion	• Recently logged or burned • Too wet for trees • Managed grazing • Organic soil • Old field regrowth	• Landslide/fire, flashflood scars • Old field or woodlot regrowth • Shale/clay substrate • Organic soil • Moisture deficiency	• Organic soil • Standing water • High ground-water table • Springs, seepage zones	• Wet depression, organic soil, • Steep slopes in agricultural areas • Flood-prone areas
Semiarid (High Plains, Central California)	• Caliche or salt pan (playa) at or very near surface • Desert pavement	• Localized water sources • Wind erosion • Overgrazing	• Overgrazing • Free from burning • Too dry for trees	• Channels with available moisture • Aquiferous substrate	• (Same as above)	• Aquiferous substrate • Seepage zone or spring • Stream valley (galleria) forest
Arid (Southwest, Great Basin, S. California)	• Rock surface • Unstable ground such as dunes or rockslides • Too dry			• Protected pockets • Favorable (moist) slopes • Logged/burned		• Plantation
Arctic and Alpine (N. Canada, Alaska, Rockies	• Rock surface • Active slopes • Persistent ice, snow, or ground frost • Ponded water during growing season	• Above tree line • Persistent ice, snow, or ground frost • Active slopes • Periglacial processes active	• Above tree line • Ice, snow, and wind pruning • Mildly active slopes • Wet depressions	• Wind/ice pruning • Avalanche, landslide scars, fire • Recent logging near tree line permafrost near surface	• (Same as above)	• Protected pockets

Beyond this sort of exercise, the particular types of plants, their densities, and physiological conditions (health) can be examined to learn about the detailed nature of the environment and its forces.

Source: From W. M. Marsh, *Environmental Analysis for Land Use and Site Planning* (New York: McGraw–Hill, 1978). Used by permission.

site and it has not grown back yet. Table 19.3 lists a number of site conditions in different bioclimatic regions that can be interpreted from vegetation.

19.7 SAMPLING VEGETATION

Whether our objective is to inventory and describe vegetation for the environmental impact and assessment studies, analyze vegetation for scientific purposes, or use vegetation as an indicator of site conditions and past events, it is usually necessary to sample vegetation in some way. Sampling is a means of selective observation that enables us to estimate various aspects of a plant population or vegetation community based on measurements of only a small portion of it. Sampling is attractive because it saves time and money; however, the proper use of sampling techniques can be difficult.

Among the techniques used in vegetation studies are quadrat sampling, stratified sampling, transect sample, systematic sampling, and windshield-survey sampling.

In any sampling problem the first task is to define the relevant population. In the case of vegetation this is usually accomplished by defining the geographic area occupied by the vegetation under study. This area, which may be a development site or particular environmental zone, is outlined on a large-scale map. The vegetation within it can then be sampled, using either the quadrat or transect method.

Quadrat sampling

Quadrats are small plots, the size of which varies with the type of vegetation being sampled. In **quadrat sampling** the first step involves subdividing the whole area into grid squares. For small study areas, individual squares may be used as a sample quadrat, whereas for large areas, it may be necessary to use some fraction of a square as a quadrat depending on the type of vegetation. Generally speaking, a 1- to 2-square meter quadrat would be used for grasslands and marshes, 4- to 6-square meters for low shrub covers, 15- to 30-square meters for brushland, and 30- to 100- square meters for woodland and forest.

Stratified sampling

Stratified sampling involves subdividing the population or study area into sub-areas or sets prior to drawing the sample. These are coherent subdivisions based on observable characteristics or prior knowledge of the area. A typical subdivision for many areas involves three strata: floodplains, valley slopes (walls), and uplands. Within these strata, quadrats are chosen on a random basis. Stratified sampling is widely used today with remote sensing imagery to define regional vegetation patterns. Indeed, aerial photographs are almost indispensable in modern vegetation studies, and more technically sophisticated remote sensing systems, such as line scanners, are showing promise for discriminating major types of vegetation. A form of stratified sampling is also used in soil mapping for land use projects where the strata are defined and sampling points assigned according to development and use zones.

Transect sampling

In **random transect sampling** the study area is divided into a number of strips, called transects. The width of the transects should vary with the type of vegetation; 5 to 10 meters would, for example, be appropriate for most forests. Of the transects selected for sampling, the entire transect may be sampled, or individual quadrats may be selected for sampling within the transect.

Systematic sampling

Systematic sampling requires no prior knowledge of the population or area under consideration. A grid is drawn over the area, and a sample is taken at each intersect in the grid. Quadrats can be used as the sample unit, and it is generally recommended that, together, the quadrats cover a minimum of 20 percent of the study area.

Windshield survey

The **windshield survey** is the quickest and least expensive sampling technique. Though more commonly employed in land use surveys than in vegetation studies, it can be helpful in gaining an overview of vegetation types. We should be aware, however, that roadside vegetation may not be representative because it may be planted, cut back in road construction or maintenance, or atypical of the area owing to the establishment of second-growth trees and weedy plants in the road right-of-way.

19.8 VEGETATION AND ENVIRONMENTAL ASSESSMENT

Finally, let us comment on vegetation as it relates to environmental assessment and impact analysis. As we mentioned earlier, few components of the landscape lend themselves to identification of environmental stress and change as does vegetation. At least five parameters or measures of impact related to vegetation can be highlighted in this context for evaluating a proposed action.

Measurement parameters

First, the sheer loss of cover, measured, for example, by the area of forest lost to development, arguable is the most visible change that can be made in the character of a landscape and often evokes a powerful emotional response in the public. *Second*, the loss of valued species, communities, and habitats is a critical measure of environmental impact as mandated by law at various levels of government. *Third* is the economic loss represented by the loss of merchantable vegetation such as timber, the loss of other economic opportunities such as maple syrup production, hunting, and fishing, and the decline of real estate value on-site, and in neighboring lands. *Fourth*, vegetation is often an integral part of larger environmental systems such as microclimate, ecosystems, soils, and hydrology, thus alteration or loss of plant cover can spell serious decline in these systems. *Fifth*, it is important to remember that natural vegetation is adjusted to a certain set of environmental conditions, and changes in these conditions, even subtle ones, are often reflected in changes in the vigor, reproduction capacity, and makeup of plant communities. Therefore, plants serve as valuable "thermometers" of environmental performance, giving us warnings when things are not working well, and when landscapes are not sustainable.

19.9 CASE STUDY

Wildlife Habitat Considerations in Residential Planning, Central Texas

Jon Rodiek and Tom Woodfin

In southcentral Texas, as in many other parts of the country, urbanization has promoted habitat fragmentation. Fragmentation is characterized by compartmentalization of otherwise large, spatially continuous areas of habitat. It has two negative impacts on wildlife: (1) reduction in total habitat area, which primarily affects population size; and (2) subdivision of the remaining area into disjunct patches, often separated by barriers, which primarily affect dispersal and migration routes. According to recent studies, temperate ecological communities appear to be more resistant to the effects of fragmentation than are tropical communities. Among the reasons cited are that temperate species tend to occur in higher densities, are more widely distributed, and have better dispersal powers. On the other hand, it can be argued that the effects of fragmentation only *seem* less severe in temperate zones because most of the habitat damage there was done long before basic documentation of original landscapes and wildlife could be recorded. Such is likely the case in the region of College Station, Texas, where much of the clearing in the floodplains and woodlands took place around the 1860–1900 period. In the past 50 years, however, the trend toward fragmentation has actually reversed in many rural areas with the regeneration of small woodland and riparian patches following the abandonment of farming. Near cities this recovery process is interrupted by conventional suburban sprawl.

In one 2800-acre drainage area, which was analyzed for cover change over a 47-year period (1940–1987), losses were measured in the following cover types: pastureland, old fields, young stands of upland hardwoods, cultivated fields, shrub–grasslands, hardwood–shrub lands, and mature bottomland hardwoods. Increases were measured in grasslands, mature upland hardwoods, savannahs, and riparian woodlands. Drainage lines changed little judging from a slight decrease in the total lengths of first-, second-, and third-order streams. Roads and houses increased significantly, however. Roads increased from 7.96 to 22.6 miles, and the number of houses and other buildings increased from 30 to 124.

On balance, the trend in land cover change can be characterized by a three-phase sequence: floodplain/riparian woodland to agrarian land to mixed suburban/urban/agrarian land. This is, of course, typical of land use changes in and near urban regions across the United States and Canada, and has serious implications for wildlife. As agrarian ecotones (or their remnants) are converted to residential tract development, they become segmented and reduced in size. The challenge for environmental planners and landscape architects is to devise more creative styles of land conversion which increase habitat areas and geographic continuity.

In this particular case, an experiment was set up to design a residential landscape that incorporates basic biogeographic principles for wildlife habitat improvement. The effort was organized around two objectives: (1) to lay out two single-family residences on 15 acres of land, and if the layout proved successful (2) to extend the site design

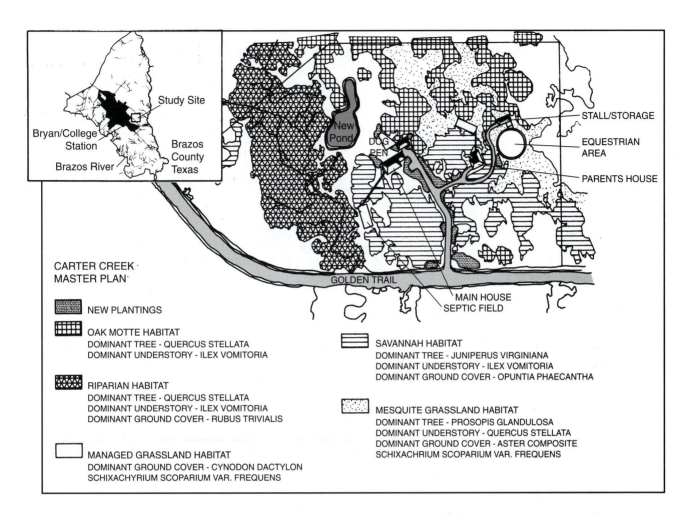

CARTER CREEK
MASTER PLAN

▨ NEW PLANTINGS

▦ OAK MOTTE HABITAT
DOMINANT TREE - QUERCUS STELLATA
DOMINANT UNDERSTORY - ILEX VOMITORIA

▩ RIPARIAN HABITAT
DOMINANT TREE - QUERCUS STELLATA
DOMINANT UNDERSTORY - ILEX VOMITORIA
DOMINANT GROUND COVER - RUBUS TRIVIALIS

☐ MANAGED GRASSLAND HABITAT
DOMINANT GROUND COVER - CYNODON DACTYLON
SCHIXACHYRIUM SCOPARIUM VAR. FREQUENS

▤ SAVANNAH HABITAT
DOMINANT TREE - JUNIPERUS VIRGINIANA
DOMINANT UNDERSTORY - ILEX VOMITORIA
DOMINANT GROUND COVER - OPUNTIA PHAECANTHA

░ MESQUITE GRASSLAND HABITAT
DOMINANT TREE - PROSOPIS GLANDULOSA
DOMINANT UNDERSTORY - QUERCUS STELLATA
DOMINANT GROUND COVER - ASTER COMPOSITE
SCHIXACHRIUM SCOPARIUM VAR. FREQUENS

concept over a larger, remaining residential tract. The ultimate goal would be to apply this approach to an entire watershed habitat reserve, a network made up of subregion habitat reserves tied together with a large number of site-scale habitat reserves.

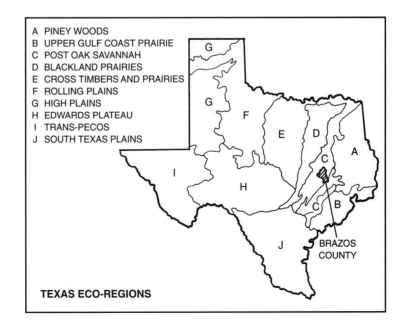

A PINEY WOODS
B UPPER GULF COAST PRAIRIE
C POST OAK SAVANNAH
D BLACKLAND PRAIRIES
E CROSS TIMBERS AND PRAIRIES
F ROLLING PLAINS
G HIGH PLAINS
H EDWARDS PLATEAU
I TRANS-PECOS
J SOUTH TEXAS PLAINS

TEXAS ECO-REGIONS

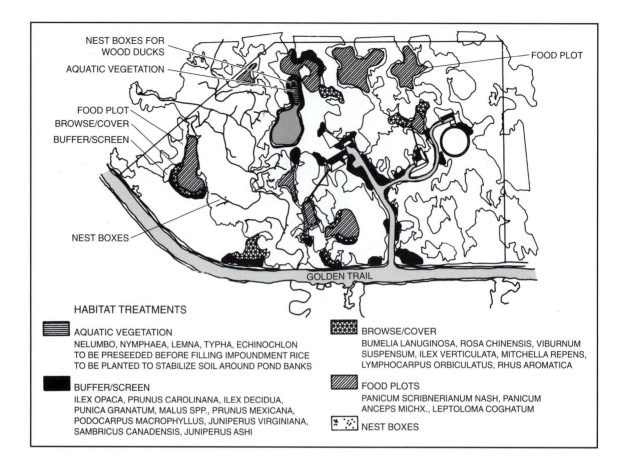

NEST BOXES FOR WOOD DUCKS

AQUATIC VEGETATION

FOOD PLOT

FOOD PLOT
BROWSE/COVER
BUFFER/SCREEN

NEST BOXES

GOLDEN TRAIL

HABITAT TREATMENTS

AQUATIC VEGETATION
NELUMBO, NYMPHAEA, LEMNA, TYPHA, ECHINOCHLON
TO BE PRESEEDED BEFORE FILLING IMPOUNDMENT RICE
TO BE PLANTED TO STABILIZE SOIL AROUND POND BANKS

BUFFER/SCREEN
ILEX OPACA, PRUNUS CAROLINANA, ILEX DECIDUA,
PUNICA GRANATUM, MALUS SPP., PRUNUS MEXICANA,
PODOCARPUS MACROPHYLLUS, JUNIPERUS VIRGINIANA,
SAMBRICUS CANADENSIS, JUNIPERUS ASHI

BROWSE/COVER
BUMELIA LANUGINOSA, ROSA CHINENSIS, VIBURNUM
SUSPENSUM, ILEX VERTICULATA, MITCHELLA REPENS,
LYMPHOCARPUS ORBICULATUS, RHUS AROMATICA

FOOD PLOTS
PANICUM SCRIBNERIANUM NASH, PANICUM
ANCEPS MICHX., LEPTOLOMA COGHATUM

NEST BOXES

In the first phase of the project, plant cover associations and related land uses were analyzed at three essential scales: watershed, subregion, and sites. Oak mottes, riparian woodlands, managed grasslands, and savannah were found to be common at all three scales. Mesquite grassland habitats were common to only the subregion and site level.

The initial planning task involved formulating a framework plan that attempted to integrate a residential complex and wildlife habitat. This plan first calls for the siting of facilities, including two residences, drives, storage facilities, and an equestrian area, in a configuration that least impacts existing woodland habitat. Vegetation was then added to provide screening, browse, and cover in selected areas, mainly along drives, open areas, and the riparian corridor. To further diversity habitat, a pond was added adjacent to the riparian habitat area. Finally, feed areas were added to the site in the form of winter and summer pasture.

The ultimate goal of this project is to mitigate the impacts of fragmentation of wildlife habitat brought on by urbanization. The strategy employed is based on the idea of establishing a skeletal corps of wildlife habitat along the riparian woodland and bottomland–hardwood landscapes. These habitats are most critical to the resident wildlife in the region. Although wooded landscapes are increasing in acreage, they currently represent only 6 percent of the subregion total. Protective zoning and enhancement of these landscapes, especially edges and linking segments, are seen as the most effective means of improving the balance between wildlife and residential development in this area.

Jon Rodiek and Tom Woodfin are landscape architects at Texas A&M University who specialize in wildlife habitat planning as a part of landscape design.

19.10 SELECTED REFERENCES FOR FURTHER READING

Carpenter, Philip L. et al. *Plants in the Landscape.* San Francisco: Freeman, 1975.

Davis, Donald D. "The Role of Trees in Reducing Air Pollution." In *The Role of Trees in the South's Urban Environment* (Symposium Proceedings), University of Georgia, 1970.

Gleason, H. A., and Cronquist, Arthur. *The Natural Geography of Plants.* New York: Columbia University Press, 1964.

Grey, Gene W., and Deneckie, F. J. *Urban Forestry.* New York: Wiley, 1978.

International Union of Forestry Organizations. *Trees and Forests for Human Settlements.* Toronto: University of Toronto Centre for Urban Forestry Studies, 1976.

McBride, J. R. "Evaluation of Vegetation in Environmental Planning." *Landscape Planning* 4, 1977, pp. 291–312.

Mooney, P. F. *Plants: Their Role in Modifying the Environment; A Selected and Annotated Bibliography.* Mississanga, Ontario: Landscape Ontario Horticultural Trades Foundation, 1981.

Schmid, J. A. *Urban Vegetation: A Review and Chicago Case Study.* Chicago: University of Chicago, Department of Geography Research Paper 161, 1975.

Thurow, Charles et al. *Performance Controls for Sensitive Lands.* Washington, DC: American Society of Planning Officials, Reports 307 and 308, 1975.

U.S. Forest Service. *Better Trees for Metropolitan Landscapes.* Washington, DC: U.S. Government Printing Office, USDA Forest Service General Technical Report NE-22, 1976.

U.S. Forest Service. *National Forest Landscape Management* (Agricultural Handbook No. 478). Washington, DC: U.S. Government Printing Office, 1974.

20

LANDSCAPE ECOLOGY, LAND USE, AND HABITAT CONSERVATION PLANNING

20.1 INTRODUCTION

Landscape fragmentation is the inevitable by-product of settlement and land use. Scholars tell us that many of the revolutions in Chinese history were actually aimed at reversing excessive farmland fragmentation and reforming rural land use. The North American landscape is juvenile compared to China's, but the rate at which it was cleared and settled is unprecedented in the world. In the continental heartland, it took only a matter of decades to fragment habitat systems such as riparian networks, wetlands, and forests as the landscape was filled with farms, railroads, highways, and settlements. Until now we have had little concern with landscape fragmentation beyond the traditional problems related to access, land ownership, water rights, and so on. But recently with recognition of the effect of fragmentation on biodiversity, landscape ecology has gained a place in the environmental planning agendas of North America and Europe.

Landscape ecology

Landscape ecology is the application of spatial (geographic) analysis to problems of habitat planning and management in mainly rural and suburban landscapes. It is a direct response to the decline in biodiversity and biological productivity as a result of habitat and ecosystem fragmentation, reduction, simplification, and contamination. From the air these landscapes have a patchy geographic aspect that some writers in the field have termed mosaics. *Mosaics* are made up of pieces of landscape systems; drainage nets, soils, woodlands, wetlands, etc., representing the disaggregated remnants of various types of habitats and ecosystems.

The study of landscape ecology focuses on three traits of landscape mosaics: structure, function, and change (or form, process, and change). The basic aim is to discover the relationships between landscape form and function in order to design landscapes that support richer and more productive mixes of plant and animal species. As a landscape planning problem, the objective is to reduce fragmentation and weld fractured landscapes back together into more functional patterns with greater ecological resilience and sustainability.

20.2 THE BIOGEOGRAPHICAL FOUNDATION

Island geography

The scientific underpinnings of landscape ecology come from biogeography, in particular island biogeography. Studies in **island biogeography** have revealed that two geographic factors, *habitat area* and *distance* between island habitats, have a strong influence on species diversity. The larger the area of an island or its terrestrial counterpart, the patch, the larger the number of species it can support. This was demonstrated in the original island biogeography studies based on freshwater bird species inhabiting islands of Southeast Asia (Fig. 20.1). It was also demonstrated that the greater the distance separating neighboring islands, the lower the genetic mixing and the lower the biodiversity.

Whether the principles of island biogeography are strictly applicable to fragmented terrestrial landscapes is debatable, but the concepts unquestionably apply. According to the area factor, island remnants of habitat, despite the fact that they look like the original habitat, are incapable of performing biologically as the former habitat once did. As a general rule, a 90 percent reduction in habitat area results in a 50 percent reduction in biodiversity. In addition, as more islands are taken down, the remaining forest islands are left farther and farther apart. This, of course, reduces mixing among isolated members of island populations (Figure 20.2).

Habitat and species loss

In tropical forest habitats, the loss of species may actually exceed 50 percent, with 90 percent habitat reduction. The reason is that the ranges of many tropical species are highly localized (i.e., endemic), often not more than a few acres or a few square

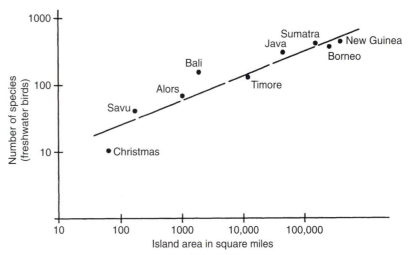

Fig. 20.1 The relationship between island size and number of species of freshwater birds from early biogeography studies in Southeast Asia. Species numbers increase with island size.

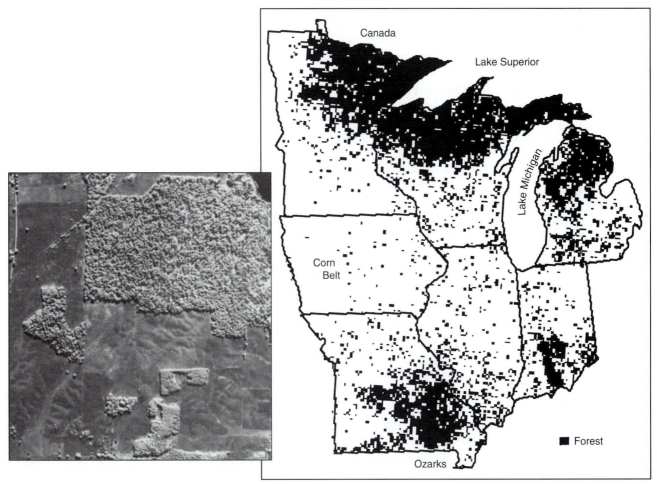

Fig. 20.2 A map of forest fragmentation in the Midwest. Michigan, Wisconsin, and Indiana were nearly fully forested only 200 years ago. Inset photograph shows fragmentation at a finer level of resolution with different sized "islands" of forest in a "sea" of farmland.

miles. This contrasts with midlatitude species, which, though fewer in total number, are often distributed in much larger ranges. Thus, in the tropics eradication of large areas of forest is more likely to destroy the entire range of many endemic species, resulting in greater than 50 percent species loss as forest loss approaches 90 percent.

In addition, studies of the short-term relationship between species decline and forest loss sometimes reveal lighter than expected losses. The reason for this, as other studies have shown, is that reduction in biodiversity does not occur suddenly with habitat reduction. Rather, species counts decline gradually over many years or many decades. Thus, short-term postclearing evaluations of biodiversity changes, based on indicator organisms such as birds, must be extended over the long-run in order to get a reliable picture of the impact of habitat loss on species numbers in general. Finally, we must recognize that our current understanding of these problems is very limited and much remains to be learned about biogeographical relations as a whole, including differences between tropical and temperate species in terms of population densities, range sizes, and fragmentation effects.

20.3 HABITAT, LAND USE, AND BIODIVERSITY

Habitat and niche

Habitat is defined as the local environment of an organism. It is sometimes referred to as the environmental address of an organism, but more frequently it is thought of as a unit of space and its environmental features—principally microclimate, soil, topography, water, available nutrition, and other organisms (Fig. 20.3). Although different organisms may occupy the same or very similar habitats—for example, different birds in the same wetland—interaction with the habitat is different for each species.

The term **niche** is used to define an organism's way of life or what it does in its habitat. Insects occupying a tree canopy habitat, for instance, may survive by acquiring food in different ways: some eat leaves, some suck nectar, and others chew bark. Niche defines an organism's functional relationship with its habitat: through its niche, each organism "sees" its habitat as unique.

Habitat replacement

Because of the special relationship each organism has with its habitat, it is virtually impossible to trade one habitat for another. Unlike humans, who have a wide range of habitat versatility, most organisms displaced from one habitat cannot simply take up life in another. In addition, when habitats are destroyed, they cannot be recreated in a manner suitable to sustain most of the organisms that once occupied them. This is especially so for complex habitats, such as many forests and wetlands, where niches are formed by many, and often subtle, interrelationships with other organisms, such as insects and microflora. Herein lies the fundamental dilemma of preserving organisms in zoos or seed banks whose habitats have been destroyed. If their habitats are gone and cannot be resurrected, then there is no place to sustain the organism and the zoo becomes an artificial life support system.

Wetland construction

Wetland construction is the most widely practiced habitat recreation enterprise in North America today. It is promoted by wetland mitigation programs that allow or require developers to create wetlands to replace wetland acreage destroyed in development projects. Although scientific evaluation of constructed wetlands is incomplete, studies to date reveal that constructed wetlands are poor replicas of the habitats they replaced because they do not support the diversity and types of species that their prototypes supported. Measured by other criteria such as hydrologic function, they may be more successful.

What do land uses do to habitat? The effects are highly variable, depending on the type and density of land use. The least disruptive land use is decidedly hunting–gathering, such as once practiced by nomadic North American native tribes,

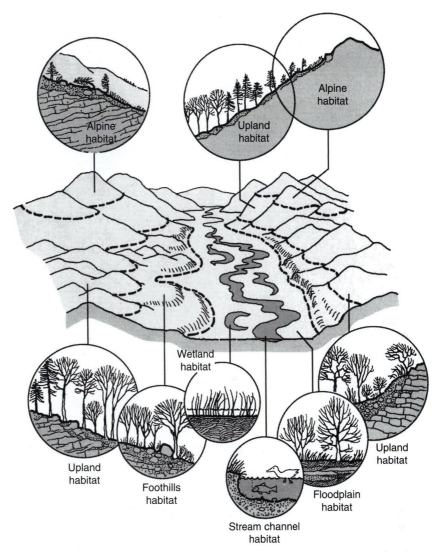

Fig. 20.3   A schematic diagram illustrating the different habitats associated with a stream valley in mountainous terrain, ranging from alpine at high elevation to aquatic and floodplains on valley floors..

Land use and habitat   because land clearing and agriculture are not involved and human populations are very light. In fact, it is not inappropriate to consider hunting–gathering societies as part of the natural landscape, because they, like other mammal populations, do relatively little to manipulate the environment. When agriculture replaces hunting and gathering, the impact on habitat increases, but initially it may not be significant because farms are usually small and widely separated. But as the landscape fills in and agriculture is organized into large-scale systems, the balance between cleared land and natural habitat shifts in favor of cleared land. The actual ratio between the two depends on the suitability of the land for agriculture and the regional pressure to use more land. As Figure 20.4 illustrates, the conversion of natural landscapes into landscapes dominated by farmland can be described more or less as a stagewise process.

Farmland development   In the early stages, land clearing follows a reasonably predictable pattern corresponding to the physiography of local watersheds. In areas of low to modest relief such as the U.S. Midwest, farmland first takes up belts of arable land lying on the

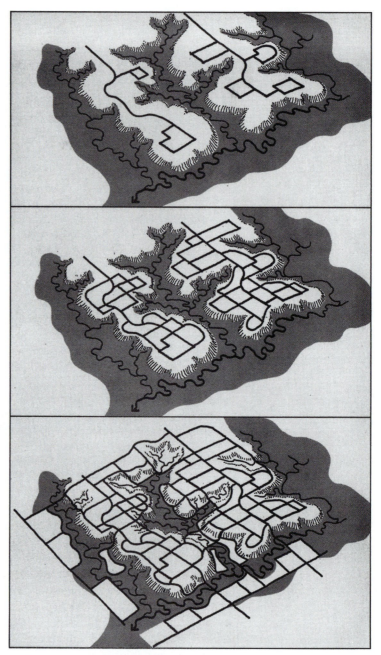

Fig. 20.4 Infilling of the landscape by agriculture resulting in reduction of open space and fragmentation of habitat corridors. Land use eventually spills over slopes and into stream valleys and wetlands.

interfluves between corridors of less arable land such as floodplains and wetlands stretching along valley floors. Next, formal roads and railways are driven through the landscape along surveyed routes that often have little relationship to local patterns of drainage, soils, and vegetation. As the landscape is cracked open, as it were, and population and agriculture increase, the pressure to clear more land often pushes farming and settlement beyond the reasonable limits of cropland into marginal lands, such as slopes and floodplains. Wetlands are drained, dams are built, streams are channelized, woodlands are removed, and the remaining habitat corridors are bro-

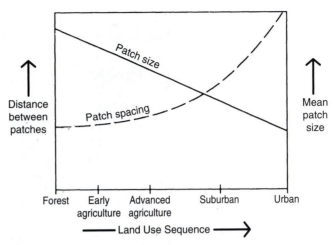

Fig. 20.5 Schematic graph illustrating the changes in patch size and spacing with land use change from wilderness to urban.

ken up or fragmented into smaller segments. The remaining uncleared landscape is left in small, widely separated patches (Fig. 20.4).

Farming impacts
Within the agriculture sector of the landscape, habitat potential is severely limited. Trees and shrubs, the habitats of many birds and insects, are reduced to field margins *or edge habitats*. Natural ground cover is largely eliminated and topsoil is reduced significantly by runoff and wind erosion associated with crop farming. Plowing greatly disrupts or destroys root, insect, and microorganism habitat and pesticide applications further limit the soil as habitat. Humus declines dramatically, earthworm, beetle, ant and other invertebrate populations decline precipitously or disappear. Wildflowers such as trillium, wild geranium, mayapples, and trout lilies, are eliminated from cropland or are forced into woodland fragments along and between fields—geographic refugia. In addition, many of the original plant and animal species are replaced by introduced species.

Biosimplification
Under intensive cultivation, the landscape becomes *biologically simplified* and often impoverished. Farmfields are reduced to biological **monocultures** dominated by several crop species with a few hardy weed species and greatly reduced and simplified surface and soil biodiversities and/or greatly imbalanced remnant populations. The remnant island patches of habitat are too small and widely spaced to support more than a fraction of the original species. As agriculture intensifies with commercialization and/or expansion into marginal lands, remnant habitat is chopped up into even smaller fragments that are subject to severe damage from surrounding cropland and farm animals (Fig. 20.5).

Whereas many species are reduced or eliminated by land clearing, agriculture, and settlement, a number of species are favored by these changes. Among the plants, **weed species** such as bull thistle, sumac, ragweed, and poison ivy have increased their geographic coverage and populations with the spread of development. These *Marginal habitats* plants have responded to (1) edge habitats such as fence lines and hedgerows along farm fields where the original species cannot sustain themselves and (2) disturbed areas such as road margins, playgrounds, and construction sites. As we noted in the previous chapter, in extreme cases, plants such as kudzu have overrun and smothered local landscapes (Fig. 20.6).

Many mammals and birds have also increased, taking up edge habitats and disturbed areas. Fragmentation of forested landscapes in the U.S. Midwest and southern Ontario, for example, has favored **opportunistic species** such as cowbirds, blue

Fig. 20.6   A landscape smothered by kudzu, a vine introduced to the American South for the purpose of arresting soil erosion in the 1930s.

Versatile opportunists jays, and crows. After being introduced into North America, European house sparrows and starlings have been very successful in agricultural areas as well as in cities and suburbs. Coyotes, rabbits, opossums, raccoons, and white-tailed deer have shown a remarkable capacity to adapt to new and changed habitats. Such versatility has enabled them to reinhabit suburban areas from which they were displaced during the clearing and construction phases of the development process. The trend to reoccupy is explained in part by the observation that as the suburban landscape ages and tree and shrub cover increase, the landscape becomes more diverse and richer as habitat. In addition, the geographical linkage within and among suburban areas is often improved as habitat corridors are created with the establishment of parks and wooded areas around schools, in vacant lots, and along streets.

20.4 ENDANGERED, THREATENED, AND PROTECTED SPECIES

The policy In 1973 the United States Congress passed the **Endangered Species Act**. This controversial law established two classes of protected species. **Endangered species** are defined as those in imminent danger of extinction in all or a significant portion of their ranges. **Threatened species** are those with rapidly declining populations that are likely to become endangered within the foreseeable future. Although the act applies to all lands and marine environments (private, state, and federal), it has focused particularly on federal construction and land use projects (e.g., military bases, interstate highways, and flood-control facilities) and on private projects using federal money or requiring federal permits. The act also makes it illegal to capture, kill, possess, buy, sell, transport, import, or export threatened or endangered species.

The central problem Although the importance of individual protected species should not be trivialized, it is not the central problem in species protection. The central problem is the much broader issue of biodiversity and habitat protection. The greatest loss of species is caused by habitat destruction, not by exploitation of individual organisms; therefore, the focus should be on the protection of entire ecosystems. In protecting ecosystems, habitat must automatically be incorporated because habitat, represented by soil, topography, water features, and so on, is the very foundation of ecosystems. The passage and enforcement of wetland laws represent a move toward habitat protection and conservation of biodiversity.

On the other hand, the protection of individual organisms is not without broader ecological benefits. By protecting individual species, their ecosystems must also be protected, and in turn a larger network of organisms and habitats also receive protection. The least prudent approach to species protection is the creation of zoolike preserves where special organisms are given showcase status without the appropriate habitat, area, and ecosystem arrangements for long-term viability.

Extinction-prone species

Some species are decidedly more prone to extinction than others. At the top of the list are **endemic species** with ranges so small that the species can be swept to extinction after only several years of land clearing for agriculture or lumber. Endemic species that are relics of once large populations and ranges are also vulnerable. Among other things, these species often have very narrow habitat requirements and are therefore subject to decline from subtle imbalances in the environment. A celebrated example is the Chinese giant panda, which depends on only a few species of bamboo for virtually its entire food supply. Should these bamboos (which are already severely limited by land use) be destroyed by development or natural change, the pandas are likely to starve. Only about 1000 pandas remain in the wild in China.

Species with small populations are also prone to extinction because loss of relatively few individuals may pull the population below the critical threshold of a breeding population. This is especially significant for animals, such as whales, accustomed to living and breeding in groups. If the group becomes too small, breeding may cease even though breeding adults are available. Small populations are also vulnerable for species, such as cheetahs, that reproduce slowly. These species average one or fewer offspring a year and the young require a number of years to reach breeding age. The recovery time is dangerously long for such populations, because they are subject to other perturbations during recovery.

Susceptible island species

Another extinction-prone class is species that have evolved in an isolated environment that has protected them from competition and predation. Many natural island species, such as the giant tortoises on the Galápagos Islands, declined when goats were introduced to the islands because goats consumed the tortoise's food sources. In other cases, island species were decimated by predators introduced by humans, or by humans themselves. Among the predators commonly associated with humans are pigs, dogs, chickens, and rats. Many bird populations in New Zealand and Hawaii, for example, evolved without serious natural predation, making them vulnerable to the new and vigorous predators introduced with early settlers. Introduced snakes have been particularly effective in decimating bird populations on a number of tropical islands. Witness that two-thirds of the birds in the Hawaiian Islands became extinct over the two hundred years following the initial settlement of the islands by Polynesians. The World Conservation Union estimates that since 1600 nearly 40 percent of the animal extinctions in the world have been caused by introduced species.

20.5 PATTERNS AND MEASURES OF LANDSCAPE FRAGMENTATION

Before the intrusion of agriculture, settlements, and associated transport systems, the biological landscape was covered by an interlocking network of ecosystems. These ecosystems were of different sizes, shapes, and compositions and despite the seem-

The physiographic framework

ingly random geographic patterns displayed by some, their distributions were actually quite structured. The basic structural framework for ecosystems was provided by physiography and since the scale, pattern, and makeup of physiographic systems were different in different parts of the continents, the landscape ecology was, accordingly, different as well.

Scale differences

In the eastern United States early settlers remarked on the patchy character of the New England forest cover, whereas settlers in the Midwest marveled at the vast uniformity of the tall grass prairie of central Illinois. Neither response is surprising in light of the physiographic character of the two locations. The physiography of Illinois favored large and continuous ecosystem areas that conformed to the broad pattern of stream corridors and the interfluves between them, whereas in New England, where the Northern Appalachian terrain was diverse and irregular with numerous rock outcrops and various glacial deposits interspersed with wetlands, physiography favored small and discontinuous ecosystems, more patches, and irregular corridors.

Fragmentation by land use

Into these landscapes were superimposed the land use systems like those we described earlier. Some land uses such as early farming honored the physiographic framework, but modern ones such as highway systems tend to treat the physiographic framework indiscriminately, especially in nonmountainous regions. Each land use system added a layer to the landscape and each fragmented the ecological patterns of the original physiographic landscape. Habitat patches became smaller and corridors became narrower and more segmented. In some instances new corridors emerged such as utility rights-of-way and systems of field hedgerows. But on balance, the landscape's ecological linkage declined as its economic (land use) linkage increased. Fewer original species could be supported, and for those remaining, populations were smaller, less productive, and less viable biologically.

Corridor types: riparian

At least five types of corridor systems can be found in most rural and/or partially developed landscapes. The most fundamental corridor system is the **riparian network**. This system has a hierarchical structure that follows networks formed by stream systems (Fig. 20.7*a*). The geographic pattern of riparian systems varies widely, but the fundamental structure is governed by the principle of *stream orders* with first-order corridors (no branches), second-order corridors (formed by at least two first-order branches), third-order corridors, and so on down the network (see Fig. 9.1). The size of the corridors in a riparian network increases with the corridor rank, and critical to their ecological function is the fact that corridors of all orders are linked together in comprehensive, self-sustaining water systems.

Interfluve

The upland counterpart to the riparian corridor system is the **interfluve corridor system** (Fig. 20.7*b*). Named for the fingers of upland terrain that lie between individual riparian corridors in a drainage network, these corridors follow drainage divides. They support distinctly different ecosystems than the adjacent stream lowlands and are less functionally integrated because they are not tied together by a flow system of drainage channels. Nevertheless, they also tend to be organized in a hierarchical fashion with main interfluves and connecting branches of progressively lower orders (also see Fig. 19.9).

Linear

Linear corridors are perhaps the least complicated corridor systems. The most glaring examples of linear corridors are rights-of-way for utilities and roads, but there are also striking natural examples, such as seashores and lakeshores. Shoreline corridors may be circular, as around an inland lake, or elongated, as along the seashore. When shoreline corridors are combined with riparian corridors, it is possible to define **corridor loops** or cells as is illustrated in Figure 20.7*c*.

Grid

Grid corridor systems are common by-products of the land survey system in the American landscape. The rectilinear pattern of land use has given rise to straight hedgerows, tree corridors, and similar border habitats (Fig. 20.7*d*). In agricultural and residential areas even drainage lines are often adjusted to the rectilinear grid. The

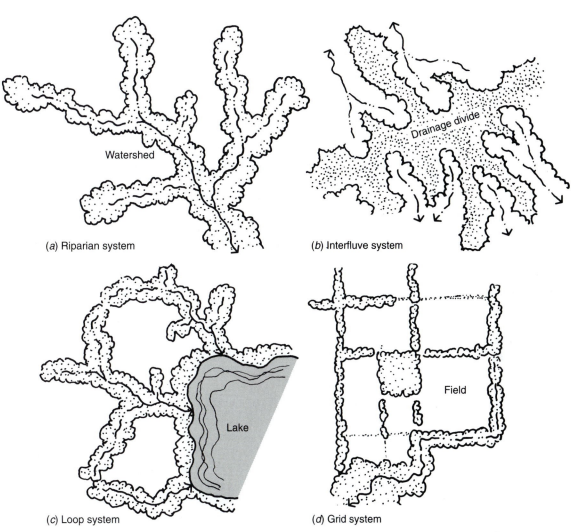

(a) Riparian system

(b) Interfluve system

(c) Loop system

(d) Grid system

Fig. 20.7 Various types of habitat corridor systems that are common to the North American landscape. The interfluve system (b) follows the divides between damage nets. The grid system (d) follows field margins, woodlots, and drainage ditches.

resultant pattern is a grid network with cells of different sizes and compositions. Most cells are farmfields, but some are woodlots, wetlands, lakes, and ponds.

Segmented

Fragmentation of any corridor leads to **segmented** or **disjointed corridor systems**. Most riparian corridors are segmented as a result of road crossings, wetland eradication, farmland clearing, urban development, and many other uses. In many coastal areas linear corridors are often so disjointed that the corridor is reduced to a series of patches, or remnant pieces protected from development by rugged terrain, wetland, or parks.

Spatial parameters

Landscape ecologists and others have attempted to quantify some of the spatial or geographic attributes of patches, corridors, and other landscape features in an effort to provide systematic measures of a landscape's ecological potential and problems. Although the effort is laudable and brings a more analytical perspective to landscape ecology, much remains unknown about the biogeographical or ecological significance of these measures, that is, about their correlations with the makeup of plant and animal populations, productivity, animal migrations, and other biological attrib-

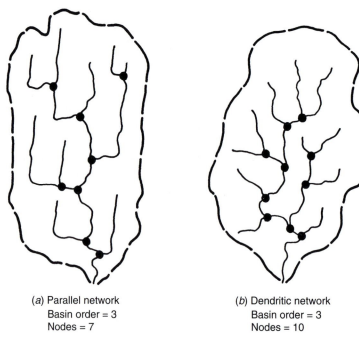

(*a*) Parallel network
Basin order = 3
Nodes = 7

(*b*) Dendritic network
Basin order = 3
Nodes = 10

Fig. 20.8   The difference in the number of intersections, or nodes, in a parallel drainage net and dendritic drainage net. Nodes increase with drainage density.

utes of landscape. Nevertheless, these measures are valuable in descriptive landscape ecology and we review several here.

Patchiness and connectivity

The density of landscape fragmentation is referred to as *patchiness*, which is simply the total number of patches of all sizes and types per unit area of land. A related measure is the *index of connectivity*, which is the ratio between the number of actual connections among patches in a study area and the maximum possible number of connections. The interaction or flow of organisms among patches appears to be influenced by the size of patches and the distance separating them. This is described by a model from geography, borrowed from physics, called the *gravity model* in which the relative attractiveness of two places (patches) to each other is a function of their sizes (masses) and the distance between them. In other words, the bigger a patch (and therefore the larger its resource base) and the closer a patch, the more attractive it will be for species.

Networks and nodes

The lines of linkage among patches and/or corridors define *networks* and the intersections within networks are *nodes*. We have already examined various types of corridor networks and it is easy to see how widely networks can vary in terms of connectivity defined by the occurrence of nodes. For example, a riparian corridor system aligned with a parallel pattern of stream channels may have relatively few nodes (low linkage) for the total length of corridors in the system, whereas a corridor system defined by a dendritic drainage pattern may have many nodes (high linkage) especially if a few intersecting land use corridors are added (Fig. 20.8).

20.6 HABITAT CONSERVATION PLANNING

Traditional conservation programs

There are several approaches to minimizing habitat reduction and species loss; **conservation programs** are one of the traditional approaches. These programs focus on open space: parks, wildlife preserves, wilderness areas, forest reserves, and landscape quality. Although parks and preserves, which are the most popular measures, have drawbacks for biodiversity preservation, they are nonetheless one important line of defense in the effort to slow habitat and species losses. Both Canada and the United States have taken leading roles in establishing national conservation programs and today have the largest areas of national parks and reserves in the world. But more conservation programs are needed, particularly in less developed countries where biodiversity is threatened by, among other things, tropical forest loss. Because most of the areas in need of conservation programs already contain local human populations, accommodations have to be made for these people.

Integrated conservation management
    Integrated conservation-development projects have been proposed as a way of reducing the land use pressure on protected areas such as parks and preserves while assisting local populations. The objective is to provide people with sustainable, income-generating opportunities. This could include (1) purchasing additional land or negotiating for the use of land for farming, and (2) setting up buffers around protected lands where certain economic activities would be encouraged. Other approaches call for *integrated landscape management*, which involves coordinating government agencies, businesses, community leaders, landowners, and others in a region to ensure that biodiversity objectives are included in the overall planning and management process.

Habitat conservation planning
    In the United States a program called **habitat conservation planning** emerged in 1982 as part of the Endangered Species Act. The objective of this program is to preserve specific areas for protected species within larger use areas such as lumbering. The selection of the habitats set aside for preservation involves both landowner and agents of the government and the plan that is worked out is binding for 50 years. For the landowner this represents a "no surprise" approach that enables him/her to develop a land use plan and use the remaining land without unexpected government interventions because of new species or habitat findings. Not surprisingly, the latter has become a point of contention, as government planners attempt to modify plans in light of new scientific findings about species ranges, migratory patterns, and so on.

Greenbelt systems
    Greenbelts or **greenway systems** are one of the oldest and most popular approaches to habitat conservation. In Europe and North America they have long been used in urban planning as a part of park and recreation planning. Although all sorts of land has been used for greenbelts in urban areas, stream corridors have been the favored settings for them because they frequently offer the only large belts of land available. Coincidentally, stream corridors are also the only remaining areas of rural type habitat in many urban regions. Today urban planners still justify greenbelts as multiuse areas for flood management, recreation, nature preservation, and other uses. Increasingly, however, habitat conservation and biodiversity are heading the list of rationale for creating greenbelts.

Rural greenbelt opportunities
    New opportunities to create greenbelts for improved biodiversity are emerging in rural areas with the decline in family farming and the shift in rural land use patterns. As small farms fall out of production, fields are abandoned, especially where the land tends to be marginal, such as in floodplains and hilly terrain. The fields fill in with woody vegetation, topsoil begins to rebuild, and habitat generally improves. This process, however, does not automatically build habitat corridors, because plowed fields, communities, and roads often remain as barriers. Thus, planning action that includes land acquisition and cooperation from private landowners is usually required to facilitate the linkage process (Fig. 20.9). Fortunately, the concept of networks and spatially integrated systems is very much a part of the traditional land use infrastructure in North American rural areas. Small roads link to larger roads; villages and towns are linked by railroads and highways; and small streams are linked to larger streams to form drainage networks. Thus, habitat restoration programs advocating development of corridor systems by linking habitat fragments together is not foreign to our way of thinking about landscape organization.

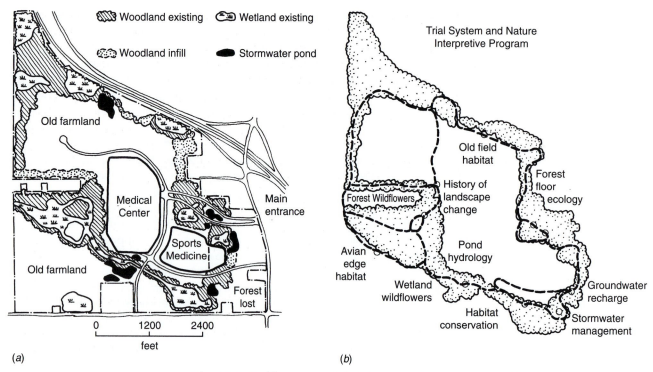

Fig. 20.9 Building a greenbelt/habitat corridor system on old farmland as part of a medical center campus. (*a*) The land use and restoration plan. (*b*) The resultant corridor with trail system and nature interpretative programs.

Counterforces Despite the increased attention to biodiversity and habitat conservation and restoration, strong counterforces are at work in the landscape. Chief among these is the emergence of major ecological barriers in the North American landscape. Consider the system of interstate highways and similar large roads, especially where development has added several tiers of land use along the highway. The resultant corridor is an imposing barrier, an ecological wasteland, that not only limits seasonal migrations and interbreeding among divided populations, but interferes with longer-term migrations as well. A restriction on regional migration may be a serious impediment to the maintenance of certain populations because they will need to shift their ranges in response to the climate changes expected in the next century. Some experts expect the ranges of many midlatitude plants and animals to be displaced northward in response to global warming trends, and there is concern that the massive regional barriers imposed by farmland, urban areas, and transportation corridors in the U.S. Midwest, for example, would greatly inhibit this biogeographical trend.

We end this chapter with some general guidelines for local and regional projects involving habitat conservation planning and design.

Some planning guidelines
- Focus on **habitat systems** and associated ecosystems rather than selected species, for without the proper habitat types and species mixes, most target species cannot be sustained.

- Focus on **critical locations** because the most promising places selected for preservation and management should be those that meet organism preferences. High preference nodes, such as lakes and wetlands in riparian corridors, are the anchor sites in a habitat system.

- Habitat systems should be founded on a firm **physiographic base**. Without a supporting physiographic system such as a fluvial corridor, habitat systems lack the resilience necessary for long-term sustainability. Accordingly, corridors made up of habitat fragments representing widely different types of physiography are less secure than corridors based on a single physiographic system or a set of interdependent physiographic systems.

■ Be mindful of **scale and connectivity**. In general, plan for fewer larger areas rather than many smaller areas, but where parcelization is natural and unavoidable the linkage between parcels should not be forced. In addition, be cognizant that critical scales or spatial thresholds may exist for certain species and ecosystems, and these may be tied to specific physiographic features, such as prairie potholes, seepage zones, cliff faces, and sinkholes.

■ **Species richness** decreases and range size increases over broad reaches of latitude. This rule holds true for birds, reptiles, amphibians, and trees as well as other groups of organisms. Therefore, the scale factor in habitat conservation planning may vary appreciably with geographic location in countries as large as the United States and Canada.

■ The changes in biodiversity, population densities, and other measures targeted in **habitat conservation planning** may show unexpected and/or disappointing results because (a) they often lag behind habitat improvements by years or even decades, and (b) the scales and configurations of planning efforts may not be adequately understood for certain habitats and species; thus, projects may inadvertently "miss the mark."

20.7 CASE STUDY

■ ### Marsh Restoration for Bird Habitat in the Fraser River Delta, British Columbia

Patrick Mooney

Iona Island is located at the mouth of the Fraser River, in the Vancouver region of British Columbia, on Canada's west coast. The primordial delta landscape here provided the native peoples with abundant food from the land, river, and sea. In the late 1880s, diking and draining of the delta for farming were begun. By 1950 much of the native vegetation had been removed, the marshes drained, and tidal flooding halted. In about 1955 the first bridges spanned the north arm of the Fraser River, linking the delta to Vancouver and opening it to urbanization. Nevertheless, the delta remains a migratory bird habitat of international significance and supports the largest wintering population of raptors in Canada.

In 1987, much of Iona Island, situated in the northwestern corner of the delta (see map) was designated as a regional park. A man-made pond in the park had been recently filled, destroying wildlife habitat. In addition, the proposed expansion of the nearby Vancouver International Airport would soon eliminate important habitat of the regionally rare Yellow-headed Blackbird (*Xanthocephalus Xanthocephalus*). In response, a wetland restoration with the goals of maximizing general avian diversity and providing Yellow-headed Blackbird habitat was implemented in 1992.

Restoration of wildlife habitat requires a two-step modeling process. Using site analysis, literature review, expert advice, field work, and laboratory testing, the restorationist seeks to understand first the habitat requirements of the intended species and then how the site conditions necessary to support the intended species can be developed.

The Yellow-headed Blackbird habitat model showed that the bird nests in open marshes, in Hardstem bulrush (*Scirpus acutis*), adjacent to open water. Transects through nearby McDonald Slough revealed that Hardstem bulrush grows at about 2 feet depth and will not grow in much deeper water (Diagram 1). Cattails (*Typha spp.*)

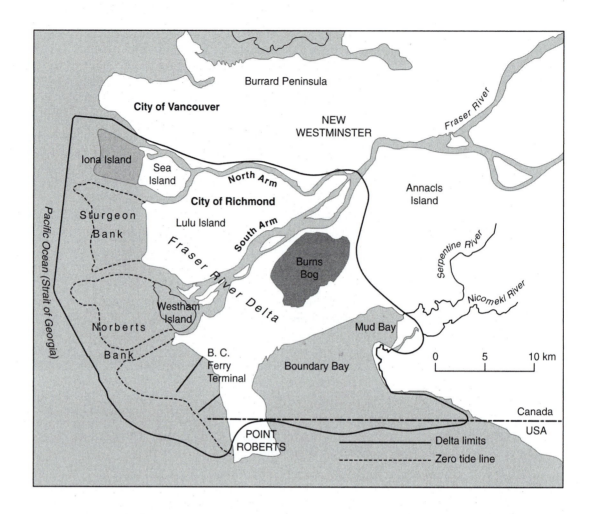

tolerate drier conditions and thus grow at higher elevations in the marsh. In the restoration, the pond bottom was contoured, and the marsh plants were placed to provide these depths while limiting the expansion of marsh vegetation into open-water areas. Soil fertility, pH, and nutrient levels between the slough and the restoration site were laboratory tested to ensure that they matched. Existing Yellow-headed

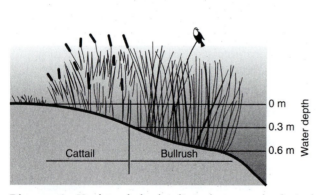

Diagram 1 Hardstem bulrush is limited to water depths in the range of 1 to 2 feet.

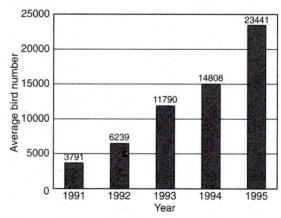

Diagram 2 Iona Island bird survey totals, 1991–1995.

headed Blackbird habitat at the airport provided emergent plants for the marsh and native shrubs and trees for the areas around the marsh.

The site was monitored for three years after completion. All the emergent plants colonized in their intended locations. In the spring of 1993, Yellow-headed Blackbirds moved into the restored pond. Two years later, 11 Yellow-headed Blackbird nests and 54 eggs were recorded. In the five years after the restoration general avian diversity increased dramatically (Disgram 2).

The project site was contaminated, however, with two exotic invasive plants: Scots Broom (*Cyticus scoparius*), and Purple Loosestrife (*Lythrum salicaria*). In subsequent years, manual cutting of the broom and the use of biological controls on the Loosestrife have controlled these plants.

The lesson of this project is that significant biodiversity can be restored even in disturbed urban habitats. It demonstrates the benefits of ecological modeling in restorations as well as the dangers of not eliminating exotic invasive plants from a restoration site. Restoration projects should be monitored so that midcourse corrections can be made and so that the successes and failures of the restoration can be used to improve future projects.

Patrick Mooney is a landscape architect at the University of British Columbia who specializes in West Coast habitat restoration. He is a fellow in the Canadian Society of Landscape Architects.

20.8 SELECTED REFERENCES FOR FURTHER READING

Beatley, Timothy. *Habitat Conservation Planning: Endangered Species and Urban Growth*. Austin: University of Texas Press, 1994.

Berger, J. J. (ed). *Environmental Restoration: Science and Strategies for Restoring the Earth*. Washington, DC: Island Press, 1990.

Butler, R. W., and Campbell, R. W., "Birds of the Fraser River Delta: Populations, Ecology, and International Significance." *Occasional Paper 65, Canada Wildlife Service, 1987.*

Cronon, William. *Changes in the Land: Indians, Colonists, and the Ecology of New England*. New York: Hill and Wang, 1983.

Collinge, S. K. "Ecological Consequences of Habitat Fragmentation: Implications for Landscape Architecture and Planning." *Landscape and Urban Planning*, 36, 1996, pp. 59–77.

Edwards, P. J., et al. (eds.). *Large-Scale Ecology and Conservation Biology*. Oxford: Blackwell Scientific, 1994.

Falk, D. A., et al. *Restoring Diversity: Strategies for Reintroduction of Endangered Species*. Washington, DC: Island Press, 1996.

Forman, R. T. T., and Godron, M. *Landscape Ecology*. New York: Wiley, 1986.

MacArthur, R. H., and Wilson, E. O. *The Theory of Island Biogeography*. Princeton, NJ: Princeton University Press, 1967.

Orians, G. H. *Blackbirds of the Americas*. Seattle: University of Washington Press, 1985.

Robinson, G. R., et al. "Diverse and Contrasting Effects of Habitat Fragmentation." *Science*, 257, 1992, pp. 524–526.

Smith, D. S., and Hellmund, P. C. (eds.). *Ecology of Greenways*. Minneapolis: University of Minnesota Press, 1993.

Vos, C. C., and Ophdam, P. *Landscape Ecology of a Stressed Environment*. London: Chapman and Hall, 1993.

21

WETLANDS, HABITAT, AND LAND USE PLANNING

21.1 INTRODUCTION

Much of environmental planning has to do with edges—the lines and ribbons in the landscape where one environment gives way to another. No edge is more important than that between land and water. More than a billion people live on this edge, and the most productive ecosystems on earth are found there. Among these ecosystems are wetlands, the organically rich shallow-water environments along streams, lakes, and seas.

Long regarded as fringe environments of marginal utility for land use, wetlands have been the object of severe misuse for centuries. In the United States (less Alaska), it is estimated that since about 1800 as much as 50 percent of the original area of wetlands has been destroyed. In the past few decades, however, wetlands have found a solid place in the environmental agendas of the United States and a few other countries. Laws have emerged calling for their protection against the pressures of land development. The environmental rationale behind these laws is basically twofold: (1) wetlands are important habitats necessary to the survival of a host of aquatic and terrestrial species; and (2) wetlands are integral parts of the hydrologic system necessary for the maintenance of water supplies and water quality.

Practical considerations   Besides the environmental quality rationale, there are a number of purely practical reasons for respecting wetlands in land use planning. *First*, the places in the landscape where wetlands form are characterized by drainage conditions that are extremely limiting to most land uses. The sources of these conditions usually extend over an area larger than the wetland itself, and in most instances the conditions do not simply disappear by scraping the wetlands away. Therefore, attempts to build in wetland sites may significantly increase overall development costs because of the need for special allowances for site drainage, flood protection, and facility maintenance. *Second*, most wetlands are usually underlain by organic soils that are unstable for most forms of development. To use such soils often requires special and often elaborate engineering schemes, or the soils must be removed by excavation and replaced with stable fill material. In either case, costs are increased significantly. *Third*, wetlands are landscape amenities and, like lakes and streams, can improve land values and design opportunities for the insightful developer. For many land uses, wetlands clearly enhance property values if they are properly integrated into land use planning schemes.

Wetlands cover such a wide spectrum of physical conditions and ecological characteristics that it is difficult to arrive at a succinct definition of them. Scientists generally agree, however, that all wetlands have three characteristics, and these serve as a general definition:

Defining traits

■ The presence of water on the surface, usually relatively shallow water, all or part of the year.

■ The presence of distinctive soils, often with high organic contents, which are clearly different from upland soils.

■ The presence of vegetation composed of species adapted to wet soils, surface water, and/or flooding.

Regulatory definitions   The regulatory agencies responsible for environmental policies have formulated various definitions of wetlands. The U.S. Fish and Wildlife Service uses the following definition, which planners have widely accepted:

Wetlands are lands transitional between terrestrial and aquatic systems where the water table is usually at or near the surface or the land is covered by shallow water. Wetlands must have one or more of the following three attributes: (1) at least periodically, the land supports predominantly hydrophytes, (2) the substrate is predominantly undrained hydric soil, and (3) the substrate is nonsoil and is saturated with water or covered by shallow water at some time during the growing season of each year.

In Canada, the following definition is used in the Canadian Wetland Registry:

> Wetland is defined as land having the water table at, near, or above the land surface or which is saturated for a long enough period to promote wetland or aquatic processes as indicated by hydric soils, hydrophylic vegetation, and various kinds of biological activity which are adapted to the wet environment.

21.2 WETLAND HYDROLOGY

Water is the most fundamental component of wetlands. Although vegetation is usually the most visible component of the wetland environment and is conventionally used to define them for regulatory purposes, water is decidedly the driving force behind the origin and maintenance of wetlands. Indeed, the traditional practice of wetland eradication almost always involved draining the problem area and limiting the influx of the normal water supply.

Hydrologic regime   Each wetland can be described as a hydrologic system with inflows and outflows of water. For most wetlands there is a particular pattern or regime to the inflows and outflows which is manifested in rises and falls in the internal water level. These fluctuations have a wide variety of periods depending on wetland setting, water sources, and climatic situation. Some are rhythmical as in the daily flux of water in tidal marshes; others are seasonal as in stream valleys that flood in the spring; and still others are sporadic, as in isolated locales where water comes anytime with the runoff from rainfall events. In all cases the hydrologic regime is essential to understanding and managing wetlands. Among other things, plant productivity and animal life cycles are adjusted to it.

Water sources   There are four possible sources of water for the wetland system: (1) direct *precipitation*, (2) *runoff* from surrounding lands including inflowing streams, (3) *groundwater* inflow, and (4) ocean *tide water* (Fig. 21.1). All wetlands receive precipitation, and virtually all receive runoff of some sort in the form of streamflow, stormwater, and/or overland flow. Depending on the soil and geologic conditions controlling subsurface water, however, only certain wetlands receive groundwater, and of course, only coastal wetlands receive ocean water or lake water. On the water loss side, wetlands lose water to evapotranspiration, seepage into the ground, stream discharge, and tidal outflow. Taken together, the inputs and outputs of water define the wetland's water balance (Fig. 21.1).

Storage water   In order to maintain most wetlands, the water balance cannot fluctuate so radically that there are long periods without substantial inputs of external water. To buffer against such deteriorating events, wetlands usually hold a large reserve of storage water in the form of soil moisture and groundwater. Organic soils, which form the substratum of most wetlands, have a very high moisture-holding capacity. At full saturation, organic matter such as muck and peat can hold more than 6 inches of water for each foot of soil. In addition, these materials have a high moisture transfer capacity whereby moisture is readily conducted from depth to the root zone by capillary action. These two hydrologic conditions account for the survival of many terrestrial wetlands—particularly those not attached to a major source of water such as a stream, or lake, or an aquifer—during prolonged periods of summer drought.

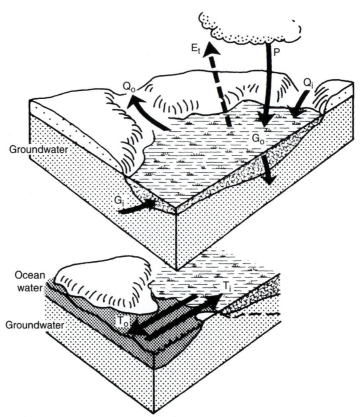

Fig. 21.1   Inflows and outflows in wetland water systems. P is precipitation, E_t is evapotranspiration, Q_i is stream inflow, Q_O is stream outflow, T_i is tidal inflow, T_O is tidal outflow, G_i is groundwater inflow, and G_O is groundwater outflow.

21.3   THE WETLAND ECOSYSTEM

Overview

Wetlands are attractive habitats for many plants and animals. The diversity of species is often higher in wetlands than in nearby upland landscapes, and where species diversity is not great, the population of individual species may be great. *Biomass* (the total weight of living matter per square meter of surface) is usually greater in wetlands than in nearby deep-water habitat or adjacent upland areas, and the productivity of the wetland vegetation, measured by the amount of new organic material produced each year, is typically greater than that of other habitats.

Wetland ecosystem

Ecological character and function are clearly the ranking attributes of wetlands in today's planning and management agendas. Among other things, wetlands are often cited as model ecosystems. An **ecosystem** is a biological energy system made up of food chains along which energy is passed from one group of organisms to another. The ecosystem's basic source of energy is the solar radiation, heat, and other essential resources taken up by plants in photosynthesis and converted into organic energy in the form of organic compounds (sugars and carbohydrates). Within the plant itself this energy is moved along two paths; some is used in respiration (plant maintenance processes) and some in growing new tissue (leaves, seeds, etc.). The total amount of new tissue manufactured in a year is termed *primary productivity*, and it is the source of energy upon which *all* other organisms in the ecosystem depend, either directly or indirectly, beginning with the herbivores and ending with the specialized predators (Fig. 21.2).

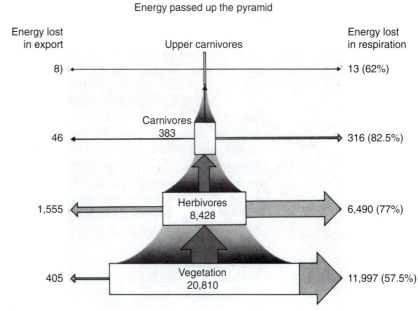

Fig. 21.2 Model of an ecosystem, called an energy pyramid, showing the four basic levels of organization, Silver Springs, Florida. All units in kilocalories per square meter per year.

Wetland productivity

Wetland **productivity**, which is measured in grams of organic matter generated per square meter of surface per year, may typically be twice that of nearby upland vegetation. A salt marsh along the Gulf of Mexico, for example, may average close to 2000 grams (per m²) per year compared to 1000 to 1500 grams per year in a neighboring subtropical forest. Many factors influence wetland productivity. Obviously, climatic conditions are important; northern bogs average only 500 to 600 grams per year because of growing season limitations. Hydrology is also important as it appar-

Nutrient supply

ently influences the supply of nutrients. Wetlands with the highest productivity are those with a moderate but continuous flow of water; stagnant wetlands, by contrast, generally have low productivities. Wastewater can also affect productivity. Wetlands receiving enriched waters from stormwater and sewage treatment facilities usually show accelerated productivities, whereas those receiving pollutants such as heavy metals and petroleum residues not only show a decline in productivity but a decline in species diversity as well.

If wetlands are capable of producing large amounts of organic matter, they must also have the means of disposing of it. The balance between productivity and dis-

Organic mass balance

posal or loss determines the **organic mass balance** of the wetland. Changes in the organic mass balance are reflected in changes in the reserve of organic matter in the wetland substratum, that is, the muck and peat deposits. When the mass balance is negative, loss exceeds productivity and this reserve declines.

The processes responsible for the loss of organic matter in wetlands are decomposition by microorganisms, consumption by herbivores, erosion by surface waters, and leaching (chemical disintegration) to groundwater. Export of organic matter by erosion and leaching varies greatly depending on the local setting and hydrology. For tidal wetlands such as mangrove swamps and salt marshes export may be as great as 40 to

Export and decomposition

50 percent of annual productivity. In wetlands that tend to be hydrologically closed, however, export is often negligible, and decomposition is virtually the sole means of loss of organic matter. Among the controls on decomposition, water depth is the most critical because it, along with the mixing motion of the water, governs the availability

of oxygen to many of the decomposing organisms. Under fully flooded conditions, little oxygen is available in the organic deposits and decomposition is slow, especially in stagnant water. If water is drained away to the point where the organic deposits are exposed to the atmosphere, decomposition rates rise dramatically. This is a principal reason for the decline in wetlands when they are artificially drained.

In a general way the ecology of a wetland is also reflected in the character of its plant cover. Traditionally, wetlands are described according to the structure (or form) and floristic composition of the plant cover. Various terms, such as swamp, marsh, and bog, have been used over the years, and although their meanings tend to vary somewhat from place to place, these terms are still meaningful and still widely used.

Marshes   **Marshes** are dominated by herbaceous vegetation, typically bladeleaf plants such as cattails, reeds, and rushes (Fig. 21.3). Although these plants may reach a height of 6 feet or more, marshes often have the look of a grassland or meadow; indeed, some marshes are called wet meadows (see Case Study 12.8). Soils are typically rich with relatively high (alkaline) pH levels, which has made marshes attractive to agriculture in many areas.

Swamps   **Swamps** are dominated by trees and shrubs (Fig. 21.3). There are many varieties of swamps in the United States and Canada. At the climatic extremes, for example, are cypress swamps in the American South (see Fig. 14.8) and northern conifer swamps in the U.S. North and Canada. Northern conifer swamps may be dominated by various tree covers: spruce, tamarack, cedar, or balsam fir, which may occur in various associations with other trees and shrubs. Owing to the short growing season and persistently wet (or flooded) soils, the trees of the northern conifer swamps are often stunted and at full maturity may reach heights of only 10 to 20 feet.

Bogs   **Bogs** are northern wetlands containing a wide diversity of vegetation. They are characterized by deep organic deposits, typically peat, and tend to be acidic. Bogs often form in ponds or small lakes where the vegetation is organized in concentric bands ranging from trees in the outer band to emergent and floating vegetation near the middle. Although bogs tend to fill in and become grown over in the long term (Fig. 21.3), many show a capacity to expand and contract with rises and falls in water level in response to changes in groundwater, streamflow, and obstructions such as beaver dams.

21.4 WETLAND TYPES AND SETTINGS

Wetlands can be classified in a variety of ways—for example, on the basis of vegetative cover (as previously described), hydrologic regime, or geographic (or physiographic) setting. The most basic control in shaping the wetland system is its *physiographic* setting. This includes the topographic situation, the proximal landscape (surrounding land use and vegetation), soils, and subsurface conditions (deposits and bedrock). Some combination of these factors produces a state of impeded (slow) drainage and/or abundant water supply that gives rise to wetland habitat.

Physiographic factors

Learning to recognize the physiographic conditions that produce wetlands is the first step in understanding their function and maintenance because setting is critical to wetland hydrology. In many instances, the relationship between setting and hydrologic function is readily apparent from casual field observation, for example, in cases of shallow waters along the shore of a pond or estuary. In other settings, however, it is not so apparent because sources of water or the conditions responsible for regulating water loss may be hidden underground or tied to sporadic hydrologic events.

We can define four general classes of wetlands based on physiographic setting and hydrologic conditions: (1) surficial, (2) groundwater, (3) riparian, and (4) composite. **Surficial wetland sites** are dependent on surface sources of water, mainly direct precipitation, and local runoff in the form of overland flow, ephemeral chan-

Surficial sites

Fig. 21.3 Photographs showing examples of major North American wetlands types: marshes, swamps, and bogs.

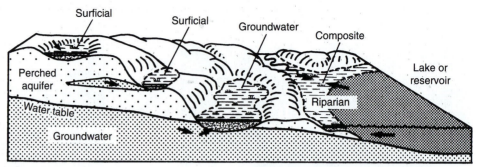

Fig. 21.4 A schematic diagram showing some types of physiographic and hydrologic conditions associated with surficial, groundwater, riparian, and composite wetland sites.

nel flow, and interflow (Fig. 21.4). These waters collect in shallow swales, closed depressions, and disturbed places where drainage has been blocked by deposits, tree throws, construction, or farming activity. The wetland begins with the establishment of a plant cover over the wet spot, which is followed by two changes that help to stabilize the site as a wetland environment: (1) accumulation of organic debris and mineral sediments leading to the formation of a wetland soil; and (2) sealing of the wetland floor with fine sediments. These changes improve the wetland's overall water balance by increasing its water-retention capacity; nevertheless, surficial wetlands typically suffer from radical seasonal variations in water supply.

Surficial wetlands also include those supported by *perched groundwater.* These are lenses of groundwater that lie near the surface above the main groundwater body (Fig. 21.4). Where perched lenses intercept the surface, water seeps out saturating the overlying soil giving rise to wetland conditions. This often accounts for the occurrence of wetlands on hillslopes and at the heads of swales and small stream channels (see Fig. 7.6). In certain northern bogs where sphagnum moss is the dominant plant cover, the wetland may actually expand upward and outward beyond the limits of the seepage zone. This is attributed to the rise of water by capillary flow within the organic mass.

Groundwater sites   **Groundwater wetland sites** are usually found at lower elevations in the landscape such as on the floors of stream valleys, sinkholes, and glacial kettles. These sites lie at or below the water table of the main groundwater body, and as low-pressure points in the groundwater system, they receive groundwater discharge. Because the supporting aquifers are often large, the water supply to the wetland is substantial and, unlike surficial wetlands, is not subject to radical fluctuations with variations in precipitation and alterations in the surrounding land cover and surface drainage patterns (Fig. 21.4). Nevertheless, some groundwater wetlands, such as northern bogs, are subject to water-level fluctuations in the range of several feet on a seasonal and longer-term basis as the water table rises and falls around them.

Riparian sites   **Riparian wetland sites** are those in and around major water features such as lakes, large streams, and estuaries (Fig. 21.4). These wetlands usually show a strong gradation in habitat with water depth from deep-water aquatic on the wet side to upland mesic on the terrestrial side. As the principal source of water, the controlling water feature governs the wetland's hydrologic regime. This is a two-sided coin, however, because the water feature is also the source of destructive processes such as storm waves, floodflows, and ice movements that can cut wetlands back and in some instances obliterate them entirely (Fig. 21.5; also see Fig. 14.12).

Composite sites   **Composite wetland sites** are those supported by two or more major sources of water. Most large, enduring wetlands fall into this class. For example, cypress

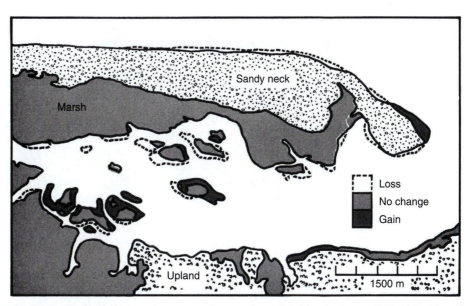

Fig. 21.5 Wetland changes in an estuarine marsh on the Atlantic Coast, 1859–1957. This illustrates a characteristic of wetlands that is not widely appreciated, namely, that wetlands can retreat as well as grow under natural conditions.

swamps in river floodplains are dependent on both floodwaters and groundwater (see Fig. 14.8). Coastal marshes are often supported by a combination of tidal water, stream discharge, and groundwater. Because each source has a different regime, the principal supply of water to the wetland may change from season to season. For example, in the Great Lakes, stream discharge peaks in spring, whereas the lakes themselves reach their highest levels in late summer. Therefore, as the streamflows feeding coastal marshes subside in late summer, the Great Lakes often augment the *Management implications* wetland's shrinking water budget. Understandably, the composite class of wetlands is often the most complex of the wetland systems and therefore the most difficult to manage because alterations may have compound effects on its various sources of water, mechanisms for internal water transfer, and water discharge. This is often illustrated with the construction of roads, canals, navigational facilities, and flood control structures where the impact on the wetland may initially appear to be minor but leads to gradual change that builds into substantial impact in the long run. One of the most controversial management dramas of this sort is currently being played out in the Florida Everglades.

21.5 COMPREHENSIVE WETLAND CLASSIFICATION SYSTEM

U.S. regulatory agencies Among the many governmental agencies concerned with wetlands, the U.S. Fish and Wildlife Service, the U.S. Environmental Protection Agency, and the U.S. Army Corps of Engineers are at the heart of the regulatory review process. In Canada, where there is no federal wetland policy comparable to that in the U.S., wetland policy and regulation are handled mainly by provincial agencies. These agencies are responsible mainly for the review of proposals for permitting projects in stream and coastal habitats where wetlands may be involved. To facilitate the review process, the regulatory agencies in the United States have adopted a wetland classification system that is comprehensive in scope and part of the U.S. National Wetlands Survey (Fig. 21.6).

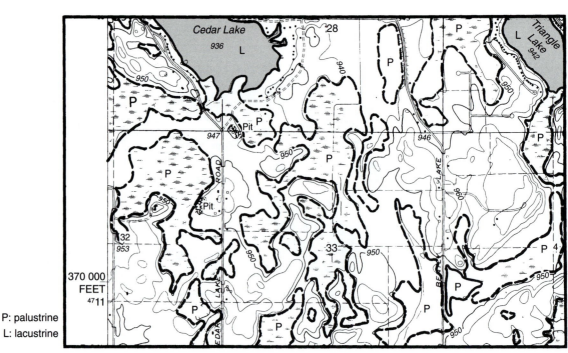

P: palustrine
L: lacustrine

Fig. 21.6   A portion of a map from the National Wetlands Survey. The base map is a standard U.S. Geological Survey topographic contour map. The code refers to wetland classification.

Deep-water limits   In this system, both wetlands and deep-water habitats are addressed. Deep-water habitats are permanently flooded lands that lie below the deep-water boundary of wetlands. In saltwater settings, deep-water habitat begins at the extreme low water mark of low spring tide. In other waters, the boundary line is at 2 meters (6.6 feet) below the low water mark. This is taken as the maximum depth of growth of emergent aquatic plants. Wetland begins landward of the deep-water line.

Wetland systems   The classification scheme is organized into three basic levels beginning with wetland systems (Table 21.1). Five **wetland systems** are recognized: marine, estuarine, riverine, lacustrine, and palustrine. The *marine system* consists of the deep-water habitat of the open ocean and the adjacent marine wetlands of the intertidal areas along the mainland coast and islands. The *estuarine system* is associated with coastal embayments and drowned river mouths and includes salt marshes, brackish tidal marshes, mangrove swamps as well as deep-water bays. The *riverine system* is limited to freshwater stream channels, and the *lacustrine system* is limited to standing waterbodies, mainly lakes, ponds, and reservoirs. Both the riverine system and the lacustrine system include deep-water habitat. By contrast, the fifth system, the *palustrine system*, includes only wetland habitat. The palustrine system is the major system because it encompasses the vast majority of North America's wetlands, namely, inland marshes, swamps, and bogs.

Wetland classes   Beyond the system level, *subsystems* are defined for all but the palustrine system. At the third level, *wetland classes* are defined, and they represent either basic habitat types or wetland types based on vegetation. Among the palustrine wetlands, the three major classes are (1) emergent wetland, (2) scrub–shrub wetland, and (3) forested wetland. *Emergent wetlands* are dominated by herbaceous vegetation including grasses, cattails, rushes, and sedges. *Scrub–shrub wetlands* are dominated by short, woody vegetation (shrubs and trees less than 20 feet high), and *forested wetlands* are dominated by trees taller than 20 feet.

Table 20.1 Three-level Classification of Wetlands and Deep-water Habitats, Showing Systems, Subsystems, and Classes

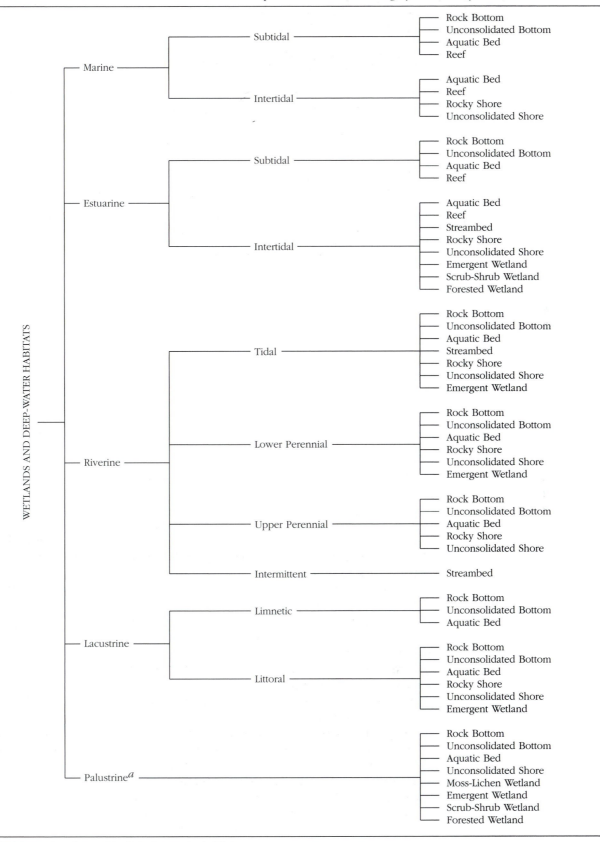

[a] The Palustrine system does not include deep-water habitats.

21.6 WETLAND MAPPING

Wetland mapping has become one of the most important inventory activities in environmental planning. No matter the nature of the project, it is essential not only to identify wetlands but to define their geographic limits as well. As with virtually all efforts to map landscape features, three problems must be faced at the outset: (1) the criteria or indicators to be used as the basis for boundary delineation; (2) the sources of data; and (3) the mapping resolution, that is, the level of geographic detail required.

Mapping criteria: vegetation

Three criteria are consistently named in regulatory policy for wetland mapping purposes: vegetation, soils, and hydrology. Depending on state and local guidelines, all three, two, or just one criterion may be required to define a wetland. Vegetation is the most commonly used criterion in wetland mapping. Many wetland edges, such as the one shown in Figure 21.7, can be identified by a relatively abrupt change in the *vegetative structure*, especially where the lowland/upland topographic transition is sharp. This border may also be marked by a third vegetative form, a belt of shrub-sized plants fringing the wetland, which is also evident in Figure 21.7. Whereas structure is useful in wetland mapping, *floristic composition* of the plant cover is considered more reliable for boundary demarcation. Certain plants can be used as wetland indicator species, and their dominance may be taken as sound evidence of wetland habitat, although there can be much debate over this. Indicator species vary from region to region, and in most locales, a preferred list of indicator species can be obtained from environmental agencies, a university botanist, or environmental organizations (also see Appendix F).

Soil

The second criterion is soil. The term *hydric* is widely used to describe wetland soils. These are mainly organic soils, such as muck or peat, but may also include saturated mineral soils. Soils are generally used as a secondary indicator after vegeta-

Fig. 21.7   Wetland border marked by the abrupt change in vegetative structure.

tion. Care should be taken in using published soil maps to corroborate wetland borders initially mapped on the basis of vegetation patterns because many of the original soil boundaries were based in part on vegetation. Thus, discovery of a strong correlation between hydric soils, as shown, for example, on U.S. Natural Resources Conservation Service maps and wetland vegetation may constitute circular reasoning. Field checks of soils are usually advisable in wetland mapping.

Flooding   The third criterion is flooding. If an area is periodically or frequently flooded and lies in a flood zone such as a designated floodplain, it is also taken as evidence of wetland. Whereas flood-prone areas such as floodplains are readily identifiable, the boundaries of such areas are often inexact because they are normally defined by hydrological calculations rather than by field measurement of floodwater coverage. Therefore, the delineation of wetlands in terms of fixing a boundary line should usually be based more on vegetation and soils than on flood zones. (Also see Section 10.4 in Chapter 10.)

Data sources: maps   The sources of data and information for wetland mapping fall into two classes: (1) published sources in the form of topographic maps, aerial photographs, and soil maps, and (2) field surveys. Large-scale *topographic contour maps*, published in the United States by the U.S. Geological Survey and in Canada by the National Mapping Branch and various provincial agencies, mark wetland areas larger than 10 acres or so. These areas are mapped from aerial photographs on the basis of visible surface water, vegetation patterns, topographic trends, and proximity to water features such as lakes and streams. *Soil maps* such as those published by the U.S. Natural Resources Conservation Service, do not necessarily show wetlands, but they do depict hydric soils. As noted, however, the original placement of the soil borders that appear on the NRCS maps was often guided by vegetation and topographic lines. Therefore, these borders cannot always be taken as reliable independent indicators at the local scale. Once again, field verification is usually necessary when using soils for wetland delineation.

Aerial imagery   *Aerial photographs* are especially helpful in wetland mapping. The U.S. Fish and Wildlife Service has used small-scale (high-altitude) infrared photographs to build crude wetland maps for much of the United States. But for project planning purposes a much higher level of mapping resolution is needed which requires large-scale aerial photographs enhanced by stereoscopic perspective. Stereoscopic models afford an exaggerated view of the vegetation structure and forms, and textural differences in forest cover and bladeleaf vegetation, for example, help in distinguishing wetland areas. In addition, infrared imagery provides enhanced scenes in which water, different vegetation types, and exposed soil contrast more sharply with each other than they do on standard aerial photographs.

Field verification   In cases calling for development or alteration of sites containing wetland, *field verification* and refinement of wetland borders are usually necessary. This involves walking the border, taking note of indicator plants, soils, and evidence of high water. The evidence of high water includes water marks on trees, debris stranded on low branches and foliage, stranded driftwood on the ground, and shallow and exposed tree roots (see Fig. 14.12). As the border is identified, it should be flagged so that it can be fixed by field survey and plotted on site maps.

21.7 MANAGEMENT AND MITIGATION CONSIDERATIONS

Wetland mitigation begins not on the development site or at the water's edge, but with programs dealing with public attitudes and information on wetlands. As with many environmental problems, the first steps toward solutions involve improved public awareness followed by understanding of the value of wetlands in the environment. In other words, society has to know enough about something and feel strongly enough about it to make a place for it in the great agenda of environmental problems. That has already happened with wetlands in a general way, and it has led to enforcement of wetland protection laws such as *Section 404* of the *U.S. Clean Water Act* as well as the enactment of new laws at the state and local levels.

Setting the agenda
     How to balance society's need to use land and the need to protect wetlands, however, requires the attention of planners and scientists working at the level of individual land use sites. This involves two levels of activity: (1) management of existing land uses and (2) planning future land uses. The principal challenge with existing land uses is with those that continue to damage and destroy wetlands. Chief among these is agriculture, which in the United States has been responsible for 70 to 90 percent of the wetland loss since the 1950s. Limiting future wetland loss to agriculture is important, but mitigation of damage from existing operations is also needed. Unfortunately, few agricultural activities are subject to wetland regulation, and until national policy is extended to include agriculture as well as forestry and mining activities, U.S. wetland protection law will remain somewhat hollow. This is not the case, however, with new urban development. Residential, industrial, and commercial land use plans in most areas are carefully scrutinized for wetland conflicts by federal, state, and local agencies.

Wetland mitigation
     The search for land use compatibility with wetlands may be approached along two lines: mitigation and management planning. **Mitigation** involves taking certain actions to counterbalance wetland losses or damage. Three types of mitigation are usually practiced: (1) restoration and enhancement of damaged environments; (2) creation of new wetlands; and (3) wetland preservation through control of potentially damaging actions. The third type usually includes various means of controlling construction activity, stormwater runoff, soil erosion, sedimentation, and building encroachment, but it may also involve various land protection arrangements such as deed restrictions and environmental easements.

Replacement
     With respect to wetland restoration, enhancement, and creation, there are two approaches: replacement mitigation (also called on-site mitigation) and mitigation banking. **Replacement mitigation**, in its simplest form, involves building new wetlands to offset acreage lost or damaged as a result of development. Normally, a replacement ratio such as 2 acres of replacement for one acre of loss or damage, is scheduled as a part of the mitigation plan. Replacement may be on the project site or in a neighboring location, but it is always tied to an individual project. **Mitigation banking**, on the other hand, is a system of compensation credits in which surplus wetland mitigation acreage (in the form of replacement, enhancement, restoration, and/or preservation) is banked against future needs. This system is not project specific, and if structured as part of a habitat conservation plan within a rational geographic framework such as a local or regional watershed, is a promising approach to compensatory wetland mitigation.

Preservation management
     The concept of **management planning** is central to wetland preservation but not widely practiced. Almost invariably management for preservation must address the larger system of which the wetland is a part. This begins with an inventory of the principal sources of water and the controls on the waterflow system, followed by an evaluation of the vulnerability of the flow system to alterations from land use that could affect the wetland. In the case of the wetland in Fig. 21.8, for example, the principal sources of water happened to be vegetated swales heading up in the surrounding upland. According to the state regulations governing the site, a buffer zone of fixed width was to be designated around the entire wetland to protect it against land use encroachment. Because of the character of the topography and drainage patterns around the wetland, however, such a buffer definition did not offer adequate protection for the sources of runoff. Therefore, an alternative buffer concept was devised based on the configuration of the swale drainage system. The total area of

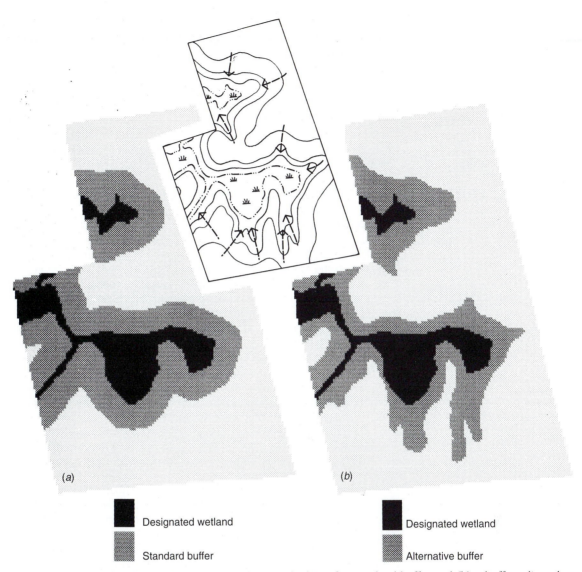

(a)

(b)

■ Designated wetland

▨ Standard buffer

■ Designated wetland

▨ Alternative buffer

Fig. 21.8 Maps showing (a) a standard 100-foot wetland buffer and (b) a buffer adjusted to the configuration of the drainage patterns around the wetland. The inset shows the topography and associated drainage pattern around the wetland.

buffer remained about the same, but the performance of the buffer in terms of the inflow of water and the long-term maintenance of the wetland improved with the functional buffer concept.

Systems approach   The main point of wetland management planning is to approach wetlands in much the same way as we would an inland lake, that is, to treat the contributing watershed as the main vehicle for preserving the waterbody. This includes understanding not only the extent and nature of the water sources, but also the role of water storage and the controls on water release in the overall maintenance of the wetland. As with lake watersheds, the sources of water have to be evaluated in terms of the directness of their linkage to the wetland, their relative contributions, and whether they tend to fall into the manageable or nonmanageable class. Nonmanageable sources are those beyond the reasonable reach of management such as precipitation and deep sources of groundwater. For manageable sources that make significant contributions to the wet-

Stormwater management   lands, care should be taken to insure that the availability of water is maintained and the delivery system is not impaired by structures, grading, or stormwater diversion.

The use of wetlands in stormwater management is widespread today, and while generally encouraged, the practice is seriously questioned for wetland habitats prone to damage from changes in water regime and water quality. Marshes polluted by many years of urban and industrial runoff, for example, have been discovered to decline sharply in species diversity, eventually reaching a state best described as a monoculture. The levels of permissible pollutant loading in wetlands are clearly a matter of much uncertainty, which at this writing awaits further research.

21.8 CASE STUDY

In Search of Better Wetland Regulation and Management

Gary F. Marx

It is a paradox that although public support is nearly unanimous for programs that protect the environment and preserve natural areas, wetlands are the only critical habitats actually protected by national law. Despite this national mandate, developers and property owners of all stripes complain publicly that stringent wetland regulations rob them of their constitutional rights to the use of their property. If one believes their rhetoric, it would appear that wetlands receive comprehensive and unbending protection from development. A closer inspection of the nuts and bolts of these programs suggests that in reality things are quite different. While some wetland areas receive protection under wetland laws, many acres of wetlands are exempt from protection, others are degraded by nearby development, while still others are lost to piecemeal encroachment by individual property owners.

First, it is commonly believed that the law prevents development in wetlands. Actually, the law just requires a permit before regulated activities can take place. While the restrictions are stringent, wetlands are routinely filled under permit for construction of roads and driveways, ponds, and large commercial developments where the requirements for permit issuance have been met. *Second*, many activities are exempt from wetlands regulation. The most significant of these activities is farming. No permits are required for farming and farming-related activities in wetlands. This includes many projects to drain, plow, and plant wetlands as part of normal farming practice. The result is that thousands of wetland acres annually are converted to farmland and their wetland value as habitat is lost. Forestry activities including logging are also exempt from wetland regulation. While cutting of trees does not completely alter the hydrology of these wetlands, habitat quality is lost or greatly reduced when the vegetation is removed. In addition, the heavy equipment used in most logging operations causes considerable disturbance to surface drainage and soils, leaving them vulnerable to serious erosion problems long after the loggers have left the scene.

This brings us to the problem of making quality distinctions among wetlands. Current wetland regulations do not distinguish between "high" and "low" quality wetlands. If enacted, such provisions could focus protection efforts on high quality ecosystems, while allowing more liberal treatment of those of lower quality. But such an allowance would be nearly impossible to administer. *First*, defining the quality of wetlands would be technically very difficult, and *second*, any scheme developed for this purpose would be open to attack from "experts" hired by the property owner. The result would be something resembling a full employment act for lawyers and wetland experts and would vastly increase the costs and time necessary to administer the program. This explains the opposition of environmental groups to include a quality factor in the wetland regulatory scheme. Consequently, all wetlands are treated as equal in the eyes of the law, requiring that as much effort is expended protecting marginal and submarginal wetlands as in protecting high quality systems.

Next is the problem of excluding wetland watersheds from protection. In some states, wetland regulations do not extend beyond the immediate boundaries of the wetland itself. This means that development can occur right up to the very edge of

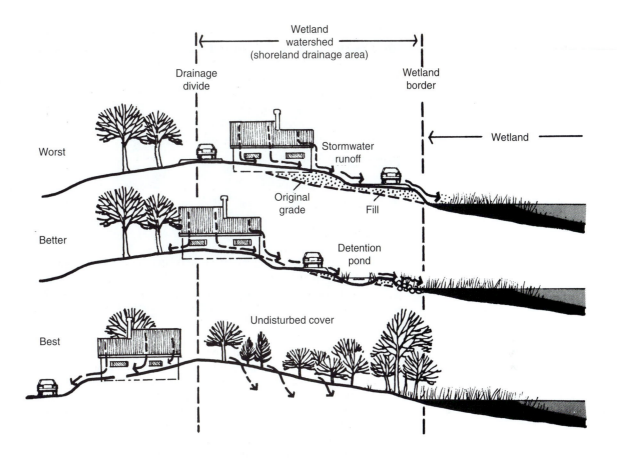

the wetland including important runoff source areas. Wetland watersheds may be cleared, graded, stormsewered, and developed without regard to effects on the hydrology and biota of adjacent wetlands. Frequently, large-scale grading associated with site preparation produces massive soil erosion resulting in heavy sedimentation of wetlands. Wetland regulations do not consider this to be a regulated situation even though wetland quality and functional capabilities are lost. The diagrams illustrate three approaches to development in wetland watersheds. The top one is clearly inappropriate whereas the lower one is clearly preferred.

Another problem is the difficulty in dealing with secondary impacts of proposed development activities, especially residential developments. For example, where a proposed subdivision is designed around existing wetland areas, only small areas of wetland fill may be required for road crossings and utilities. Issuance of wetland fill permits for such projects is routine, but often fails to address the larger impacts that occur when individual homes are built in the development. Many of the lots in these subdivisions extend into the regulated wetland with a building site provided in the upland portion of the lot. Many owners assume the developer handled all the necessary permitting as part of the original planning process and believe they can do as they please within their own lot, including draining and filling of the wetland.

In one particularly troublesome case, an application was filed to place fill in less than one tenth of an acre of wetland for access to a development area of 30 lots on a peninsula in a freshwater lake. A belt of wetland fringed the entire peninsula and the lots were platted so that each lot extended from interior upland across the wetland to the lake shore. A permit was issued for the road crossing, with the stipulation that lots would be deed restricted against the placement of fill material and the cutting of trees larger than 4-inch diameter.

The first purchaser of a lot in the subdivision cut every tree smaller than 12 inches in diameter, placed stone riprap shore protection along his entire shoreline, and graded the site level between the remaining few trees to facilitate lawn devel-

opment. The purchaser of the next lot also removed all but the largest trees and graded the site down to the water's edge. While enforcement actions will be brought against these owners, the damage is already done, and for the most part, it is irreversible. Twenty-eight lots remain to be sold in this subdivision, and if each owner is similar to the first two, the enforcement effort to keep up with the situation—given the huge caseload of field officers—will be overwhelming.

This raises an issue that is at the heart of many compliance problems with private property owners: education. While people hear about wetlands and are aware that they should be protected, most do not recognize the variety of wetlands that are protected, and further are resistant to regulation when it affects their activities on their own property. "What effect can my little infringement possibly have; it's the developers who do the real damage." But of course they fail to see the big picture and the cumulative impacts of numerous small wetland encroachments in a drainage basin. Such impacts can have devastating effects on the fish and wildlife of such watersheds, as well as significant losses of water quality, groundwater recharge, flood damage amelioration, and other wetland values which can only be replaced at great expense if they can be replaced at all.

While some of these situations can be addressed by more aggressive enforcement and attention to detail, others require changes in the law to reduce the loss of wetland area that continues to occur under the current laws. Although the current political climate makes such changes to the law unlikely, it is important to the long-term vitality of our natural resources to hold the line on any changes in the law that would further weaken the government's ability to protect the wetlands that remain.

Gary F. Marx is district supervisor in charge of wetland permitting and regulation for the Michigan Department of Environmental Quality.

21.9 SELECTED REFERENCES FOR FURTHER READING

Cowardin, L. M., et al. *Classification of Wetlands and Deepwater Habitats of the United States.* Washington, DC: U.S. Government Printing Office, 1979.
Environmental Law Institute. *Wetland Mitigation Banking.* Washington, DC: Environmental Law Institute, 1993.

Frayer, W. E., et al. *Status and Trends of Wetlands and Deepwater Habitats in the Coterminous United States, 1950s to 1970s.* St. Petersburg, FL: Fish and Wildlife Service, U.S. Department of the Interior, 1983.

Mitsch, W. J., and Gosselink, J. G. *Wetlands.* New York: Van Nostrand, 1986.

National Wetlands Policy Forum. *Protecting America's Wetlands: An Action Agenda.* Washington, DC: Conservation Foundation, 1988.

Redfield, A. C. "Development of a New England Salt Marsh." *Ecological Monographs* 42:2, 1972.

Salvesen, David. *Wetlands: Mitigating and Regulating Development Impacts.* Washington, DC: Urban Land Institute, 1990.

Sather, J. H., and Smith, R. D. *An Overview of Major Wetland Functions and Values.* Washington, DC: Fish and Wildlife Service, U.S. Department of the Interior, 1984.

Tiner, R. W., Jr. *Wetlands of the United States: Current Status and Recent Trends.* Washington, DC: U.S. Government Printing Office, 1984.

U.S. Fish and Wildlife Service. *National Wetlands Priority Conservation Plan.* Washington, DC: U.S. Department of the Interior, 1989.

GLOSSARY

Acid rain   Precipitation with pH levels much below average as a result of the integration of oxides in polluted air with moisture.

Active layer   The surface layer in a permafrost environment, which is characterized by freezing and thawing on a seasonal basis.

Aggradation   Filling in of a stream channel with sediment, usually associated with low discharges and/or heavy sediment loads.

Albedo   The percentage of incident radiation reflected by a material. Usage in earth science is usually limited to shortwave radiation and landscape materials.

Alluvial fan   A fan-shaped deposit of sediment laid down by a stream at the foot of a slope; very common features in dry regions, where streams deposit their sediment load as they lose discharge downstream.

Alluvium   Any material deposited by running water; the soil material of floodplains and alluvial fans.

Angiosperm   A flowering, seed-bearing plant. The angiosperms are presently the principal vascular plants on earth.

Angle of repose   The maximum angle at which a material can be inclined without failing. In civil engineering the term is used in reference to clayey materials.

Aquifer   Any subsurface material that holds a relatively large quantity of groundwater and is able to transmit that water readily.

Atterberg Limits Test   A test used to determine a soil's response to the addition of water based on its changes in the physical state, as from plastic to liquid.

Backscattering   That part of solar radiation directed back into space as a result of diffusion by particles in the atmosphere.

Backshore   The zone behind the shore—between the beach berm and the backshore slope.

Backshore slope   The bank or bluff landward of the shore that is comprised of *in situ* materials.

Backswamps   A low, wet area in the floodplain, often located behind a levee.

Bankfull discharge   The flow of a river when the water surface has reached bank level.

Barrier dune   A large mass or ridge of coastal sand dunes that parallels the shoreline in areas of extensive dune building.

Baseflow   The portion of streamflow contributed by groundwater. It is a steady flow that is slow to change even during rainless periods.

Base level   In a stream system, this represents the lowest elevation, such as sea level, controlling downcutting.

Bay-mouth bar   A ribbon of sand deposited across the mouth of a bay.

Bed shear stress   The force of flowing water dissipated against the streambed; a product of two main variables, channel slope and water depth.

Berm   A low mound that forms along sandy beaches; also used to describe elongated mounds constructed along water features and site borders.

Best management practices   Measures used to prevent or reduce the detrimental, environmental impacts caused by development, especially impacts from stormwater runoff.

Bifurcation ratio   The branching ratio in stream networks; the number of streams in one order to that in the next, higher order.

Biological oxygen demand   Oxygen requirements by aquatic microorganisms that deprive fish and other aquatic animals of essential oxygen supplies.

Biomass   The total weight of organic matter within a prescribed surface area, usually 1 square meter.

Bioremediation   The use of communities of plants, animals, and microorganisms to treat wastewater; for example, wetlands are used to remediate stormwater.

BMPs   *See* **Best management practices**

BOD   *See* **Biological oxygen demand.**

Bog   A cool/cold climate wetland characterized by peat deposits, an acidic pH, and often a diverse plant cover.

Boreal forest   Subarctic conifer forests of North America and Eurasia; floristically homogeneous forests dominated by fir, spruce, and tamarack. In Russia, it is called *taiga*.

Boundary layer   The lower layer of the atmosphere; the lower 300 meters of the atmosphere where airflow is influenced by the earth's surface.

Boundary sublayer   The stratum of calm air immediately over the ground that increases in depth with the height and density of the vegetative cover.

Bowen Ratio   The ratio of sensible-heat flux to latent-heat flux between a surface and the atmosphere.

Buffer   The zone around the perimeter of a wetland or lake where land use activities are limited in order to protect the water features.

Buildable land units   Parcels of various size within a designated project area that are suitable for development as defined by a prescribed development program.

CAFE standards   Corporate average fuel economy standards for automobiles as mandated by the U.S. Clean Air Act as part of the National Ambient Air Quality Standards.

Capillarity   The capacity of a soil to transfer water by capillary action; capillarity is greatest in medium-textured soils.

Carrying capacity   The level of development density or use an environment is able to support without suffering undesirable or irreversible degradation.

Chaos theory   A concept of landscape change in which the environment experienced by humans behaves more or less erratically.

Choropleth map   A map comprised of areas of any size or shape representing qualitative phenomena (e.g., soils) or quantitative phenomena (e.g., population); often has a patchwork appearance.

Climate   The representative or general conditions of the atmosphere at a place on earth. It is more than the average conditions of the atmosphere, for climate may also include extreme and infrequent conditions.

Closed forest   A forest structure with multiple levels of growth from the ground up; a forest in which undergrowth closes out the area between the canopy and the ground.

Clustering   A land use development concept in which facilities are grouped closely together with large areas of surrounding open space.

Coastal dune   A sand dune that forms in coastal areas and is fed by sand from the beach.

Coefficient of runoff   A number given to a type of ground surface representing the proportion of a rainfall converted to overland flow. It is a dimensionless number between 0 and 1.0 that varies inversely with the infiltration capacity; impervious surfaces have high coefficients of runoff.

Collection zone   The central, upper part of a small watershed where runoff from the contributing zone concentrates and forms channel flow.

Colluvium   Any material made up of a mixture of runoff and mass wasting (e.g., landslide) deposits.

Concentration time   The time taken for a drop of rain falling on the perimeter of a drainage basin to go through the basin to the outlet.

Conditional stability   A condition in the landscape in which stability is dependent on one or two essential factors, such as plants holding an oversteepened slope in place; also called metastability.

Conduction   A mechanism of heat transfer involving no external motion or mass transport. Instead, energy is transferred through the collision of vibrating molecules.

Constraint   Any feature or condition of the built or natural environment that poses an obstacle to land use planning.

Conveyance zone   The central route of drainage, usually a channel and valley, in a drainage basin.

dBA   Decibel scale that has been adjusted for sensitivity of the human ear.

Decibel   Unit of measurement for the loudness of sound based on the pressure produced in air by a noise; denoted dB.

Declination of the sun   The location (latitude) on earth where the sun on any day is directly overhead; declinations range from 23.27°S latitude to 23.27°N latitude.

Degradation   Scouring and downcutting of a stream channel, usually associated with high discharges.

Density   *See* **Development density**.

Depression storage   Rainwater and overland flow held in shallow, low spots in the terrain.

Design storm   A rainstorm of a given intensity and frequency of recurrence used as the basis for sizing stormwater facilities such as stormsewers.

Detention   A strategy used in stormwater management in which runoff is detained on site to be released later at some prescribed rate.

Development density A measure of the intensity of development or land use; defined on the basis of area covered by impervious surface, population density, or building floor area coverage, for example.

Discharge The rate of waterflow in a stream channel; measured as the volume of water passing through a cross-section of a stream per unit of time, commonly expressed as cubic feet (or meters) per second.

Discharge zone An area where groundwater seepage and springs are concentrated.

Disturbance An impact on the environment characterized by physical alteration such as forest clearing.

Disturbance theory An alternative to the community-succession concept in plant ecology in which the principal habitat change agent is external (mainly abiotic) forces such as fire, floods, and land use.

Diurnal damping depth The maximum depth in the soil that experiences temperature change over a 24-hour (diurnal) period.

Drainage basin The area that contributes runoff to a stream, river, or lake.

Drainage density The number of miles (or km) of stream channels per square mile (or km^2) of land.

Drainage divide The border of a drainage basin or watershed where overland separates between adjacent areas.

Drainage network A system of stream channels usually connected in a hierarchical fashion. *See also* **Principle of stream orders.**

Drainfield The network of pipes or tiles through which wastewater is dispersed into the soil.

Earth mound A type of soil-absorption system for residential development in which the drainfield is constructed above ground and covered with a soil medium.

Ecosystem A group of organisms linked together by a flow energy; also a community of organisms and their environment.

Ecotone The transition zone between two groups, or zones, of vegetation.

Edge city A type of urban center characterized by business and commercial land uses that has developed at selected intersections in the U.S. interstate highway system near large urban centers.

Effective impervious cover Impervious cover that is connected by a ditch or pipe to a drainage network and natural water features.

Emergent wetland Wetland dominated by herbaceous vegetation growing in shallow water.

Endangered species According to the U.S. Endangered Species Act, a species in imminent danger of extinction in all or a significant portion of its range.

Energy balance The concept or model that concerns the relationship among energy input, energy storage, work, and energy output of a system such as the atmosphere or oceans.

Environmental assessment A preliminary study or review of a proposed action (project) and the influence it could have on the environment; often conducted to determine the need for more detailed environmental impact analysis.

Environmental impact statement A study required by U.S. federal law for projects (proposed) involving federal funds to determine types and magnitudes of impacts that would be expected in the natural and human environment and the alternative courses of action, including no action.

Environmental inventory Compilation and classification of data and information on the natural and human features in an area proposed for some sort of planning project.

Ephemeral stream A stream without baseflow; one that flows only during or after rainstorms or snowmelt events.

Erodibility The relative susceptibility of a soil to erosion.

Erodibility factor A value used in the universal soil loss equation for different soil types representing relative erodibility; called the *K*-factor by the U.S. Soil Conservation Service.

Erosion The removal of rock debris by an agency such as moving water, wind, or glaciers; generally, the sculpting or wearing down of the land by erosional agents.

Estuarine wetland Coastal wetlands associated with bays and estuaries of the ocean.

Eutrophication The increase of biomass of a waterbody leading to infilling of the basin and the eventual disappearance of open water; sometimes referred to as the aging process of a waterbody.

Evapotranspiration   The loss of water from the soil through evaporation and transpiration.

Exchange time   *See* **Residence time.**

Facility   Any part of the built environment, especially structures and mechanical systems.

Facility planning   Planning for facilities such as power-generating stations or sewage treatment plants; usually carried out by engineers.

Feasibility study   A type of technical planning aimed at identifying the most appropriate use of a site.

Fetch   The distance of open water in one direction across a waterbody; one of the main controls of wave size.

Filtration   A term generally applied to the removal of pollutants, such as sediment, with the passage of water through a soil, organic, and/or fabric medium.

Floodway fringe   The zone designated by U.S. federal flood policy as the area in a river valley that would be lightly inundated by the 100-year flood.

Floristic system   The principal botanical classification scheme in use today. Under this scheme the plant kingdom is made up of divisions, each of which is subdivided into smaller and smaller groups arranged according to the apparent evolutionary relationships among plants.

Formation   A structural unit of vegetation that may be considered a subdivision of a biochore; a formation may be made up of several communities. In the traditional terminology, it is called a physiognomic unit; in geology, a major unit of rock.

Fragmentation   The process by which the landscape and its various habitats are broken down into smaller and smaller parcels as a result of land use development.

Frequency   The term used to express how often a specified event is equaled or exceeded.

Frost wedging   A mechanical weathering process in which water freezes in a crack and exerts force on the rock, which may result in the breaking of the rock; a very effective weathering process in alpine and polar environments.

Geographic Information System (GIS)   Computer mapping system designed for ready applications in problems involving overlapping and complex distributional patterns. Two classes of GIS are vector and raster.

Geomorphic system   A physical system comprised of an assemblage of landforms linked together by the flow of water, air, or ice.

Geomorphology   The field of earth science that studies the origin and distribution of landforms, with special emphasis on the nature of erosional processes; traditionally, a field shared by geography and geology.

Global coordinate system   The network of east-west and north-south lines (parallels and meridians) used to measure locations on earth; the system uses degrees, minutes, and seconds as the units of measurement.

Gradient   The inclination or slope of the land; often applied to systems such as streams and highways.

Grafting   The practice of attaching additional channels to a drainage network. In agricultural areas, new channels appear as drainage ditches; in urban areas, as stormsewers.

Gravity water   Subsurface water that responds to the gravitational force; the water that percolates through the soil to become groundwater.

Greenbelt   A tract of trees and associated vegetation in urban and rural areas; may be a park, nature preserve, or part of a transportation corridor.

Green infrastructure   Infrastructure that relies on soft or green measures rather than structural or hard measures in stormwater management. Green infastructure includes grass-lined swales, infiltration galleries, and porous pavers.

Groin   A wall or barrier built from the beach into the surf zone for the purpose of slowing down longshore transport and holding sand.

Gross sediment transport   The total quantity of sediment transported along a shoreline in some time period, usually a year.

Ground frost   Frost that penetrates the ground in response to freezing surface temperatures.

Ground sun angle   The angle formed between a beam of solar radiation and slope of the surface that it strikes in the landscape.

Groundwater   The mass of gravity water that occupies the subsoil and upper bedrock zone; the water occupying the zone of saturation below the soil-water zone.

Gullying   Soil erosion characterized by the formation of narrow, steep-sided channels etched by rivulets or small streams of water. Gullying can be one of the most serious forms of soil erosion of cropland.

Habitat   The local environment of an organism from which it gains its resources. Habitat is often variable in size, content, and location, changing with the phases in an organism's life cycle.

Habitat conservation planning   A program under the U.S. Endangered Species Act designed to protect habitat areas based on plans involving both government agents and landowners.

Habitat corridor   A belt or zone representing a habitat system (one or several related habitat types), such as a stream valley.

Hardpan   A hardened soil layer characterized by the accumulation of colloids and ions.

Hazard assessment   Study and evaluation of the hazard to land use and people from environmental threats such as floods, tornadoes, and earthquakes.

Heat island   The area or patch of relatively warm air that develops over urbanized areas.

Heat syndrome   Various disorders in the human thermoregulatory system brought on by the body's inability to shed heat or by a chemical imbalance from too much sweating.

Heat transfer   The flow of heat within a substance or the exchange of heat between substances by means of conduction, convection, or radiation.

Hillslope processes   The geomorphic processes that erode and shape slopes; mainly mass movements, such as soil creep and landslides, and runoff processes, such as rainwash and gullying.

Horizon   A layer in the soil that originates from the differentiation of particles and chemicals by moisture movement within the soil column.

Hydraulic gradient   The rate of change in elevation (slope) of a groundwater surface such as the watertable.

Hydraulic radius   The ratio of the cross-sectional area of a stream to its wetted perimeter.

Hydric soil   Soil characterized by wet conditions; saturated most of the year; often organic in composition.

Hydrograph   A streamflow graph that shows the change in discharge over time, usually hours or days. *See also* **Hydrograph method**.

Hydrograph method   A means of forecasting streamflow by constructing a hydrograph that shows the representative response of a drainage basin to a rainstorm; the use of "normalized" hydrograph for flow forecasting in which the size of the individual storm is filtered out. *See also* **Hydrograph**.

Hydrologic cycle   The planet's water system, described by the movement of water from the oceans to the atmosphere to the continents and back to the sea.

Hydrologic equation   The amount of surface runoff (overland flow) from any parcel of ground is proportional to precipitation minus evapotranspiration loss, plus or minus changes in storage water (groundwater and soil water).

Hydrometer method   A technique used to measure the clay content in a soil sample that involves dispersing the clay particles in water and drawing off samples at prescribed time intervals.

Hypothermia   A physiological disorder associated with cold conditions and characterized by the decline of body temperature, slowed heartbeat, lowered blood pressure, and other symptoms.

Impervious cover   Any hard surface material, such as asphalt or concrete, that limits infiltration and induces high runoff rates.

Infiltration beds   A general term applies to beds underlain by gravel and/or covered by vegetation designed to infiltrate stormwater. Also called infiltration galleries or biomediation beds.

Infiltration capacity   The rate at which a ground material takes in water through the surface; measured in inches or centimeters per minute or hour.

Inflooding   Flooding caused by overland flow concentrating in a low area.

Infrared film   Photographic film capable of recording near infrared radiation (just beyond the visible to a wavelength of 0.9 micrometer), but not capable of recording thermal infrared wavelengths.

Infrared radiation Mainly longwave radiation of wavelengths between 3.0–4.0 and 100 micrometers, but also includes near infrared radiation, which occurs at wavelengths between 0.7 and 3.0–4.0 micrometers.

In situ A term used to indicate that a substance is in place as contrasted with one, such as river sediment, that is in transit.

Interception The process by which vegetation intercepts rainfall or snow before it reaches the ground.

Interflow Infiltration water that moves laterally in the soil and seeps into stream channels. In forested areas this water is a major source of stream discharge.

Island biogeography The study of biodiversity as related to island size and separation distance.

Isopleth map A map comprised of lines, called isolines, that connect points of equal value.

Lacustrine wetland Wetland associated with standing waterbodies such as ponds, lakes, and reservoirs.

Land cover The materials such as vegetation and concrete that cover the ground. *See also* **Land use**.

Landfill A term applied to various managed waste disposal sites involving ground burial.

Landscape The composite of natural and human features that characterize the surface of the land at the base of the atmosphere; includes spatial, textural, compositional, and dynamic aspects of the land.

Landscape design The process of laying out land uses, facilities, water features, vegetation, and related features and displaying the results in maps and drawings.

Landscape ecology The application of geographic (spatial) analysis to problems of habitat planning and management in natural landscapes. A field of biogeography concerned with habitat fragmentation and biodiversity.

Landscape planning The decision making, technical, and design processes associated with the determination of land uses and the utilization of terrestrial resources.

Landslide A type of mass movement characterized by the slippage of a body of material over a rupture plane; often a sudden and rapid movement.

Land use The human activities that characterize an area, for example, agricultural, industrial, and residential.

Latent heat The heat released or absorbed when a substance changes phase as from liquid to gas. For water at 0°C, heat is absorbed or released at a rate of 2.5 million joules per kilogram (597 calories per gram) in the liquid/vapor phase change.

Leachate Fluids that emanate from decomposing waste in a sanitary or chemical landfill.

Leaching The removal of minerals in solution from a soil; the washing out of ions from one level to another in the soil.

Levee A mound of sediment that builds up along a river bank as a result of flood deposition.

LID *See* **Low impact development.**

Life form The form of individual plants or the form of the individual organs of a plant. In general, the overall structure of the vegetative cover may be thought of as life form as well.

Lineament Straight features in the landscape marked by slopes, segments of stream channels, soil patterns, or vegetation.

Line scanner A remote sensing device that records signals of reflected radiation in scan lines that sweep perpendicular to the path (flight line) of the aircraft.

Littoral drift The material that is moved by waves and currents in coastal areas.

Littoral transport The movement of sediment along a coastline. It is comprised of two components: longshore transport and onshore-offshore transport.

Load *See* **Sediment load**.

Loess Silt deposits laid down by wind over extensive areas of the midlatitudes during glacial and postglacial times.

Longshore current A current that moves parallel to the shoreline. Velocities generally range between 0.25 and 1 m/sec.

Longshore transport The movement of sediment parallel to the coast.

Low impact development Land use development designed specifically to minimize environmental impact in terms of energy use, air pollution, stormwater runoff, and land consumption. Applies to architecture, landscape architecture, and landscape planning.

Magnetic declination The deviation in degrees east or west between magnetic north and true north.

Magnitude and frequency concept The principle that large, landscape-changing events, such as major floods, occur at low frequencies, whereas small events with little capacity for landscape change occur at high frequencies.

Manning formula A formula used to determine the velocity of streamflow based on the gradient, hydraulic radius, and roughness of the channel; an empirical formula widely used in engineering for sizing channels and pipes.

Marsh A wetland dominated by herbaceous plants, typically cattails, reeds, and rushes.

Mass balance The relative balance in a system, based on the input and output of material such as sediment or water; the state of equilibrium between the input and output of mass in a system.

Mass movement A type of hillslope process characterized by the downslope movement of rock debris under the force of gravity; includes soil creep, rock fall, landslides, and mudflows; also termed *mass wasting*.

Meander A bend or loop in a stream channel.

Meander belt The corridor formed by a stream's meander system; defined by a set of lines drawn along the outer edge of active meanders.

Meander belt axis A line drawn down the center of a stream's meander belt.

Metastability *See* **Conditional stability**.

Microclimate The climate of small spaces such as an inner city, residential area, or mountain valley.

Misfit streams Streams that are either too large or too small for their valleys, such as small urban streams overloaded with stormwater discharge.

Mitigation A measure used to lessen the impact of an action on the natural or human environment.

Mitigation banking In wetland mitigation planning, the practice of building surplus acreage of compensation credits through replacement, enhancement, restoration, and/or preservation of wetland.

Model Any device, including conceptual constructs, mathematical formulas, or hardware apparatus, used in problem solving and analysis.

Monoculture An ecosystem dominated by few species with large populations, such as a farmfield or plantation forest.

Montmorillonite A type of clay that is notable for its capacity to shrink and expand with wetting and drying.

Moraine The material deposited directly by a glacier; also, the material (load) carried in or on a glacier. As landforms, moraines usually have hilly or rolling topography.

Morphogenetic region The concept that global landscapes are the product of different climatic regimes.

Mosaic A term used in landscape ecology to describe the patchy character of habitat as a result of fragmentation by land use.

Mudflow A type of mass movement characterized by the downslope flow of a saturated mass of clayey material.

National Pollution Discharge Elimination System U.S. federal program requiring a permit for the release of pollutants into natural water, including contaminated stormwater from communities over 100,000 population.

Nearshore circulation cell The circulation pattern of water and sediment formed by the combined action of rip currents, waves, and longshore currents.

Net sediment transport The balance between the quantities of sediment moved in two (opposite) directions along a shoreline.

Niche A term used to define an organism's way of life within its habitat; what it does to survive and reproduce.

Nonpoint source Water pollution that emanates from a spatially diffuse source such as the atmosphere or agricultural land.

NPDES *See* **National Pollution Discharge Elimination System.**

Nutrients Various types of materials that become dissolved in water and induce plant growth. Phosphorus and nitrogen are two of the most effective nutrients in aquatic plants.

Ogallala Aquifer The largest aquifer in North America, stretching from Nebraska to Texas.

Open forest A forest structure with a strong upper one or two stories and limited undergrowth; a forest that is largely open at ground level.

Open space Term applied to underdeveloped land, usually land designated for parks, greenbelts, water features, nature preserves and the like.

Open system A system characterized by a throughflow of material and/or energy; a system to which energy or material is added and released over time.

Opportunities and constraints A type of study often carried out in planing projects to determine the principal advantages and drawbacks to a development program proposed for a particular site.

Outflooding Flooding caused by a stream or river overflowing its banks.

Outwash plain A fluvioglacial deposit comprised of sand and gravel with a flat or gently sloping surface; usually found in close association with moraines.

Overdraft A condition of groundwater withdrawal in which the safe aquifer yield has been exceeded and the aquifer is being depleted.

Overland flow Runoff from surfaces on which the intensity of precipitation or snow melt exceeds the infiltration capacity; also called Horton overland flow, for hydrologist Robert E. Horton.

Oxbow A crescent-shaped lake or pond in a river valley formed in an abandoned segment of channel.

Ozone One of the minor gases of the atmosphere; a pungent, irritating form of oxygen that performs the important function of absorbing ultraviolet radiation.

Palustrine wetland Wetlands associated with inland sites that are not dependent on stream, lake, or oceanic water.

Parallels The east-west running lines of the global coordinate system. The equator, the Arctic Circle. and the Antarctic Circle are parallels; all parallels run parallel to one another.

Parent material The particulate material in which a soil forms. The two types of parent material are *residual* and *transported*.

Partial area concept A stormwater runoff model based on the observation that only a fraction of the area of a drainage basin contributes overland flow to stream discharge.

Passive solar collector A solar collector that operates without the aid of powered machinery.

Peak annual flow The largest discharge produced by a stream or river in a given year.

Peak discharge The maximum flow of a stream or a river in response to an event such as a rainstorm or over a period of time such as a year.

Peak flow *See* **Peak discharge**.

Pedon The smallest geographic unit of soil defined by soil scientists of the U.S. Department of Agriculture.

Percolation rate The rate at which water moves into soil through the walls of a test pit; used to determine soil suitability for wastewater disposal.

Percolation test A soil-permeability test performed in the field to determine the suitability of a material for wastewater disposal; the test most commonly used by sanitarians and planners to size soil-absorption systems.

Perennial stream A stream that receives inflow of groundwater all year; a stream that has a permanent baseflow.

Performance concept The concept of setting standards on how an environment or land use is expected to perform; includes formulation of goals, standards, and controls.

Periglacial environment An area where frost-related processes are a major force in shaping the landscape.

Permafrost A ground-heat condition in which the soil or subsoil is permanently frozen; long-term frozen ground in periglacial environments.

Permeability The rate at which soil or rock transmits groundwater (or gravity water in the area above the watertable); measured in cubic feet (or meters) of water transmitted through a specified cross-sectional area when under a hydraulic gradient of 1 foot per 1 foot (or 1 m per 1 m).

Photopair A set of overlapping aerial photographs that are used in stereoscopic interpretation of aerial photographs.

Photosynthesis The process by which green plants synthesize water and carbon dioxide and, with the energy from absorbed light, convert it into plant materials in the form of sugar and carbohydrates.

Physiography A term from physical geography traditionally used to describe the composite character of the landscape over large regions; today used mainly in planning and landscape architecture to describe the physical, mainly natural, character of a site or planning region.

Piping The formation of horizontal tunnels in a soil due to sapping, that is, erosion by seepage water. Piping often occurs in areas where gullying is or was active and is limited to soils resistant to cave-in.

Plane coordinate system A grid coordinate system designed by the U.S. National Ocean Survey in which the basic unit is a square measuring 10,000 feet on a side.

Planned unit development (PUD) A planning strategy aimed at reducing urban sprawl and related impacts by clustering development into carefully planned units.

Plant production The rate of output of organic material by a plant; the total amount of organic matter added to the landscape over some period of time, usually measured in grams per square meter per day or year.

Plume The stream of exhaust (smoke) emanating from a stack or chimney.

Point source Water pollution that emanates from a single source such as a sewage plant outfall.

Pollutant loading The amount of pollution released to runoff from different land uses measured in pounds or kilograms per acre or square kilometer.

Pollution Contamination of the environment from foreign substances or from increased levels of natural substances.

Porosity The total volume of pore (void) space in a given volume of rock or soil; expressed as the percentage of void volume to the total volume of the soil or rock sample.

Primary productivity The total amount of organic matter added to the landscape by plant photosynthesis, usually measured in grams per square meter per day or year.

Principal point The center of an aerial photograph, located at the intersection of lines drawn from the fiducial marks on the photo margin.

Principle of limiting factors The biological principle that the maximum obtainable rate of photosynthesis is limited by whichever basic resource of plant growth is in least supply.

Principle of stream orders The relationship between stream order and the number of streams per order. The relationship for most drainage nets is an inverse one, characterized by many low-order streams and fewer and fewer streams with increasingly higher orders. *See also* **Stream order**.

Progradation The process of seaward growth of a shoreline.

Pruning In hydrology the cutting back of a drainage net by diverting or burying streams; usually associated with urbanization or agricultural development.

Quadrat sampling A field sampling technique in which small plots, called quadrats, are laid out in the landscape and from which the sample is drawn.

Radiation The process by which radiant (electromagnetic) energy is transmitted through free space; the term used to describe electromagnetic energy, as in infrared radiation or shortwave radiation.

Radiation beam The column of solar radiation flowing into or through the atmosphere.

Rainfall erosion index A set of values representing the computed erosive power of rainfall based on total rainfall and the maximum intensity of the 30-minute rainfall.

Rainfall intensity The rate of rainfall measured in inches or centimeters of water deposited on the surface per hour or minute.

Rainshadow The dry zone on the leeward side of a mountain range of orographic precipitation.

Rainsplash Soil erosion from the impact of raindrops.

Rainwash Soil erosion by overland flow; erosion by sheets of water running over a surface; usually occurs in association with rainsplash; also called *wash*.

Rating curve A graph that shows the relationship between the discharge and stage of various flow events on a river. Once this relationship is established, it may be used to approximate discharge using stage data alone.

Rational method A method of computing the discharge from a small drainage basin in response to a given rainstorm. Computation is based on the coefficient of runoff, rainfall intensity, and basin area.

Recharge   The replenishment of groundwater with water from the surface.

Recharge zone   An area where groundwater recharge is concentrated.

Recurrence   The number of years on the average that separate events of a specific magnitude, for example, the average number of years separating river discharges of a given magnitude or greater.

Regime   The characteristic pattern of a process or system over time as in the seasonal regime of streamflow or the wet/dry regime of precipitation in California.

Regulatory floodway   A zone designated by the U.S. federal flood policy as the lowest part of the floodplain where the deepest and most frequent floodflows occur.

Relief   The range of topographic elevation within a prescribed area.

Replacement mitigation   The practice of building new wetland to offset wetland acreage lost or damaged because of land use development.

Residence time   The time taken to exchange the water in an aquifer or a lake for new water. Also called exchange time.

Restoration planning   An area of planning that addresses damaged environments, such as degraded wetland habitats and disturbed stream channels.

Retention   A strategy used in stormwater management in which runoff is retained on site in basins, underground, or released into the soil.

Retrogradation   Shoreline retreat due to erosion, water level rise, land subsidence, or some combination of the three.

Riffle   A short segment of stream channel characterized by rapid and often rough flow.

Riparian   A reference to the environment along the banks of a stream; often more broadly applied to the larger lowland corridor on the stream valley floor.

Riparian wetland   Wetlands that form on the edge of a major water feature such as a lake or stream.

Rip current   A relatively narrow jet of water that flows seaward through the breaking waves. It serves as a release for water that builds up near shore.

Riprap   Rubble such as broken concrete and rock placed on a surface to stabilize it and reduce erosion.

Risk management   An area of planning that involves preparation and response to hazards, such as floods, hurricanes, and toxic waste accidents.

Riverine wetland   Wetlands associated with streams and stream channels.

Runoff   In the broadcast sense the flow of water from the land as both surface and subsurface discharge; in the more restricted and common use, surface discharge in the form of overland flow and channel flow.

Safe well yield   The maximum pumping rate that can be sustained by a well by lowering the water level below the pump intake.

Sand bypassing   The transfer of sand around an obstacle, such as a harbor breakwater, by artificial means, for example, dredging and barging.

Sapping   An erosional process that usually accompanies gullying in which soil particles are eroded by water seeping from a bank.

SAS   *See* **Soil-absorption system**.

Scatter diagram   A graph characterized by a series of plotted points showing the relationship between two quantitative variables.

Scattering   The process by which minute particles suspended in the atmosphere diffuse incoming solar radiation.

Scouring   The principal process of channel erosion in streams characterized by heavy particles bumping and skidding against the streambed.

Scrub-scrub wetland   Wetland dominated by shrubs and short trees.

Secure landfill   A class of landfills designed and constructed expressly for the disposal of hazardous waste.

Sediment load   The material transported by streams: bed load, suspended load, and dissolved load.

Seepage   The process by which groundwater or interflow water seeps from the ground.

Sensible heat   Heat that raises the temperature of a substance and thus can be sensed with a thermometer. In contrast to latent heat, it is sometimes called the heat of dry air.

Sensitive environment   Special environments, such as wetlands or coastal lands, that require protection from development because of their aesthetic and ecological value.

Septic system   Specifically, a sewage system that relies on a septic tank to store and/or treat wastewater; generally, an on-site (small-scale) sewage disposal system that depends on the soil to dispose of wastewater.

Septic tank   A vat, usually placed underground, used to store wastewater.

Setback   A term used in site planning to indicate the critical distance that a structure or facility should be separated from an edge, such as a backshore slope or lake shore.

Shoreland   The discontinuous belt of land around a waterbody that is not drained via stream basins.

Side-looking airborne radar (SLAR)   The radar system used in remote sensing; so named because the energy pulse is beamed obliquely on the landscape from the side of the aircraft.

Sieve method   A technique used to separate the various sizes of coarse particles in a soil sample.

Siltation   A term generally applied to the deposition of sediment in natural waters due to soil erosion and stormwater runoff.

Sink   Places or features in both terrestrial and aquatic environments, such as wetlands, reservoirs, and coastal embayments, where debris both natural and human collects and is stored.

Sinuousity   A measure of the curviness of a stream channel; the ratio of channel length to the length of the meander belt axis.

Site adaptive planning   Site planning that gives careful consideration to landscape features and systems, both natural and man-made.

SLAR   *See* **Side-looking airborne radar**.

Slope failure   A slope that is unable to maintain itself and fails by mass movement, such as a landslide, slump, or similar movement.

Slope form   The configuration of a slope, for example, convex, concave, or straight.

Sluiceway   A large drainage channel or spillway for glacier meltwater.

Slump   A type of mass movement characterized by a back rotational motion along a rupture plane.

Small circle   Any circle drawn on the globe that represents less than the full circumference of the earth. Thus the plane of a small circle does not pass through the center of the earth. All parallels except the equator are small circles.

Soil-absorption system   Small-scale wastewater disposal system that relies directly on soil to absorb and disperse sewage water from a house or larger building.

Soil creep   A type of mass movement characterized by a very slow downslope displacement of soil, generally without fracturing of the soil mass. The mechanisms of soil creep include freeze-thaw activity and wetting and drying cycles.

Soil-forming factors   The major factors responsible for the formation of a soil: climate, parent material, vegetation, topography, and drainage.

Soil-heat flux   The rate of heat flow into, from, or through the soil.

Soil material   Any sediment rock, or organic debris in which soil formation takes place.

Soil profile   The sequence of horizons, or layers, of a soil.

Soil structure   The term given to the shape of the aggregates of particles that form in a soil. Four main structures are recognized: blocky, platy, granular, and prismatic.

Soil texture   The cumulative sizes of particles in a soil sample; defined as the percentage by weight of sand, silt and clay-sized particles in a soil.

Solar constant   The rate at which solar radiation is received on a surface (perpendicular to the radiation) at the edge of the atmosphere. Average strength is 1372 joules/m^2 • sec, which can also be stated as 1.97 cal/cm^2 • min.

Solar gain   A general term used to indicate the amount of solar radiation absorbed by a surface or setting in the landscape.

Solar heating   The process of generating heat from absorbed solar radiation; a widely used term in the solar energy literature.

Solifluction   A type of mass movement in periglacial environments, characterized by the slow flowage of soil material and the formation of lobeshaped features; prevalent in tundra and alpine landscapes.

Solstice   The dates when the declination of the sun is at 23.27°N latitude (the Tropic of Cancer) and 23.27°S latitude (the Tropic of Capricorn)—June 21-22 and December 21-22, respectively. These dates are known as the winter and summer solstices, but which is which depends on the hemisphere.

Source control   Managing stormwater at its place of origin using various on-site techniques.

State plane coordinate system *See* **Plane coordinate system**.

Stereoscope A viewing device used to gain a three-dimensional image from a photopair.

Stormflow The portion of streamflow that reaches the stream relatively quickly after a rainstorm, adding a surcharge of water to baseflow.

Stormwater Surface runoff in response to heavy rainfall and/or snowmelt that rushes over the land to stream channels. Also used to refer to surface runoff or overland flow from developed areas.

Stormwater garden A green infrastructure facility designed to hold and remediate stormwater by passing it through a constructed complex of wetland plants and soils.

Stratified sampling A sampling technique in which the population or study area is divided into sets or subareas before drawing the sample.

Stream order The relative position, or rank, of a stream in a drainage network. Streams without tributaries, usually the small ones, are first-order; streams with two or more first-order tributaries are second-order, and so on.

Subarctic zone The belt of latitude between 55° and the Arctic and Antarctic circles.

Subbasin A small drainage basin within the watershed of a lake or impoundment.

Sun angle The angle formed between the beam of incoming solar radiation and a plane at the earth's surface or a plane of the same altitude anywhere in the atmosphere.

Sun pocket A small space designed especially to take advantage of solar radiation and heating.

Surficial wetland A wetland originating from impaired local drainage, usually surface or near surface runoff.

Surge A large and often destructive wave caused by intensive atmospheric pressure and strong winds.

Suspended load The particles (sediment) carried aloft in a stream of wind by turbulent flow; usually clay- and silt-sized particles.

Sustainability planning An area of planning in which the objective is to achieve long-term and productive balance between land use and the environment.

Swamp A wetland dominated by trees and/or shrubs. There are many varieties including cypress, mangrove, and conifer.

Taxon Any unit (category) of classification system, usually biological.

Technical planning Data collection, analysis, and related activities used in support of the decision-making process in planning.

Temperate forest A forest of the midlatitude regions that could be described as climatically temperate, for example, broadleaf deciduous forests of Europe and North America, comprised of beeches, maples, and oaks.

Temperature inversion An atmospheric condition in which cold air underlies warm air. Inversions are highly stable conditions and thus not conducive to atmospheric mixing.

Texture The term used to describe the composite sizes of particles making up a soil sample, such as loam, sandy loam, and clay loam, based on the percentage by weight of sand, silt, and clay particles.

Thermal gradient The change in temperature over distance in a substance; usually expressed in degrees Celsius per centimeter or meter.

Thermal infrared system Line scanner capable of recording thermal infrared energy at wavelengths of 3–5 and 8–14 micrometers.

Thermokarst Sinkholelike depressions that result from differential melting of permafrost.

Threatened species According to the U.S. Endangered Species Act, a species with a rapidly declining population that is likely to become endangered.

Threshold The level of magnitude of a process at which sudden or rapid change is initiated.

Tolerance The range of stress or disturbance a plant is able to withstand without damage or death.

Toposequence Changes in soil composition and related features with topographic gradients on landforms.

Topsoil The uppermost of the soil, characterized by a high organic content; the organic layer of the soil.

Township and range A system of land subdivision in the United States that uses a grid to classify land units. Standard subdivisions include townships and sections.

Transect sampling A field sampling technique in which the sample is drawn from strips or transects laid out across the study area.

Transmission   The lateral flow of groundwater through an aquifer; measured in terms of cubic feet (or meters) transmitted through a given cross-sectional area per hour or day.

Transpiration   The flow of water through the tissue of a plant and into the atmosphere via stomatal openings in the foliage.

Transported soil   Soil formed in parent material comprised of deposits laid down by water, wind, or glaciers.

Tree line   The upper limit of tree growth on a mountain where forest often gives way to alpine meadow.

Tundra   Landscape of cold regions, characterized by a light cover of herbaceous plants and underlain by permafrost.

Turbidity   A measure of the clearness or transparency of water as a function of suspended sediment.

Turbulent flow   Flow characterized by mixing motion in which the primary source of flow resistance is the mixing action between slow-moving and faster-moving molecules in a fluid.

Universal Soil Loss Equation   A formula for estimating soil erosion by runoff based on rainfall, plant cover, slope, and soil erodibility.

Urban boundary layer   A general term referring to the layer of air over a city that is strongly influenced by urban activities and forms.

Urban canyon   City street lined with tall buildings; an urban terrain feature that has a pronounced effect on airflow, radiation, and microclimate as a whole.

Urban climate   The climate in and around urban areas; it is usually somewhat warmer, foggier, and less well lighted than the climate of the surrounding region.

Urban design   An area of professional activity by architects, landscape architects, and urban planners dealing with the forms, materials, and activities of cities.

Urbanization   The term used to describe the process of urban development, including suburban residential and commercial development.

Variable source concept   *See* **Partial area concept.**

Variance   An allowable exception or alternative interpretation to a land use or environmental ordinance.

Vascular plants   Plants in which cells are arranged into a pipelike system of conducting, or vascular, tissue. Xylem and phloem are the two main types of vascular tissue.

Watertable   The upper boundary of the zone of groundwater. In fine-textured materials it is usually a transition zone rather that a boundary line. The configuration of the watertable often approximates that of the overlying terrain.

Wave base depth   The depth offshore where wave motion first touches bottom; roughly equal to 1.0 to 2.0 times wave height.

Wavelength   The distance from the crest of one wave to the crest of the next wave.

Wave refraction   The bending of a wave, which results in an approach angle more perpendicular to the shoreline.

Wellhead protection   Land use planning and management to control contaminant sources in the area contributing recharge water to community wells.

Wetland   A term generally applied to an area where the ground is permanently wet or wet most of the year and is occupied by water-loving (or tolerant) vegetation, such as cattails, mangrove, or cypress.

Wetland disposal system   A means of sewage disposal and treatment in which the waste is dispersed into and filtered through natural or constructed wetlands.

Wetted perimeter   The distance from one side of a stream to the other, measured along the bottom.

Windshield survey   A rapid and general sampling method for vegetation and land use based on observations from a moving automobile.

Zenith   For any location on earth, the point that is directly overhead to an observer. The zenith position of the sun is the one directly overhead.

Zenith angle   The angle formed between a line perpendicular to the earth's surface (at any location) and the beam of incoming solar radiation (on any date).

APPENDIX A

U.S. AND CANADIAN SOIL CLASSIFICATION SYSTEMS

Table A.1 USDA Comprehensive System (Orders and Suborders)

Order	Outline Description	Rapid Recognition Characteristics	Suborders	
Alfisols	Argillic horizon present; base content moderate to high	All mineral horizons present except one	With gleying Others in cold climates Others in humid climates Others in subhumid climates Others in subarid climates	Aqualfs Boralfs Udalfs Ustalfs Xeralfs
Aridisols	Semidesert and desert soils	Ochric or argillic horizon present, no oxic or spodic horizon; usually dry	With argillic horizon Others	Argids Orthids
Entisols	Weakly developed, usually azonal	No diagnostic horizon except ochric, anthropic, albic, or agric	With gleying With strong artificial disturbance On alluvial deposits With sandy or loamy texture Others	Aquents Arents Fluvents Psamments Orthents
Histosols	Developed in organic materials	30% or more organic matter	Rarely saturated, 75% fibric Usually saturated, 75% fibric Usually saturated, partly decomposed Usually saturated, highly decomposed	Folists Fibrists Hemists Saprists
Inceptisols	Moderately developed; not listed elsewhere	Cambic or histic horizon present; no argillic, natric oxic, or petrocalcic horizon no plinthite	With gleying On volcanic ash In tropical climates With umbric epipedon With plaggen epipedon Others	Aquepts Andepts Tropepts Umbrepts Plaggepts Ochrepts
Mollisols	With dark A horizon and high base status	Mollic horizon present; no oxic horizon	With albic argillic horizon With gleying On highly calcareous materials Others in cold climates Others in humid climates Others in subhumid climates Others in subarid climates	Albolls Aquolls Rendolls Borolls Udolls Ustolls Xerolls
Oxisols	With oxic horizon	Oxic horizon present	With gleying With humic A horizon Others in humid climates Others in drier climates Usually dry	Aquox Humox Orthox Ustox Torrox
Spodosols	With spodic horizon	Spodic horizon present	With gleying With little humus in spodic horizon With little iron in spodic horizon With iron and humus	Aquods Ferrods Humids Orthods
Utisols	Argillic horizon present; base status low	Mean annual temperature 8°C or above; soils not listed elsewhere	With gleying With humic A horizon In humid climates Others in subhumid climates Others in subarid climates	Aquults Humults Udults Ustults Xerults

(Continued)

Table A.1 *(Continued)*

Order	Outline Description	Rapid Recognition Characteristics	Suborders	
Vertisols	Cracking clay soils	30% or more clay, with gilgai or other signs of up-and-down movement	Usually moist Dry for short periods Dry for long periods Usually dry	Uderts Usterts Xererts Torrerts
Mountain Soils		These vary greatly over short distances; with many steep slopes.		

Table A.2 Canadian Soil Classification System (Orders and Great Groups)

Order, Great Group	General Description
Chernozemic Brown Dark brown Black Dark gray	Soils of the semiarid and subhumid grasslands of the Manitoba, Saskatchewan, and Albert. Typically, rich organic accumulation in A horizon and parent material made up principally of silt and clay deposits. Calcium carbonate content is high, and pH is consistantly above 7. Soils support the great wheat-growing area of Canada.
Solonetzic Solonetz Solodized solonetz Solod	Soils developed in salinized materials mainly within the area of chernozemic soils of the Interior Plains. Very limited geographic coverage (less than 1 percent of Canada) and low agricultural potential owing to the abundant salt.
Luvisolic Gray brown luvisol Gray luvisol	Soils with silicate clay accumulation in the B horizon and fairly strong organic accumulation in A horizon. Found in forested areas and in loamy glacial deposits originally of calcareous composition. In southern Onterio and Quebec, they have been extensively cultivated.
Podzolic Humic podzol Ferro-humic podzol Humo-ferric podzol	Soils with a strong B horizon comprised of iron and aluminum oxides. Characterized by heavy leaching favored by humid climate and sandy parent material. A horizon usually contains significant organic accumulation. Found throughout Canada, covering a total of 15.6 percent of the country.
Brunisolic Melanic brunisol Eutric brunisol Sombric brunisol Dystric brunisol	Soils with a brownish-colored B horizon and a substantial organic accumulation in the A horizon. They form under forest covers in the humid subarctic of northern British Columbia and southern Yukon. Often rocky and thin with a pH around 5-6.
Regosolic Regosol Humic regosol	Soils of recent sandy deposits and active geomorphic environments. Horizons are absent or very weak. They are found with all sorts of climates and vegetation.
Gleysolic Humic gleysol Gleysol Luvic gleysol	Soils of mineral composition that are saturated all or part of the year and characterized by reducing conditions. Gleyed horizons of gray or bluish color are a common trait. Occur in all regions as poorly drained counterpart of other soils.
Organic Fibrisol Mesisol Humisol Folisol	Soils composed of largely organic matter. Found in areas of prolonged saturation and swamp and bog vegetation. Most are underlain by permafrost; located in large area on southern Hudson Bay and prominent within areas of luvisols.
Cryosolic Turbic cryosol Static cryosol Organic cryosol	Soils with permafrost within 1-2m of surface and mean annual soil temperature under 0°C. Vegetation and texture are highly variable; surface drainage is often poor, and frost action induces mechanical mixing of the active layer. The most extensive soil order in Canada, covering 40 percent of the country.

Source: The Canadian System of Soil Classification, Ottawa, 1978.

Table A.3   Unified Soil Classification System

Letter	Description	Criterion	Further criteria
G	Gravel and gravelly soils (basically pebble size, larger than 2 mm diameter)	Texture	Based on uniformity of grain size and the presence of smaller materials such as clay and silt
S	Sand and sandy soils	Texture	W Well graded (uniformly sized grains) and clean (absence of clays, silts, and organic debris) C Well graded with clay fraction, which binds soil together P Poorly graded, fairly clean
M	Very fine sand and salt (inorganic)	Texture; composition	Based on performance criteria of compressibility and plasticity
C	Clays (inorganic)	Texture; composition	L Low to medium compressibility and low plasticity
O	Organic silts and clays	Texture; composition	H High compressibility and high plasticity
P_t	Peat	Composition	

APPENDIX B

LANDFORMS AND SOIL MATERIALS AND THEIR DRAINAGE CHARACTERISTICS

Feature	Composition	Drainage
Alluvial fan	Complex: sand, silt, clay with fraction pebbles and cobbles, markedly stratified and highly heterogeneous	Variable: upper portions may be well drained, lower portions may be very poor owing to groundwater seepage
Arroyo (also gulch or wash)	Complex: silt, sand, and pebbles in bed and valley of stream in arid setting	Poor: subject to seasonal and flash flooding
Barrier beach	Sand and pebbles	Good, but water table often within several feet of surface
Beach	Variable; typically sand and pebbles but may also be clayey and silty or bedrock and rock rubble	Good if sandy, but water table usually within several feet of surface
Beach ridge	Mainly sand, but pebbles common in lower portion	Excellent, especially in those with high elevation
Bog	Organic (muck, peat) with fraction mineral clay	Very poor
Cuesta	Bedrock often with partial coverage of thin soil and talus footslope	Good, but groundwater seepage common along footslope
Cusp or cuspate foreland	Sand and pebbles	Good, but water table often within several feet of surface
Delta	Complex: usually clay, silt, and sand with local concentrations of organic material in stratified mass	Very poor to poor; high water table; subject to frequent flooding
Drumlin	Clayey with admixture of coarser fractions as large as boulders	Good to poor
Escarpment (see Cuesta)		
Esker	Stratified sand and pebble mixture (gravelly) in the form of a sinuous ridge	Excellent
Floodplain	Complex: all varieties of soil possible including organic; assortment of stratified channel deposits with fraction flood and colluvial deposits	Poor to very poor; subject to high water tables and flooding
Ground moraine	Often sand, silt, clay admixture; but may be highly variable ranging from compacted clays to sand, pebbles, cobbles, boulders; usually gently rolling	Good to poor
Kame	Mainly stratified sand and gravel in the form of a conical-shaped hill	Excellent
Lake plain	Clayey with local concentrations of beach and dune sand	Poor to fair
Lake terrace	Usually sand and pebbles but may be bedrock or clay and silt	Excellent to good
Levee	Sand, silt, and clay deposits resting on floodplain (channel) sediments	Poor; slightly better than adjacent floodplain
Marsh	Organic (muck, peat) with fraction mineral clay ranging	Very poor
Moraine	Often sand, silt, clay mixture, but may be highly variable ranging from compacted clays to sand, pebbles, cobbles, boulders; usually in form of irregular hilly terrain	Good to poor
Outwash plain	Sandy	Usually excellent, but high water table in some locales
Pediment	Thin layer of sand and gravel over bedrock	Good, but infiltration capacity may be poor
River terrace	Variable; stratified clays, silts, sand	Excellent to fair
Sand dune (barchans, seifs, parabolic, hairpin, transverse, or coastal)	Pure sand	Excellent
Scarp (see Cuesta)		
Scree slope	Cobbles and pebbles (30°–40° slope)	Excellent
Spit	Sand and pebbles	Good, but water table often within several feet of surface
Swamp	Organic (muck, peat) with fraction mineral clay	Very poor
Talus slope	Boulders in slabs, sheets, or blocks (30°–40° slope)	Excellent
Tidal flat	Sand, silt, or clay with local concentrations of organic material	Very poor
Till plain	Often sand, silt, clay admixture, but may be highly variable ranging from compacted clays to sand, pebbles, cobbles, boulders (usually gently rolling)	Good to poor

Source: Adapted from William M. Marsh, *Environmental Analysis for Land Use and Site Planning* (New York: McGraw–Hill, 1978).

APPENDIX C

U.S. RAW SURFACE WATER STANDARDS FOR PUBLIC WATER SUPPLIES

	Surface Water Criteria, mg/liter	
Substance	Permissive Criteria	Desirable Criteria
Coliforms (MPN)	10,000	<100
Fecal Coliforms (MPN)	2,000	<20
Inorganic chemicals (mg/l)		
Ammonia-N	0.5	<0.01
Arsenic[a]	0.05	Absent
Barium[a]	1.0	Absent
Boron[a]	1.0	Absent
Cadmium[a]	0.01	Absent
Chloride[a]	250	250
Chromium[a] (hexavalent)	0.05	Absent
Copper[a]	1.0	Virtually absent
Dissolved oxygen	≥4	Near saturation
Iron	0.3	Virtually absent
Lead[a]	0.05	Absent
Manganese[a]	0.05	Absent
Nitrate[a]—N	10	Virtually absent
Selenium[a]	0.01	Absent
Silver[a]	0.05	Absent
Sulfate[a]	250	<50
Total dissolved solids[a]	500	<200
Urany ion[a]	5	Absent
Zinc[a]	5	Virtually absent
Organic chemicals (mg/l)		
ABS		
Carbon chloroform extract[a]	0.15	<0.04
Cyanide[a]	0.20	Absent
Herbicides		
2,4 – D + 2,4,5 – T + 2,4 – TP[a]	0.1	Absent
Oil and gases[a]	Virtually absent	Absent
Pesticides[a]		
Adrian	0.017	Absent
Chlordane	0.003	Absent
DDT	0.042	Absent
Dieldrin	0.017	Absent
Endrin	0.001	Absent
Heptachlor	0.018	Absent
Lindane	0.056	Absent
Methoxychlor	0.035	Absent
Toxaphene	0.005	Absent
Phenols[a] 0.001 Absent		

[a]Substances that are not significantly affected by the following treatment process: coagulation (less than about 50 mg/liter of alum, ferric sulfate, or copperas, with alkali addition as necessary but without coagulant aids or activated carbon), sedimentation (6 hours or less), rapid sand filtration (3 gpm/ft^2 or less), and disinfection with chlorine (without consideration to concentration or form of chlorine residual).

Source: "Raw Water Quality Criteria for Public Supplies," National Technical Advisory Committee Report (U.S. Department of the Interior, issued by the Federal Water Pollution Control Administration, 1968).

APPENDIX D

U.S. NATIONAL AIR QUALITY STANDARDS

Pollutant	Primary Standard, Micrograms Per Cubic Meter	Secondary Standard, Micrograms Per Cubic Meter
Particulate Matter		
Annual geometric mean	75	60
Maximum 24-hour concentration[a]	260	150
Sulfer Oxides		
Annual arithmetic mean	80 (0.03 ppm)	—
Max, 24-hour concentration[a]	365 (0.14 ppm)	—
Max, 3-hour concentration[a]	—	1,300 (0.5 ppm)
Carbon Monoxide		
Max, 8-hour concentration[a]	10 (9 ppm)	10
Max, 1-hour concentration[a]	40 (35 ppm)	40
Ozone		
Max. hourly avg. concentration[a]	235 (0.12 ppm)	235
Nitrogen Dioxide		
Annual arithmetic mean	100 (0.05 ppm)	100
Hydrocarbons		
Max, 3-hour concentration[a]	160 (0.24 ppm)	160
(6–9 A.M.)	160 (0.24 ppm)	160
Lead		
Max. arithmetic means (average over calendar quarter)	1.5	1.5

[a]Not to be exceeded more than once a year per site.

Note: ppm indicates parts of pollutant per million parts of air.

Source: Environmental Protection Agency Regulations on National Primary and Secondary Ambient Air Quality Standards.

APPENDIX E

U.S. NOISE STANDARDS

Table E.1 EPA Noise Criteria and Standards and Risks

Sound Levels (decibels)	Source	Risk from Exposure
140	Jet engine (25 m distance)	Harmful to hearing
130	Jet takeoff (100 m away)	
	Threshold of pain	
120	Propeller aircraft	
110	Live rock band	Chance of hearing loss
100	Jackhammer/pneumatic chipper	
90	Heavy-duty truck	
	Los Angeles, 3rd floor apartment next to freeway	
	Average street traffic	
80	Harlem, 2nd floor apartment	Damage possible with prolonged exposure
70	Private car	—
	Boston row house on major avenue	
	Business office	
	Watts–8 mi. from touchdown at major airport	
60	Coversational speech or old residential area	
50	San Diego–wooded residential area	—
40	California tomato field	
	Soft music from radio	
30	Quiet whisper	—
20	Quiet urban dwelling	
10	Rustle of leaf	
0	Threshold of hearing	

Source: U. S. Environmental Protection Agency

Table E.2 OSHA Noise Exposure Limits

Noise (dB$_A$)	Permissible Exposure (hours and minutes)
85	16 hrs
87	12 hrs 6 min
90	8 hrs
93	5 hrs 18 min
96	3 hrs 30 min
99	2 hrs 18 min
102	1 hr 30 min
105	1 hr
108	40 min
111	26 min
114	17 min
115	15 min
118	10 min
121	6.6 min
124	4 min
127	3 min
130	1 min

Exposures above or below the 90 dB limit have been "time weighted" to give what OSHA believes are equivalent risks to a 90 dB eight-hour exposure. From U. S. Federal Register.

APPENDIX F

COMMON AND SCIENTIFIC NAMES OF NORTH AMERICAN WETLAND PLANTS

Scientific name	Common name
Acer rubrum L.	Red maple
Alisa plantago-aquatica L.	(Water plantain)
Alnus spp.	Alders
A. rugosa (DuRoi) Spreng.	Speckled alder
A. tenuifolia Nutt.	Thinleaf alder
Alopecurus aequalis Sobol.	Foxtail
Andromeda glaucophylla Link	Bog rosemary
Arctophilia fulva (Trin.) Anderss.	Pendent grass
Aristida stricta Michx.	(Three-awn)
Ascophyllum spp.	(Rockweeds)
A. nodosum (L.) LeJol	Knotted wrack
Aulacomnium palustre (Hedw.) Schwaegr.	(Moss)
Avicennia germinans (L.) L.	Black mangrove
Azolla spp.	Mosquito ferns
Baccharis halimifolia L.	Sea-myrtle
Beckmannia syzigachne (Steud.) Fernald	Slough grass
Betula nana L.	Dwarf birch
B. pumila L.	Bog birch
Brasenia schreberi J. F. Gmel.	Water shield
Calamagrostis canadensis (Michx.) Beauv.	Bluejoint
Calopogon spp.	Grass pinks
Caltha palustris L.	Marsh marigold
Campylium stellatum (Hedw.) C. Jens	(Moss)
Carex spp.	Sedges
C. aquatilis Wahlenb.	(Sedge)
C. atherodes Spreng.	Slough sedge
C. bipartita All.	(Sedge)
C. lacustris Willd.	(Sedge)
C. lasiocarpa Ehrh.	(Sedge)
C. lyngbyei Hornem.	(Sedge)
C. paleacea Schreb. ex Wahlenb.	(Sedge)
C. pluriflora Hulten	(Sedge)
C. ramenskii Kom.	(Sedge)
C. rariflora (Wahlenb.) J. E. Smith	(Sedge)
C. rostrata J. Stokes	Beaked sedge
Cassiope tetragona (L.) D. Don	Lapland cassiope
Caulerpa spp.	(Green algae)
Cephalanthus occidentalis L.	Buttonbush
Ceratophyllum spp.	Coontails
Chamaecyparis thyoides (L.) B. S. P.	Atlantic white cedar
Chamaedaphne calyculata (L.) Moench	Leatherleaf
Chara spp.	(Stoneworts)
Chenopodium glaucum L.	(Goosefoot)
Chiloscyphus fragilis (Roth) Schiffin	(Liverwort)
Chrondus crispus Stackhouse	Irish moss
Cladina spp.	Reindeer mosses
C. rangiferina (L.) Harm	(Reindeer moss)
Cladium jamaicense Crantz	Saw grass
Colocasia esculenta (L.) Scott	Taro
Conocarpus erectus L.	Buttonwood
Cornus stolonifera Michx.	Red osier dogwood
Cymodocea filiformis (Kuetz) Correll	Manatee grass
Cyperus spp.	Nut sedges
Cyrilla racemiflora L.	Black ti-ti
Decodon verticillatus (L.) Elliott	Water willow

(Continued)

Scientific name	Common name
Dendranthema articum (L.) Tzvel	Artic daisy
Dermatocarpon fluviatile G. H. Web) Th. Fr.	(Lichen)
Distichlis spicata (L.) Greene	(Salt grass)
Drepanocladus spp.	(Moss)
Dryas integrifolia Vahl	(Dryas)
Echinochloa crusgalli (L.) Beauv.	Barnyard grass
Eichhornia crassipes (Mart.) Solms	Water hyacinth
Eleocharis sp.	(Spike rush)
E. palustris (L.) Roem. & J. A. Schultes	(Spike rush)
Elodea spp.	Water weeds
Elymus arenarius L.	(Lyme grass)
Empetrum nigrum L.	Crowberry
Enteromorpha spp.	(Green algae)
Eriophorum spp.	Cotton grasses
E. russeolum Fr.	(Cotton grass)
E. vaginatum L.	(Cotton grass)
Fissidens spp.	(Moss)
F. julianus (Mont.) Schimper	(Moss)
Fontinalis spp.	(Moss)
Fraxinus nigra Marshall	Black ash
F. pennsylvanica Marshall	(Red ash)
Fucus spp.	Rockweeds
F. spiralis L.	(Rockweed)
F. vesiculosus L.	(Rockweed)
Glyceria spp.	Manna grasses
Gordonia lasianthus (L.) J. Ellis	Loblolly bay
Habenaria spp.	(Orchids)
Halimeda spp.	(Green algae)
Halodule wrightii Aschers.	Shoal grass
Hallophila spp.	(Sea grass)
Hippuris tetraphylla L.f.	(Mare's tail)
Hydrilla verticillata Royle	(Hydrilla)
Ilex glabra (L.) Gray	Inkberry
I. verticillata (L.) Gray	Winterberry
Iva frutescens L.	Marsh elder
Juncus spp.	Rushes
J. gerardii Loiseleur	Black grass
J. militaris Bigel.	Bayonet rush
J. roemerianus Scheele	Needlerush
Kalmia angustifolia L.	Sheep laurel
K. polifolia Wangenh.	Bog laurel
Kochia scoparia (L.) Schrad.	Summer cypress
Languncularia racemosa (L.) C. F. Gaertn.	White mangrove
Laminaria spp.	(Kelps)
Larix laricina (DuRoi) K. Koch	Tamarack
Laurencia spp.	(Red algae)
Ledum decumbens (Ait.) Small	Narrowleaf Labrador tea
L. groenlandicum Oeder	Labrador tea
Lemma spp.	(Duckweeds)
L. minor L.	Common duckweed
Leucothoe axillaris (Lam.) D. Don	Coastal sweetbells
Ligusticum scothicum L.	Beach lovage
Lithothamnion spp.	Coralline algae
Lycopodioum alopecuroides L.	Foxtail clubmoss
Lyonia lucida (Lam.) K. Koch	Fetterbush

(Continued)

Scientific name	Common name
Lythrum salicaria L.	Purple loosestrife
Macrocystis ssp.	(Kelps)
Magnolia virginiana L.	Sweet bay
Marsupella spp.	(Liverworts)
M. emarginata (Ehrenberg) Dumortier	(Liverwort)
Myrica gale L.	Sweet gale
Myriophyllum spp.	Water milfoils
M. spicatum L.	(Water milfoil)
Najas spp.	Naiads
Nelumbo lutea (Willd.) Pers.	American lotus
Nitella spp.	(Stoneworts)
Nuphar luteum (L.) Sibth. & J. E. Smith	(Yellow water lily)
Nymphaea spp.	(Water lilies)
N. odorata Soland. in Ait.	(White water lily)
Nyssa aquatica L.	Tupelo gum
N. sylvatica Marshall	Black gum
Oncophorus wahlenbergii Brid.	(Moss)
Panicum capillare L.	Old witch grass
Pedicularis sp.	(Lousewort)
Peltandra virginica (L.) Kunth	Arrow arum
Pelvetia spp.	(Rockweeds)
Penicillus spp.	(Green algae)
Persea borbonia (L.) Spreng.	Red bay
Phragmites australis (Cav.) Trin. ex Steud	Reed
Phyllospadix scouleri Hook.	(Surfgrass)
P. torreyi S. Wats.	(Surfgrass)
Picea mariana (Mill.) B. S. P.	Black spruce
P. sitchensis (Bong.) Carriere	Sitka spruce
Pinus contora Dougl. *ex* Loudon	Lodgepole pine
P. palustris Mill.	Longleaf pine
P. serotina Michx.	Pond pine
Pistia stratiotes L.	Water lettuce
Plantago maritima L.	Seaside plantain
Podostemum ceratophyllum Michx.	Riverweed
Polygonum spp.	Smartweed
P. amphibium L.	Water smartweed
P. bistorta L.	Bisort
Pontederia cordata L.	Pickerelweed
Potamogeton spp.	Pondweeds
P. gramineus L.	(Pondweed)
P. natans L.	Floating-leaf pondweed
Populus balsamifera L.	Balsam poplar
P. deltoides W. Bartram *ex* Marshall	Cottonwood
Potentilla anserina L.	Silverweed
P. fruticosa L.	Schrubby cinquefoil
P. palustris (L.) Scop.	Marsh cinquefoil
Puccinellia grandis Swallen	(Alkali grass)
Quercus bicolor Willd.	Swamp white oak
Q. lyrata Walter	Overcup oak
Q. michauxii Nutt,	Basket oak
Ranunculus pallasii Schlecht.	(Crowfoot)
R. trichophyllus D. Chaix	White water crowfoot
Rhizophora mangle L.	Red mangrove
Rhododendron maximum L.	Great laurel
Rhynchospora spp.	Beak rushes

(Continued)

Scientific name	Common name
Rubus chamaemorus L.	Cloudberry
Rumex maritimus L.	Golden dock
R. mexicanus Meisn.	(Dock)
Ruppia spp.	Ditch grasses
R. maritima L.	Widgeon grass
Sagittaria spp.	Arrowheads
Salicornia spp.	Glassworts
S. europaea L.	(Samphire)
S. virginica L.	(Common pickleweed)
Salix spp.	Willows
S. alaxensis (Andrss.) Coville	Feltleaf willow
S. fuscescens Anderss.	Alaska bog willow
S. ovalifolia Trautv.	Ovalleaf willow
S. planifolia Pursh	Diamondleaf willow
S. reticulata L.	Netleaf willow
Salvinia spp.	Water ferns
Sarcobatus vermiculatus (Hook.) Torr.	Greasewood
Scirpus spp.	Bulrushes
S. acutus Muhl. *ex* Bigel.	Hardstem bulrush
S. americanus Pers.	Common threesquare
S. robustus Pursh	(Bulrush)
Scolochloa festucacea (Willd.) Link	Whitetop
Solidago sempervirens L.	Seaside goldenrod
Sparganium hyperboreum Laest.	(Bur-reed)
Spartina alterniflora Loiseleur	Saltmarsh cordgrass
S. Cynosuroides (L.) Roth	Big cordgrass
S. foliosa Trin.	California cordgrass
S. patens (Ait.) Muhl.	Saltmeadow cordgrass
Sphagnum spp.	Peat mosses
Spiraea beauverdiana C. K. Schneid.	Alaska spiraea
S. douglasii Hook.	(Spiraea)
Spirodela spp.	Big duckweeds
Stellaria spp.	(Chickweed)
Suaeda californica S. Wats.	(Sea blite)
Tamarix gallica L.	Tamarisk
Taxodium distichum (L.) L. C. Rich	Bald cypress
Thalassia testudinum K. D. Koenig	Turtle grass
Thuja occidentalis L.	Northern white cedar
Tolypella spp.	(Stoneworts)
Trapa natans L.	Water nut
Triglochin maritimum L.	Arrow grass
Typha spp.	Cattails
T. angustifolia L.	Narrow-leaved cattail
T. latifolia L.	Common cattail
Ulmus americana L.	American elm
Ulva spp.	Sea lettuce
Utricularia spp.	Bladderworts
U. macrorhiza LeConte	(Bladderwort)
Vaccinium corymbosum L.	Highbush blueberry
V. oxycoccos L.	Small cranberry
V. uliginosum L.	Bog blueberry
V. vitis-idaea L.	Mountain cranberry
Vallisneria americana Michx.	Wild celery
Verrucaria spp.	(Lichens)
Wolffia spp.	Watermeals

(Continued)

Scientific name	Common name
Woodwardia virginica (L.) J. E. Smith	Virginia chain-fern
Xanthium strumarium L.	(Cocklebur)
Xyris spp.	Yellow-eyed grasses
Xyris smalliana Nash	(Yellow-eyed grass)
Zannichellia palustris L.	Horned pondweed
Zenobia pulverulenta (W. Bartram) Pollard	Honeycup
Zizania aquatica L.	Wild rice
Zizaniopsis miliacea (Michx.) Doell & Aschers.	Southern wild rice
Zostera marina L.	Eelgrass
Zosterella dubia (Jacq.) Small	Water stargrass

Source: U.S. Department of Agriculture, National List of Scientific Names, 1982.

Note: Common names that refer to a higher taxon (category) and common names that are not widely used or generally agreed upon are given in parentheses.

CREDITS

Introduction Figure 0.1: Courtesy White Mountains Attractions Association. Figure 0.2: Alexander Svirsky.

Chapter 1 Figure 1.2: The New York Public Library. Figure 1.3: ©Susan McCartney/Photo Researchers. Figure 1.4: ©Greenpeace/Calhoun. Figure 1.8: Courtesy U. S. Army Engineer Waterways Experiment Station.

Chapter 2 Figure 2.3: ©Charlie Ott/Photo Researchers. Page 32: ©Kenneth Murray/Photo Researchers. Figure 2.6: Vivian Stockman/www.ohvec.org. Figure 2.10: ©Bruce Coleman, Inc. Figure 2.11: Courtesy U. S. Army Corps. of Engineers. Figure 2.12: ©John Buitenkant/Photo Researchers. Figure 2.14: Courtesy Royal Canadian Air Force. Figure 2.15: W. M. Marsh. Figure 2.19: ©Jim Hughs/Visuals Unlimited.

Chapter 3 Figure 3.1: Courtesy U. S. Air Force. Figure 3.3 (left): ©Douglas Faulkner/Photo Researchers. Figure 3.3 (right): W. M. Marsh. Figure 3.4: ©Bohemian Nomad Picturemakers/Corbis. Figure 3.9: Courtesy Harvard Archives. Figure 3.10(top): ©Roth Stein/Library of Congress. Figure 3.10(center): Courtesy R Battalion, U. S. Air Force. Figure 3.10(bottom): Courtesy R. P. Hoblitt, U. S. Geological Survey.

Chapter 4 Figure 4.1: ©Tom McHugh/Photo Researchers. Figure 4.7: J. G. Marsh and W. M. Marsh. Figure 4.8: N. L. Marsh. Figure 4.9: Courtesy Robert L. Schuster/U. S. G. S. Figure 4.11: Courtesy U. S. Natural Resources Conservation Service.

Chapter 5 Figure 5.9: N. L. Marsh. Figure 5.10(left): Erwin W. Cole/W. M. Marsh. Figure 5.10(right): W. M. Marsh.

Chapter 6 Figure 6.4: ©Jack Dermid/Photo Researchers. Figure 6.8: ©Robert P. Carr/Bruce Coleman, Inc. Page 123: Corel Corporation.

Chapter 7 Figure 7.7: ©Bruce Clarke/Index Stock. Page 140: N. L. Marsh.

Chapter 8 Figure 8.9: Courtesy Abrams Aerial Survey Corporation. Figure 8.13: Courtesy U. S. D. A. Soil Conservation Service.

Chapter 9 Figure 9.10 and Page 183: W. M. Marsh.

Chapter 10 Figure 10.3: ©Warren Winter/Zuma Press. Page 200: Charles Schlinger.

Chapter 11 Figure 11.5: W. M. Marsh. Figure 11.6: N. L. Marsh. Figure 11.10a: ©Shirley Richards/Photo Researchers. Figure 11.10b: ©Jeff Greenberg/Photo Researchers.

Chapter 12   Figure 12.5: Courtesy U. S. D. A. Page 238: W. M. Marsh. Figure 12.11: N. L. Marsh. Page 242: Courtesy S. Goldman.

Chapter 13   Figure 13.6: W. M. Marsh.

Chapter 14   Figure 14.5a: Courtesy Canadian Royal Air Force. Figure 14.5b: W. M. Marsh. Figure 14.11: Courtesy N. O. A. A. Restoration Center. Page 279: B. K. Ferguson, W. M. Marsh and K. King.

Chapter 15   Figure 15.3(a): W. M. Marsh. Figure 15.11a: N. L. Marsh. Figure 15.12: Courtesy North Carolina Division of Marine Fisheries. Page 300: W. M. Marsh.

Chapter 16   Figure 16.7: Courtesy Williams and Woo, Inc. Figure 16.8a: W. M. Marsh. Figure 16.8b: Courtesy J. Dozier.

Chapter 17   Figure 17.8: ©Greg Dimijian/Photo Researchers.

Chapter 18   Figure 18.9b, Figure 18.12, and Page 353: Troy L. Péwé.

Chapter 19   Figure 19.2: Courtesy Long Island State Park and Recreation Commission. Figure 19.3: N. L. Marsh. Figure 19.4: Courtesy Korab Photography. Figure 19.5: W. M. Marsh.

Chapter 20   Figure 20.2: Courtesy F. R. Thompson, North Central Forest Experiment Station, U. S. Forest Service. Figure 20.6: ©G. Carleton Ray/Photo Researchers.

Chapter 21   Figure 21.3: Courtesy Frank C. Golet, University of Rhode Island. Figure 21.7: K. King and W. M. Marsh. Page 412: W. M. Marsh and G. F. Marx.

INDEX